Orion Nebula
Stocktrek Images

Infinity

New Research Frontiers

This interdisciplinary study of infinity explores the concept through the prism of mathematics and then offers more expansive investigations in areas beyond mathematical boundaries to reflect the broader, deeper implications of infinity for human intellectual thought. More than a dozen world-renowned researchers in the fields of mathematics, physics, cosmology, philosophy, and theology offer a rich intellectual exchange among various current viewpoints, rather than a static picture of accepted views on infinity.

The book starts with a historical examination of the transformation of infinity from a philosophical and theological study to one dominated by mathematics. It then offers technical discussions on the understanding of mathematical infinity. Following this, the book considers the perspectives of physics and cosmology: Can infinity be found in the real universe? Finally, the book returns to questions of philosophical and theological aspects of infinity.

Rev. Dr. Michael Heller is the Founder and Director of the Copernicus Center for Interdisciplinary Studies, Cracow. He is a Professor of Philosophy at the Pontifical University of John Paul II, Cracow, Poland; an Adjunct Member of the Vatican Observatory; and Ordinary Member of the Pontifical Academy of Sciences. Heller is the author of more than thirty books and nearly two hundred scientific papers.

Dr. W. Hugh Woodin is a Professor of Mathematics and the Chair of the Mathematics Department at the University of California, Berkeley. Professor Woodin has published numerous articles and books and is the managing editor of the *Journal of Mathematical Logic* and editor of *Mathematical Research Letters*, *Mathematical Logic Quarterly*, and *Electronic Research Announcements* (American Mathematical Society).

Infinity

New Research Frontiers

Edited by

Michael Heller

Pontifical University of John Paul II

W. Hugh Woodin

University of California, Berkeley

CAMBRIDGE UNIVERSITY PRESS
Cambridge, New York, Melbourne, Madrid, Cape Town, Singapore,
São Paulo, Delhi, Dubai, Tokyo, Mexico City

Cambridge University Press
32 Avenue of the Americas, New York, NY 10013-2473, USA

www.cambridge.org
Information on this title: www.cambridge.org/9781107003873

First published 2011

Printed in the United States of America

A catalog record for this publication is available from the British Library.

Library of Congress Cataloging in Publication data

Infinity : new research frontiers / edited by Michael Heller, W. Hugh Woodin.
   p.   cm.
Includes index.
ISBN 978-1-107-00387-3 (hardback)
1. Infinite – History.   2. Mathematics – Philosophy.
I. Heller, Michael.   II. Woodin, W. H. (W. Hugh)   III. Title.
QA9.I46   2011
111′.6–dc22   2010049374

ISBN 978-1-107-00387-3 Hardback

The infinite! No other question has ever moved so profoundly the spirit of man; no other idea has so fruitfully stimulated his intellect; yet no other concept stands in greater need of clarification than that of the infinite. . . .

David Hilbert (1862–1943)

Contents

Contributors

Wolfgang Achtner
Campus Minister and Assistant
Professor (Privatdozent) of Systematic
Theology, Justus Liebig University,
Giessen; Assistant Professor
(Privatdozent) of Systematic Theology,
Johann Wolfgang Goethe University,
Frankfurt; and Founder and Director
of the Transscientia Institute for
Interdisciplinary Research, Philosophy
and Religion, Germany

Anthony Aguirre
Associate Professor of Physics,
Department of Physics, University
of California, Santa Cruz,
United States

Marco Bersanelli
Professor of Astronomy and
Astrophysics, Department of Physics,
University of Milan, Italy

Enrico Bombieri
IBM von Neumann Professor of
Mathematics, School of Mathematics,
Institute for Advanced Study (New
Jersey), United States

Harvey M. Friedman
Distinguished University Professor of
Mathematics, Philosophy, and Computer
Science, Department of Mathematics,
The Ohio State University, Columbus,
United States

David Bentley Hart
Visiting Professor and Robert J. Randall
Chair in Christian Culture, Department
of Theology, Providence College
(Rhode Island), United States

Michael Heller
Founder and Director, Copernicus
Center for Interdisciplinary Studies,
Cracow; Professor of Philosophy, The
Pontifical University of John Paul II,
Cracow, Poland; Adjunct Member,
Vatican Observatory, and Ordinary
Member, Pontifical Academy of
Sciences

Edward Nelson
Professor of Mathematics, Department
of Mathematics, Princeton University
(New Jersey), United States

Graham Oppy
Professor and Head, School of
Philosophical, Historical and
International Studies, Monash
University at Clayton, Victoria,
Australia

Carlo Rovelli
Senior Member, Academic Institute of
France; Professor of Physics, University
of the Mediterranean; Center for
Theoretical Physics, National Center for
Scientific Research (CNRS)-Luminy,
Marseille, France

Rudy Rucker
Professor Emeritus, Mathematics and
Computer Science, Departments of
Mathematics and Computer Science,
San José State University (California),
United States

Robert John Russell
The Ian G. Barbour Professor of
Theology and Science in Residence,
The Graduate Theological Union;
Founder and Director, The Center
for Theology and the Natural
Sciences, Berkeley (California),
United States

Denys A. Turner
Horace Tracy Pitkin Professor of
Historical Theology, Yale Divinity
School, New Haven (Connecticut),
United States

W. Hugh Woodin
Professor of Mathematics, Department
of Mathematics, University of
California, Berkeley, United States

Preface

Infinity: *New Research Frontiers* was developed to explore new research domains involving concepts of infinity in a technically rigorous, as well as interdisciplinary, context. It is the culmination of a creative research initiative that began with the international conference "New Frontiers in Research on Infinity," held August 18 to 20, 2006, in San Marino. The conference was co-organized by the Center for Theology and the Natural Sciences (CTNS), Berkeley, California,[1] and the John Templeton Foundation (JTF), Philadelphia, Pennsylvania,[2] and it was funded by a generous grant to CTNS from JTF with assistance from Euresis, Milan,[3] and the Republic of San Marino.[4] The invitation-only conference, whose theme was the concept and meaning of infinity in the multifaceted contexts of mathematics, physics, cosmology, philosophy, and theology, laid the intellectual groundwork for this volume.

Esteemed researchers in these diverse fields were invited to contribute to this book with the purpose of pursuing one of the "biggest questions" facing humankind: the notion of infinity. As the great German mathematician David Hilbert stated, as quoted in the epigraph at the beginning of this volume, "No other question has ever moved so profoundly the spirit of man." Infinity has usually been regarded as a "limiting concept," that is, a concept to which we arrive by extrapolating from what we know and what is limited and finite. However, some thinkers, following the example of Descartes, claim that infinity is a "primordial concept" and that all other concepts are derived from it. Thus, in this book, we ask the basic question: Is our world of everyday experience embedded in something transcending it?

Within, we offer our intellectual adventures with the concept of infinity, showing its multiform, interdisciplinary nature. The perspective adopted in the book is to show the tension between different viewpoints and opinions, presenting an academic "dispute in action" rather than a static panorama of accepted views. Previously residing in the realm of philosophical and theological speculations, considerations about infinity are

---

[1] "Promoting the creative mutual interaction between contemporary *theology and the natural sciences*": http://www.ctns.org/.

[2] "Supporting science, investing in the big questions": http://www.templeton.org/.

[3] The Association for the Promotion of the Development of Culture and Scientific Work: http://www.euresis.org/.

[4] See http://www.ctns.org/infinity/ for information about the symposium on which this book is based.

currently dominated by mathematics, and it is mathematics that influences our thinking about infinity in other domains.

Thus, first viewed through the prism of mathematics, infinity is then investigated more expansively in areas beyond mathematical boundaries to reflect the broader, deeper implications of the concept for human intellectual thought and explorations of our lives in the vastness of the universe: infinity in mathematics, in physics and cosmology, and in philosophy and theology.

Following the comprehensive introduction, which takes us through some of the aspects of infinity from the vantage points of various human pursuits, Part I presents a chapter showing the historical process of the expansion of investigations of infinity into other areas of human endeavor – infinity as a transformative concept in science and theology. Mathematics is powerfully represented in the book in a series of five mathematical chapters (divided into two parts) by prominent mathematicians, who present modern views of infinity. Part II contains two somewhat less technical discussions, and Part III contains three advanced studies contributing to our understanding of mathematical infinity. These include how infinity fits into current research in the foundations of mathematics and range from a prediction of imminent collapse to a prediction of complete vindication in the conception of the infinite.

Can infinity be found in the real universe? The question is far from trivial, and it is considered by prominent scientists in physics and cosmology in the four chapters presented in Part IV. Given the profundity of the subject matter, the editors devoted the remaining four chapters to exploring the inevitable historical, philosophical, and theological aspects of the vast notion of infinity in Part V. These reflections are based on Western metaphysics and the Judeo-Christian tradition and have no agenda other than to explore developments in these areas of human concern.

The integration of the somewhat disparate disciplines represented in this volume has resulted in an exploration of the concept and meaning of infinity that would not otherwise have been possible. Given the recent surge in activity among researchers trying to expand the mathematical and conceptual foundations of our understanding of the universe, we believe this book will appeal to a broad, educated readership. In addition, a volume of this sort makes it possible to present in one place a rather high level of technical information while rounding out the discussion of a profound conceptual and foundational question in a multidisciplinary format that would be inappropriate in a technical mathematics research volume. Thus, we hope this book will be of benefit to those without advanced mathematics and scientific backgrounds who are deeply interested in a topic that is typically explored only by specialists and who may find new inspiration in contemplating a boundless universe within these pages, which provide fascinating reading for all those who are not afraid of intellectual challenge.

***

Michael Heller, Founder and Director, Copernicus Center for Interdisciplinary Studies, Cracow; Professor of Philosophy, The Pontifical University of John Paul II, Cracow, Poland; Adjunct Member, Vatican Observatory, and Ordinary Member, Pontifical Academy of Sciences

and

W. Hugh Woodin, Professor of Mathematics, Department of Mathematics, University of California, Berkeley, United States

Acknowledgments

The editors wish to thank the co-organizers of the 2006 conference, "New Frontiers in Research on Infinity," on which this volume is based: the Center for Theology and the Natural Sciences (CTNS), Berkeley, California, and the John Templeton Foundation (JTF), Philadelphia, Pennsylvania. In addition, we wish to thank JTF, Euresis, and the Republic of San Marino for co-sponsoring the conference.

While much of the conceptual structure of, and resources for, the conference and this book came initially from JTF senior staff and consultants, Robert John Russell, Director of CTNS, and his staff worked closely with JTF to develop and host the conference. Dr. Russell is also a contributor to this volume and, in addition, he worked with the co-editors during the writing and review process.

We also wish to thank Charles L. Harper, Jr., currently Chancellor for International Distance Learning and Senior Vice President, Global Programs, of the American University System, as well as President of Vision-Five.com Consulting, who served as one of the original project developers and the conference convener in his former role as Senior Vice President and Chief Strategist of JTF, working with Robert Russell of CTNS.

Hyung S. Choi, currently Director of Mathematical and Physical Sciences at JTF, assumed an integral role in developing the academic program for the symposium in conjunction with Charles Harper and Robert Russell.

Pamela M. Contractor, President and Director of Ellipsis Enterprises, Inc., working in conjunction with JTF and the volume editors, served as developmental editor of this book, along with Matthew P. Bond, Assistant Editor and Manager of Client Services at Ellipsis.

Finally, the editors acknowledge Lauren Cowles, Senior Editor, Mathematics and Computer Science at Cambridge University Press–New York, for supporting and overseeing this book project.

Introduction

Rudy Rucker

A stimulating factor in discussions of infinity is that the concept arises in many different contexts: mathematics, physics, metaphysics, theology, psychology, and even the arts. The founder of modern set theory, Georg Cantor, was well aware of these distinctions, and he collapses them into three domains.

> The actual infinite arises in three contexts: *first* when it is realized in the most complete form, in a fully independent other-worldly being, *in Deo*, where I call it the Absolute Infinite or simply Absolute; *second* when it occurs in the contingent, created world; *third* when the mind grasps it *in abstracto* as a mathematical magnitude, number, or order type. I wish to make a sharp contrast between the Absolute and what I call the Transfinite, that is, the actual infinities of the last two sorts, which are clearly limited, subject to further increase, and thus related to the finite.[1]

Mathematical infinities occur as, for instance, the number of points on a continuous line, the size of the endless natural number sequence 1, 2, 3, . . . , or the class of all sets.

In physics, we encounter infinities when we wonder if there might be infinitely many stars, if the universe might last forever, or if matter might be infinitely divisible.

In metaphysical discussions of the Absolute, we can ask whether an ultimate entity must be infinite, whether lesser things can be infinite as well, and how the infinite relates to our seemingly finite lives.

The metaphysical questions carry over to the theological realm, and with an added emotional intensity. Theologians might, for instance, speculate about how a finite, created mind experiences an infinite God's love.

In the psychological domain, some might argue that it's impossible to talk coherently about infinity at all, whereas others report meditative mental perceptions of the infinite.

---

[1] Georg Cantor. 1980. *Gesammelte Abhandlungen,* p. 378. Berlin: Springer.. This translation is taken from my book, *Infinity and the Mind*, p. 9. Princeton: Princeton University Press, 2004. Robert John Russell also mentions this quote in his chapter in the present volume, "God and Infinity: Theological Insights from Cantor's Mathematics."

And, finally, in the arts, we seek to find representations of our looming intimations of the infinite, perhaps in paintings, or in music, poetry, or prose.

In the following remarks, I'll say a bit about the different kinds of infinity, with special attention to the contents of the essays gathered in this volume. I'll divide my remarks into four sections: (1) mathematical infinities, (2) physical infinities, (3) metaphysical and theological infinities, and (4) psychological and artistic infinities.

Mathematical Infinities

Enrico Bombieri's genial, discursive chapter, "The Mathematical Infinity," gives us a historical survey of many areas in which infinity has cropped up in mathematics, running from the Pythagoreans to the P and NP problem in computer science. Wolfgang Achtner's chapter, "Infinity as a Transformative Concept in Science and Theology," describes how the evolution of mathematical and physical notions of infinity has advanced in concert with our theological notions of infinity. I'll say more about Achtner's chapter in the section on metaphysical and theological infinities.

For now, I'll describe a high point of the history of mathematical infinity in my own words. Set theory, or the mathematical theory of infinity, was in large part created by Georg Cantor in the late 1800s. Cantor distinguishes between a specific set and the abstract notion of its size. In Cantor's theory, there's no contradiction or incoherence in having, say, two times a transfinite cardinal be the same transfinite cardinal. And, unlike finite sets, an infinite set can have the same cardinality as a proper subset of itself. Cantor calls these infinite number sizes "transfinite cardinals."

Cantor's celebrated theorem of 1873 shows that there are transfinite cardinals of strictly different sizes. Using a so-called diagonal argument, Cantor proved that the size of the set of whole numbers is strictly less than the size of the set of all points on a line. More generally, he showed that the cardinality of any set must be less than the cardinality of its power set, that is, the set that contains all the given set's possible subsets. Along with a principle known as the axiom of choice, the proof method of Cantor's theorem can be used to ensure an endless sequence of ever-larger transfinite cardinals.

The transfinite cardinals include aleph-null (the size of the set of whole numbers), aleph-one (the next larger infinity), and the continuum (the size of the set of points on a line). These three numbers are also written as $\aleph_0$, $\aleph_1$, and c. By definition, $\aleph_0$ is less than $\aleph_1$, and by Cantor's theorem, $\aleph_1$ is less than or equal to c. And we can continue on past $\aleph_1$ to such numbers as $\aleph_2$ and $\aleph_{\aleph_0}$.

The continuum problem is the question of which of the alephs is equal to the continuum cardinality c. Cantor conjectured that $c = \aleph_1$; this is known as Cantor's continuum hypothesis, or CH for short. The continuum hypothesis can also be thought of as stating that any set of points on the line must either be countable (of size less than or equal to $\aleph_0$) or have a size as large as the entire space (be of size c).

In the early 1900s a formalized version of Cantor's theory of infinite sets arose. This theory is known as ZF, which stands for "Zermelo-Fraenkel set theory." Informally speaking, the theory is often taken to include the axiom of choice.

The continuum hypothesis is known to be undecidable on the basis of the axioms in ZF. In 1940, the logician Kurt Gödel was able to show that ZF can't disprove CH, and in 1963, the mathematician Paul Cohen showed that ZF can't prove CH. Set theorists continue to explore ways to extend the ZF axioms in a reasonable way so as to resolve CH. In the early 1970s, Kurt Gödel suggested that CH may be false and that the true size of c could be the larger infinity $\aleph_2$. In 2001, the mathematician W. Hugh Woodin seemingly espoused this view as well.[2]

In one of his two chapters for the present volume, "The Realm of the Infinite," W. Hugh Woodin mounts farther than ever into the pinnacles of the infinite. Woodin and like-minded set theorists feel that Cantor's continuum problem may in fact be solvable, and that the answer is likely to have some relation to large cardinal axioms, which posit higher and higher levels of infinity.

Near the start of his chapter, Woodin makes the point that asserting the consistency of set theory is equivalent to asserting that certain destructive types of proofs will never be found to exist in the physical world. Thus, in this sense there is a direct, albeit subtle, connection between set theory and the physical world. Woodin feels that this connection lends some validity to the belief that the universe of set theory is real.

Some skeptics maintain that Cantor's continuum problem is, in fact, a meaningless question, akin to asking about, say, the fictional Frodo Baggins's precise height. In "The Realm of the Infinite," Woodin deploys a number of refined arguments for the reality and the objectivity of the continuum problem.

In particular, he wants to undermine what he calls the "generic-multiverse position," which suggests that, although we have many diverse models of set theory, there is really no one true model wherein something like the continuum hypothesis is definitely true or false. In the multiversal kind of view, CH is true in some models, false in others, and that's the end of it.

The perennial hope among set theorists is that if we can attain still higher conceptions of infinity, these insights may end up by shedding light on even such relatively low-level questions as the continuum problem. These novel kinds of infinity are collectively known as large cardinals. It may be that by extending set theory with some new axioms about large cardinals, we could narrow in on a more complete and satisfying theory.

Let me remark in passing that there is some similarity between a contemplative monk's quest for God and a set theorist's years-long and highly focused study of large cardinals.

Woodin formulates his analysis in terms of such highly advanced modern notions as his Ω Conjecture, the Inner Model Program, and his Set Theorist's Cosmological Principle. This is twenty-first-century mathematics, and a delight to behold, even if the details will lie beyond many of us.

Woodin is arguing for a maximally rich universe of set theory, in which new levels of surprise and creativity can be found at arbitrarily large levels. In his words, "It is a fairly common (informal) claim that the quest for truth about the universe of sets is analogous to the quest for truth about the physical universe. However, I am claiming an important distinction. While physicists would rejoice in the discovery that the conception of the physical universe reduces to the conception of some simple fragment or model, the set theorist rejects this possibility. By the very nature of its conception, the set of all truths

[2] W. Hugh Woodin. 2001. The continuum hypothesis, part I and part II. *Notices of the American Mathematical Society* (June/July and August).

of the transfinite universe (the universe of sets) cannot be reduced to the set of truths of some explicit fragment of the universe of sets."

Infinity isn't the only concept that lies on the interface between mathematics and philosophy. In his other startling contribution to this volume, "A Potential Subtlety Concerning the Distinction between Determinism and Nondeterminism," Woodin uses his brand of intellectual legerdemain to argue that there really is no coherent distinction between free will and determinism. The proofs are rigorous and subtle, drawing in deep results from recursion theory and nonstandard model theory.

Woodin's line of thought is similar to the following argument, which is drawn from computer science. Philosophers often suppose that there are only two options. *Either* we are deterministic and all of our decisions are predictable far in advance *or* our behavior is utterly capricious and essentially random. But there is a third way. A human being's behavior may indeed be generated by something like a mathematical algorithm, but it may be that the workings of the algorithm do not admit for any kind of shortcuts or speedups. That is, human behavior can be deterministic without being predictable.

Those not well versed in mathematical set theory sometimes imagine there to be a strong likelihood that our formal science of the infinite may contain an inconsistency. If, for instance, ZF set theory were to be inconsistent, then at some point we'd learn that the theory breaks down and begins "proving" things like 0=1. In this case, the theory would be useless.

Most professional set theorists develop a kind of sixth sense, whereby they feel themselves to be proving things about a Platonic world of actually infinite objects. One of my thesis advisors, Gaisi Takeuti, used to say, "Why would you believe in electrons or in a tiny village in Russia that you've never seen – yet deny the reality of $\aleph_1$ or of the set of all real numbers?" To a mathematician who's "looking" at the class of all sets every day, it seems quite evident that set theory is consistent – in much the same way that a physicist is sure that the laws of physics are consistent. A theory has a concrete model if and only if it is consistent.

In his chapter, "Concept Calculus: Much Better Than," Harvey Friedman, who has often worked in the field of proof theory, takes a novel approach to questions about the consistency or inconsistency of the standard theory or ZF set theory. How do we win the confidence of someone who hesitates to believe in a world of actually infinite sets? Friedman's new idea is that we might possibly model set theory in terms of the ordinary informal concepts of "better than" and "much better than." The chapter is an interesting tour de force, quite technical and demanding in its details. Friedman's intended program is to find further deep connections between logic and common sense, and this is to be commended.

Cantor was well aware that some people are in some sense blind to the possibility of infinity. As Cantor puts it: "The fear of infinity is a form of short-sightedness that destroys the possibility of seeing the actual infinite, even though it in its highest form has created and sustains us, and in its secondary transfinite forms occurs all around us and even inhabits our minds."[3]

---

[3] I quote and translate this remark in my book, *Infinity and the Mind*, p. 43. The original quote appears in Georg Cantor, *Gesammelte Abhandlungen*, p. 374.

In his chapter, "Warning Signs of a Possible Collapse of Contemporary Mathematics," Edward Nelson speaks of "the strong emotions of loathing and oppression that the contemplation of an actual infinity arouses in me. It is the antithesis of life, of newness, of becoming – it is finished." Nelson proceeds to argue that the notion of infinity is somehow inconsistent. He takes an extreme finitist tack and presents a case that even the existence of very large finite sets is questionable. But, in the end, there's a certain circularity to any a priori arguments for or against the possibility of infinity.

Physical Infinities

The science of physical infinities is much less developed than the science of mathematical infinities. The main reason is simply that the status of physical infinities is quite undecided. In physics, one might look for infinities in space, time, divisibility, or dimensionality, and I'll discuss this in this section.

It is worth noting, however, that we are still conspicuously lacking in any physical application for the transfinite numbers of set theory. Along these lines, Georg Cantor hoped he could find an application for transfinite set theory in the realm of physics – at one time he proposed that ordinary matter might be made of aleph-null particles and that aether (which we might now term electromagnetic fields) might be made of aleph-one particles. Cantor conjectured that the matter and fields might be decomposable into meaningful pieces based on his notions of the accumulation points of infinite series. It would be a great day for set theorists if anything along the lines of such theories ever reached fruition.

Some of the Greeks speculated that space had to be infinite because the notion of an edge to space is incoherent. But, as Carlo Rovelli mentions in "Some Considerations on Infinity in Physics," if we view the space of our three-dimensional surface as curved into the hypersurface of a hypersphere, we're able to have a space that is both finite and unbounded.

Is our universe really shaped like that? In "Infinity and the Nostalgia of the Stars," Marco Bersanelli discusses how recent measurements of the cosmic microwave background indicate that the overall curvature of our space is very close to being that of a flat, Euclidean space, although the possibility still remains that our space might after all be a large hypersphere, or even a negatively curved space. Bersanelli couches the result in an amusing way:

> It is as if Eratosthenes in his famous measurement of the radius of the Earth in 250 BC was not able to measure any curvature: then his conclusion would have been that the Earth might be flat and infinite, or that its radius is greater than a given size compatible with the accuracy of his observation.
>
> (Bernaselli, Chapter 9)

Bersanelli makes another point that is not so well known. Even if we knew our space were to be precisely flat, we could not inevitably conclude that it was as infinite as an endless plane, for we might allow for the possibility that space might be multiply connected – like the surface of a torus. It is possible to find finitely large and multiply connected spaces that are, in fact, flat or negatively curved.

But suppose our physical universe were indeed infinite in its spatial extent. What then? In his chapter, "Warning Signs of a Possible Collapse of Contemporary Mathematics," Edward Nelson speaks of his discomfort with the concept of an actually infinite universe, and he remarks that in such a universe, every possible variation of our planet Earth would appear infinitely many times. Anthony Aguirre mentions this possibility as well in "Cosmological Intimations of Infinity." In "Infinity and the Nostalgia of the Stars," Marco Bersanelli makes the point that spatial infinity may be boring: "It seems that spatial infinity, in order to be perceived as a fascinating concept, has to maintain some kind of genuine element of variety and surprise."

It is worth remarking that repetition is not inevitable in an infinite universe. Put differently, the mere fact that a collection is infinite does not entail that it's exhaustive. As a very simple counterexample, consider an infinite set of integers that has only *one* odd member, the number 3. Someone who starts at 3 and looks for another odd number is going to be disappointed. As a slightly more sophisticated counterexample, think of a nonperiodic tessellation of a plane, for instance, by Penrose tiles. Although the same few tiles reoccur infinitely often, there is no one particular pattern that can be repeated to obtain the whole. It is possible for each location in an infinite universe to have its own unique qualities.

This being said, if the physical universe really were to be infinite, then there really might be other people exactly like us out there somewhere. Simply working through the number crunching needed to formalize such an argument gives us a little taste of how big infinity really is. Is it discouraging to imagine a copy of oneself on another world? Perhaps not – perhaps it's liberating. Even if you do something wrong here, maybe one of your copies will get it right!

In his chapter, "Infinities in Cosmology," Michael Heller also mentions the question of repetition in an infinite space. As Heller remarks, the physicist Max Tegmark has observed that, in some senses, the notion of a spatially infinite universe is close to the notion of a multiverse of many mutually inaccessible spacetimes.[4] Like me, Heller is unwilling to grant that spatial infinity entails endless duplication. As he puts it, ". . . in the set of real numbers, each number is an individual entity that is never repeated in the entire uncountable infinite set of reals. The 'individualization principle,' in this case, consists in both peculiar properties of a given real number and the ordering properties of the whole set of reals. If such a principle works with respect to such apparently simple entities as real numbers, should we not expect that something analogous could be at work at much higher levels of complexity?"

And what of temporal infinities? In the light of the Big Bang theory, cosmologists think of our universe as having a finitely long past; whether it might have an endless future is an open question.

Under the infinite-future view we might suppose that the space of our universe will continue much as it is now, with the galaxies drifting farther and farther apart, the stars burning to dust, and the remaining particles possibly decaying into radiation. In the finite-future view, we suppose that at some definite future time a cosmic catastrophe will destroy our universe: it might be that our space collapses to a point, or perhaps

---

[4] Max Tegmark. 2003. Parallel universes. *Scientific American* (May). Also available online at http://space.mit. edu/home/tegmark/crazy.html.

it might be that a parallel sheet of space moves through ours, annihilating everything. In any of the catastrophic finite-future scenarios, we can still wonder if the end of our universe might be followed by the birth of a new one, in which case the future might, in some sense, be infinite after all.

In his stimulating chapter, "Cosmological Intimations of Infinity," Anthony Aguirre takes up some deep considerations relating to the possible infinitude of space and time. To begin with, he points out that, even if we adopt the Big Bang scenario under which the universe in some sense sprang into being at some finitely removed past time, it is possible for a Big Bang universe to be infinitely large. Even more heartening for those infinitistically inclined, Aguirre remarks that, even though it *appears* as if a Big Bang occurred, it may also be that our past time line is, in fact, infinite.

Along these lines, in their popular exposition, *The Endless Universe*,[5] physicists Paul Steinhardt and Neil Turok envision, as hinted at earlier, two sheets of hyperspace passing through each other and, at a stroke, filling each other with energy and light.

Aguirre prefers the popular inflationary universe scenario, under which our universe, whether finite or infinite, has at some point expanded very much more rapidly than the speed of light. Aguirre points out that, mathematically speaking, one of these inflationary universes can be infinite instead of finite – it takes only a touch of mathematical trickery to make the bubbles infinitely large, at least as seen from the inside. Even more intriguing, we can have an infinite number of inflationary "pocket universes" coexisting, and these pocket universes may themselves be infinite. One might think of them as bubbles spontaneously forming in a pot of boiling water.

A variation on the theme of multiple pocket universes is the notion of the multiverse, wherein, as mentioned previously, many versions of our universe may all exist, nestled together in some quantum mechanical, infinite-dimensional Hilbert space. As Marco Bersanelli remarks in "Infinity and the Nostalgia of the Stars," the notion of an exhaustive multiverse lacks a certain aesthetic appeal. In "Infinities in Cosmology," Michael Heller points out that the multiverse model in some sense neuters the interesting philosophical question regarding why our world is the particular way it is. Heller adds that the multiverse idea, although it seems to make superfluous the notion of God as Great Designer, gives added force to the image of God as Creator.

Continuing in the vein of finding subtler kinds of physical infinities, Anthony Aguirre's "Cosmological Intimations of Infinity" closes with an argument that if our universe had only finitely many states, then our lives would be dominated by random large-scale thermodynamic fluctuations. From this, Aguirre concludes, "... the reasoning may indeed be telling us something profound: that the very coherence of our experience means that the universe has infinite possibilities."

In his chapter, "Infinities in Cosmology," Michael Heller distinguishes between two kinds of infinities in cosmology. On the one hand, he talks about "infinitely distant" regions such as one might find beyond the reaches of an endless space or an endless time. On the other hand, he talks about the "infinitely divergent" regions called singularities, where "the standard structure of spacetime breaks down: when one approaches such regions, some physical magnitudes tend to infinity."

---

[5] Paul J. Steinhardt and Neil Turok. 2008. *Endless Universe: Beyond the Big Bang.* New York: Broadway Books.

Heller describes how mathematically oriented cosmologists have come to terms with the infinitely distant regions by using a trick from differential geometry. They apply specialized representations of distance that allow them to draw an infinite spacetime as a tidy diamond shape known as a Penrose diagram, In these diagrams, the infinities lie along the corners and edges of the diamond, in much the same style that the artist M. C. Escher uses in some of his infinitely regressing images.

The singularities of spacetime are more difficult to deal with. Heller strikingly describes one approach in terms of geodesics, which are spacetime paths that objects might naturally follow, and in terms of apophatic theology, which is the technique of describing God by using negations, that is, by describing the things that *cannot* be said about God. As he puts it, "We collect information from inside a given spacetime (by following the behavior of geodesics in it) to learn something about the way its structure breaks down. The *apophatic* character of our knowledge is mitigated by tracing vestiges of what we do not know in the domain open for our investigation."

Heller makes the interesting point that both kinds of cosmological infinity can have global effects, in that their presence in a given universe can affect all of its spacetime. Here again he makes an interesting connection to theology. "Both 'infinitely distant' and 'infinitely divergent' transcend the regular parts of spacetime and at the same time are, as nonlocal elements of the model, somehow present everywhere in the model. Analogously, God transcends the world and at the same time is present within it."

What of the possibility of infinities in the small? Might matter or space itself be infinitely subdivisible? If this were the case, then each object would, in principle, contain a potentially infinite collection of particles. Of course, that perennial spoilsport quantum mechanics bids to rule this out, but perhaps there's a way around the barrier.

In "Some Considerations on Infinity in Physics," Carlo Rovelli discusses the fact that several modern theories of physics propose that space itself may be quantized. Perhaps quantum mechanics does pose an unbreachable lower bound on size – often this is identified with the so-called Planck length. Or it may be that there's some underlying deeper structure to the universe that also resists endless subdivision. In any of these cases, it seems that neither matter nor space can be viewed as infinitely divisible.

Even so, as I hinted at earlier, there remains a possibility that there may still be some kind of physics that operates below the quantum level. I recently came across a passage in Michio Kaku's book, *Parallel Worlds*, in which he discusses a 1984 theory of "string duality" ascribed to Keiji Kikkawa and Masami Yamasaki. The string duality theory also allows for interesting physics below the Planck length. The Planck length becomes something like an interface between two worlds, one that is, so to say, "inside" the Planck length, and another world that is "outside." As Kaku puts it:

> This means that the physics within the Planck length is identical to the physics outside the Planck length. At the Planck length, spacetime may become lumpy and foamy, but the physics inside the Planck length and the physics at very large distances can be smooth and are in fact identical.[6]

I'll leave the last word on the topic of physical infinities to Carlo Rovelli. In "Some Considerations on Infinity in Physics," he recognizes that all of our current speculations

---

[6] Michio Kaku. 2006. *Parallel Worlds: A Journey through Creation, Higher Dimensions and the Future of the Cosmos*, p. 237. New York: Anchor.

are inconclusive. As he puts it, "I think that what is truly infinite may just be the abyss of our ignorance."

Metaphysical and Theological Infinities

Although Plato thought of the Absolute as finite, all theologians and metaphysicians from Plotinus on have supposed the Absolute to be infinite. What is meant by "the Absolute" depends, of course, on the philosopher in question; it might be taken to mean God, an overarching universal mind, or simply the class of all possible thoughts.

As mentioned earlier, Wolfgang Achtner's chapter, "Infinity as a Transformative Concept in Science and Theology," gives us a very rich and interesting historical overview, which dovetails nicely with the survey in Enrico Bombieri's "The Mathematical Infinity." Working very much in the spirit of our volume, Achtner looks for the ways in which mathematical, physical, and theological attitudes toward infinity have advanced hand in hand. Achtner sees four steps in this advance.

1. The passage from what the Greeks called the *peiron* (limited, clearly defined, having a simple form) to the *apeiron* (unlimited, indescribable, chaotic). An early example of something *apeiron* was the irrational number length of the diagonal of a square. For the early thinkers, being infinite was a privation, a lack of structure, and it seemed natural to deny that God or the One would have such an unpleasant property.

2. Aristotle's realization that the *apeiron* could be represented in a logical form by using the notion of potential infinity. Rather than throwing up our hands in horror because an irrational number like pi or the square root of two can't be represented as a simple ratio of two whole numbers, we've learned to write our irrational numbers as endless sequences of decimal digits. If we view these sequences as approximation schemata, we are viewing them as potential infinities. As mentioned previously, the mystic philosopher Plotinus was one of the first to view being infinite as a positive attribute, and it seems fair to say that he sometimes thought of God as a potential infinity, a goal toward which a human soul might strive.

3. Gregory of Nyssa, Nicholas of Cusa, and, much later, Georg Cantor all came to think that a truly divine being might have an actually infinite nature, rather than being a potentially infinite process of endless growth. I'll say more about Gregory later. It's important to note that, before Cantor, *apeiron* notions of logical incoherence were still mixed in with the concept of the actual infinite. Although Galileo paved the way, it was Cantor's great achievement to demonstrate that we could, in fact, discuss the infinite in scientific as well as in theological terms.

4. It was Cantor as well who saw that there exists an Absolute Infinity that lies beyond the mathematical transfinites that set theorists discuss. As Achtner mentions, when we reach out to this realm, it's easy to fall into paradoxes. If the Absolute Infinite is the set of all transfinite numbers, then we need to be careful not to call the Absolute Infinite a transfinite number itself, for then it takes on the contradictory quality of being a number that is less than itself.

In closing, Achtner remarks that "the step from finiteness (*peiron*) to potential infinity and the transfinite is associated with the liberation from a purely sensual encounter of the world in favor of a rational relation. The step from the transfinite to absolute infinity is the liberation from merely rational quantitative thinking to the intuitive insight of the unity and infinity of the all-encompassing infinity of God." However, in defense of mathematics, I would add that Gödel-Bernays class-set theory or Gaisi Takeuti's more esoteric nodal transfinite type theory make it clear that mathematics, too, can speak about the Absolute.[7]

Graham Oppy's chapter, "God and Infinity: Directions for Future Research," outlines some possible connections between theology and transfinite set theory, or, as he puts it, the study of the "ultimate source of everything" and the mathematics of "limits and bounds."

One of the salient questions here is whether we might speak of God as infinite, and most theologians would answer this in the affirmative. If, for instance, God is to be omniscient, then it seems as if God might well know infinitely many propositions. But once we learn about Cantor's transfinite numbers, we find ourselves on a slippery slope. It would seem odd to say that there's some respect in which God is as big as aleph-one, but not as big as aleph-two. Hence, as Oppy points out, we're more likely to end up saying that God is "Absolutely Infinite," in Cantor's phrase.

In line with what I said earlier, and despite what Oppy remarks in one of his footnotes, most set theorists would be comfortable with identifying the Absolute Infinite with the proper class of all transfinite ordinals, a class that is variously called On or Ω. This identification is, to repeat a point, an example of how transfinite set theory is a kind of mathematical metaphysics, that is, an exact science of the Absolute.

In "God and Infinity: Theological Insights from Cantor's Mathematics," Robert John Russell further pursues the connections between theology and Cantorian set theory, bringing up two points relating to Cantor's formulation of the ordinal numbers and to the reflection principle used in the foundations of set theory. These points seem rather central to the aims of this book, and I'll summarize them in some detail.

Russell's first point has to do with the fact that Cantor distinguishes between cardinal and ordinal numbers. Cardinality has to do with the size of a number, whereas ordinality has to do with the linear order pattern in which the number elements are arranged. In the finite realm, these notions are equivalent, but in the transfinite realm we can have two numbers of the same cardinality that differ in their ordinality: for instance, the ordinals ω and $\omega + \omega$ have the same cardinality, but they represent different linear order patterns and are different ordinal numbers. A cardinal is simply an ordinal that doesn't have the same size as any previous ordinals. For Russell, what is significant in this context is that set theorists formally represent an ordinal α as the set of all ordinals less than α. Thus, aleph-null is the set of all finite ordinals, aleph-one is the set of all finite or countable ordinals, and so on.

Russell now applies this idea to a certain dilemma faced by theologians, which he describes as follows: "1. The Infinite is the negation of the finite. Yet if it is nothing more than this negation, the Infinite too is finite. 2. To avoid being merely finite through this negation, the Infinite transcends the negation by uniting itself with the finite without destroying their difference."

---

[7] Rudy Rucker. 1977. The one/many problem in the foundations of set theory. In *Logic Colloquium '76*, R. O. Gandy and J. M. E. Hyland (eds.), pp. 567–93. Amsterdam: North-Holland.

The idea of having an infinite ordinal contain all of the lesser ordinals supplies a very pretty solution to the problem. And the God-like Absolute Infinity, although too large to be a normal ordinal, is the class of all ordinals that can possibly exist.

Russell's second point relates to a somewhat abstract thought schema that is used by set theorists – the reflection principle. In my formulation in the next paragraph, we might take "the Absolute" to be the class of all ordinals, which is, again, known as the Absolute Infinite. Or we might take the Absolute to be the class of all sets. Or, going a little further, we might take the Absolute to be the ineffable ground of all being. Or, going right to the end of the line, we might suppose that the Absolute is God.

The reflection principle: If P is any simply describable property enjoyed by the Absolute, then there must be something smaller than the Absolute that also has property P.

The motivation for the reflection principle is that, if it were to fail for some property P, then we could define the Absolute as the unique thing that has property P, thus violating the principle that the Absolute should transcend any human description of it.

Putting it in theological terms, we might echo a quote from the fourth-century theologian Gregory of Nyssa: "No matter how far our mind may have progressed in the contemplation of God, it does not attain to what He is, but to what is beneath Him."[8]

The useful thing about the reflection principle is that it turns what seems like a privation into a creative force. Our inability to fully conceive of the Absolute means that each of our intuitions about it can lead to evidence for smaller, fully formed, and conceivable concepts. Thus, in set theory, we use the reflection principle to argue for the existence of infinite sets: the Absolute universe of all sets is infinite; therefore, by reflection there must be an ordinary set that is also infinite. Mathematicians such as W. Hugh Woodin have pushed the reflection principle much, much farther than this humble start.

Russell focuses on a less-often-discussed aspect of the reflection principle. As well as being a generative force, the reflection principle does, after all, give us a certain amount of insight into the Absolute itself.

> But this [the reflection principle] means that we do, in fact, know something about the Absolutely Infinite: all of the properties it possesses must be shared with and disclosed to us through the properties of the transfinites. The Absolute Infinite is in this sense knowable, comprehensible; each of its properties must be found in at least one transfinite number. The Absolute is disclosed through the relative, Absolute Infinity through the transfinite, and yet it is precisely through this same disclosure that Absolute Infinity remains hidden, ineffable, incomprehensible... it is as though the transfinites form an endless veil surrounding Absolute Infinity. The veil is all we can ever know... Yet, genuine knowledge about Absolute Infinity is forever revealed in the veil that hides it....
>
> (Russell, Chapter 13)

Russell recasts this into theological terms: "What God has chosen to disclose to us... God's existence as Creator, Gods' goodness, love and beauty – is a veil behind which the Reality of God is endlessly hidden precisely as it is endlessly revealed."

---

[8] I discuss this remark of St. Gregory's in my *Infinity and the Mind*, p. 44. The quote itself comes from Allan B. Wolter. 1941. Duns Scotus on the nature of man's knowledge of God. *Review of Metaphysics* I.2: 9. More on Gregory of Nyssa's thinking can be found in the chapters by David Bentley Hart (Chapter 12) and Denys Turner (Chapter 14) in the present volume.

In its own way, David Bentley Hart's chapter, "Notes on the Concept of the Infinite in the History of Western Metaphysics," is as intensely technical as W. Hugh Woodin's "The Realm of the Infinite" or Harvey Friedman's "Concept Calculus: Much Better Than."

Hart feels that the starting point of the metaphysical notion of infinity is the notion of absolute indeterminacy. He also remarks that the metaphysical infinite is a domain in which the principle of noncontradiction fails, and both A and the negation of A can be true. Waxing a bit territorial, he says that things such as numbers, matter, space, or time can never really attain to a truly metaphysical infinitude. As he puts it, "We see here, then, that between the mathematical and the metaphysical senses of 'infinite' there exists not merely a distinction, but very nearly an opposition... any possible analogy is at best pictorial, affective, and immeasurably remote."

Given the richness of the metaphysical thoughts that Hart goes on to describe, I would hope that he's overstating his case that the metaphysical infinity bears no fruitful relations to the infinites of mathematics and physics. He describes some wonderfully juicy and intricate distinctions to be found in the metaphysicians' writings, and I'd like to think that some mathematical logician of the order of Harvey Friedman or W. Hugh Woodin might find some formal applications of these arcane metaphysical notions after all.

Hart views the Christian "Logos" as being like an intermediate divinity between God and the mundane world. He speaks of there being a certain pathos to the notion of a higher One that lies completely outside of our world – this notion is tragic because, in order to reach the divine, ordinary mortals need to discard or at least forget their ordinary lives.

In discussing Erich Przywara's *analogia entis*, Hart presents a possible resolution to our distancing from the divine Infinite. Here we regard of God as living within us and we view our behaviors as in some way mirroring those of the Creator. As Hart puts it, "Every creature exists in a state of tension between essence and existence, in a condition of absolute becoming, oscillating between what it is and that it is, striving toward its essence and existence alike, receiving both from the movement of God's grace while possessing nothing in itself... sharing in the fullness of being that God enjoys in infinite simplicity, and so infinitely other than the source of its being."

Marco Bersanelli's "Infinity and the Nostalgia of the Stars" includes a somewhat similar remark: "The belief that each human being is in a direct relationship with the Infinite provides a solid foundation to irreducible dignity of the single person, a foundation that is hard to maintain otherwise."

Although Cantor himself seems to set metaphysics apart from mathematics and physics, I still have hopes that theological notions can enrich mathematical thought and vice versa. I was heartened to notice that Hart writes quite a bit about Gregory of Nyssa, whom I mentioned earlier for his anticipation of the set-theoretic reflection principle. Hart refers to Gregory as one of the first early thinkers to "attribute to God, or to develop a philosophical description of, positive infinity," that is, of an infinity that is discussed in terms of "fullness of the whole" rather than in terms of the "fecundity of chaos."

Denys Turner's chapter, "A (Partially) Skeptical Response to Hart and Russell," takes issue with the two chapters that I just discussed. His issue seems to be that saying that God is infinite is not really the same kind of thing as saying that God is good. Roughly speaking, he feels that "being good" has some positive content that we can understand, while "being infinite" is in the nature of a negative statement akin to "not being finite," with no positive content of its own. In his words:

> To get at the infinity of God, we would have to "get at" some intelligible content to the notion of an infinite *being*. And that we cannot get at. For we possess no concepts and no language, except that of negation, for doing so. Such a notion is utterly incomprehensible, beyond any power of expression. . . .
>
> (Turner, Chapter 14)

The best response to this kind of remark is not so much a counterargument, but a demonstration. My feeling is that the mathematical science of set theory has indeed shown that we can form positive and meaningful ideas about infinite entities. And to deny this is perhaps only to reveal one's unfamiliarity with the discipline in question.

On the other hand, I myself have but the slightest familiarity with the fine points of theological discourse. Reading these chapters, I felt new kinds of thought patterns emerging in my mind. I'm still not quite sure what to make of them, but I'll be happy to leaf through the final form of this book and ponder it all again.

Psychological and Artistic Infinities

Can we, in fact, think about infinite things? In the nineteenth century, the mathematician Bernard Bolzano formulated an argument for the infinitude of the class of all possible thoughts. If T is a thought, suppose we write T* to stand for the thought, "T is a thought." T and T* are distinct thoughts, so if we start with any thought T, we can get an endless sequence of possible thoughts: T, T*, T**, T***, and so on. Some, such as the philosopher Josiah Royce, have viewed this as evidence that the mind has access to infinity. One might say that the T series is only potentially infinite, that is, we don't literally think it all at once. But we do understand the idea of the T series all at once.

We can present something like the T series as a mental loop, along the lines of a meditation such as "I know that I know that I know . . ." But, rather than trying to handle an infinitely long sentence, why not just think, "I know," or perhaps "I am"? People who meditate on these kinds of things may experience moments of illumination. Perhaps they'll even have a vision of a blinding white light. Are they seeing infinity? Or seeing God? Are these visions real?

Perhaps. Those who've had a strong mystical experience are often very insistent that their vision had some objective content. But how, one might ask, could a finite brain hold infinity? Well, perhaps it is not that the mind actually *contains* the vision of infinity – any more than an eyeball contains a mountain. Perhaps the vision of infinity lies somehow outside the brain, let us say, in another dimension or at another level of being.

Another possibility is that matter really is infinitely divisible. Then the brain is actually infinite, and we have infinite thoughts all the time without really noticing it! Certainly one's thoughts do often seem to have infinitely subtle shadings and intricacies. It's not as if our thoughts feel like the abrupt machine-language instructions of a computer program.

In the early 1970s, when I was working on my Ph.D. dissertation in set theory at Rutgers University, I was fortunate enough to meet with the great logician Kurt Gödel at his office in Princeton. I was intrigued by his remarks about infinite sets being objectively existing objects that he could in some sense perceive. I asked him how to see the infinite sets. I'll quote his answer from my book *Infinity and the Mind*:

> He said three things. (i) First one must close off the other senses, for instance, lying down in a quiet place. It is not enough, however, to perform this negative action, one must actively seek with the mind. (ii) It is a mistake to let everyday reality condition possibility, and only to imagine the combinings and permutations of physical objects – the mind is capable of directly perceiving infinite sets. (iii) The ultimate goal of such thought, and of all philosophy, is the perception of the Absolute.[9]

Many of us have some immediate feeling for the infinite. We may express it in mathematics, in philosophy – or in art.

Originally I'd planned to contribute an article of my own to this present anthology. I was trained as a set theorist, and I've authored a book about infinity – but that was nearly thirty years ago. In the intervening years, my preferred means of self-expression has changed to the writing of fiction – and not even reputable literary fiction! Mostly I write science fiction now.

Rather than being a formal essay on mathematical philosophy, my initially proposed contribution to this anthology took the form of a short story entitled "Jack and the Aktuals, or, Physical Applications of Transfinite Set Theory." Partly inspired by W. Hugh Woodin's work, my story was constructed to illustrate the notion that our physical space might be transfinitely subdivisible, and that the generalized continuum hypothesis could fail in the sense that the cardinalities of the successive power sets might be distributed with no readily discernible pattern. And, of course, the story was also created as a means to achieving some artistic representations of infinity.

Quite reasonably, the publishers felt uneasy about including a science-fiction story in their academic volume, but the general editors generously suggested that I write this introduction for the volume instead. I'm glad to have had this opportunity; it's been fun to think all these things over yet again.

As a bonus, you can find my story "Jack and the Aktuals, or, Physical Applications of Transfinite Set Theory" online as an Adobe Acrobat PDF file on my Web site. The link is http://www.rudyrucker.com/pdf/jackandtheaktuals.pdf.

Marco Bersanelli includes a discussion of art and infinity in his chapter, "Infinity and the Nostalgia of the Stars." He makes the point that infinity is often suggested by the motifs of pattern repetition, vast landscape, presence of a hedge or limiting horizon, an element of indefiniteness, presence of a human figure. In some sense, these themes can be found in our scientific cosmologies as well.

---

[9] *Infinity and the Mind*, p. 169.

Bersanelli points out that when artists and philosophers speak of the infinitely great, they are often thinking of our deep desires for happiness and love, rather than of any endless arrays of objects. As he concludes, "perhaps the personal experience in which reference to an Infinity is most indisputable is love: the boundless intergalactic space becomes like nothing compared to the true love for a single person."

PART ONE

Perspectives on Infinity from History

CHAPTER 1

Infinity as a Transformative Concept in Science and Theology

Wolfgang Achtner

1.1 Introduction

As finite human beings in space, time, and matter, our thinking and feeling about infinity have always been associated with religious awe, whether in terms of wondering, being frightened, or being inspired. In this sense, infinity has a very strong affective-emotional component. At the same time, infinity is a scientific concept, and it has a strong intellectual-rational component that is mostly intertwined with a kind of intuitive and even ecstatic grasp of the infinite God. The affective-emotional component is emphasized in many records. One of the most famous is the text of Blaise Pascal (1623–62). He was writing when modern Europe, with its drive for infinite scientific progress, was about to emerge.

> If I consider the short period of time in which my life is intertwined with the eternity preceding and following it, and if I consider the small space in which I dwell, and if I consider just that which I see, which is lost in infinite space, of which I know nothing, and which knows nothing of me, then I am struck with wonder that I live here and not there.
>
> (Pascal 1978, pp. 114–15)

However, the rational structure of infinity still needs to be clarified. Many philosophers and mathematicians have struggled to give a rational account of infinity. The most famous quotation derives from the German mathematician David Hilbert (1862–1943). In an address given in 1925 in Münster, he said, "The infinite has always stirred the emotions of mankind more deeply than any other question; the infinite has stimulated and fertilized reason as few other ideas have" (Hilbert 1925).

Religious awe and wonder are just the beginning of a process of deeper religious and rational *transformation*. It is based on both affective-emotional involvement and intellectual-rational clarification. First, I want to argue in this chapter that religious and scientific – especially mathematical – concepts of infinity are deeply intertwined. Secondly, I want to say that at least three levels of infinity can be discerned. Jumping from one level to another is always associated with an act of transformation, and that results creatively in new religious and scientific insights. The chapter will be divided

19

into three parts according to these three levels: (1) from πέρας to ἄπειρον, (2) from ἄπειρον to potential infinity, and (3) from potential infinity to actual infinity. As an outlook the chapter concludes with (4) infinity in modern theology and the evolution of the religious Self.

1.2 From πέρας to ἄπειρον: Anaximander

As early as the pre-Socratics of early Greek philosophy, infinity was a matter for philosophical and religious consideration. Anaximander (610–546 BC), who lived in the period of transition from mythology to rationality, was the first to speak of it. He coined a koan-like saying about infinity, and it definitely expressed the pessimistic mood of the early Greeks concerning the destiny of human life. In Simplicius's physics (Simplicius 1954, pp. 13, 22–9), it is preserved as the only sentence of Anaximander that we have.[1]

The ἄπειρον is the opposite of πέρας, which means both a frontier or border and also the definite and clear. On the other hand, ἄπειρον had a negative connotation, and it referred to something to be afraid of and to be avoided. The latter derived from an understanding that human life was thought to be possible and healthy only in a confined and rationally ordered area.

From Anaximander's point of view, the ἄπειρον was not only vague and indefinite, but also showed an abyss in which the ordered life and ordered cosmos would dissolve. Perhaps the ancient Greek philosophers experienced a kind of existential barrier for thinking appreciatively about the ἄπειρον. The idea of infinity was born, and now people could think about it. However, they could think of it neither in a mathematical way nor in a scientific way. They could only think about things that were definite, clear, and certain; that is, they thought about objects with a clear πέρας that were accessible to mathematical description. These thoughts were expressed with the idea of mathematical commensurability. At this stage of philosophical discussion, the ἄπειρον could not be a subject of mathematical research.

From the perspective of the transition from mythology to rationality, this phenomenon can be interpreted as the struggle of rationality (the form of thinking the clear, definite, and certain) to stabilize itself against the disordering powers of a mythological (and thus archaic) way of living. Finding order, certainty, clearness – πέρας – then must be understood as the drive of rationality in seeking to rule out mythology. As a struggle against a mythological perception of the κόσμος and life, this struggle also becomes apparent in the concept of time, which is associated with Anaximander's ἄπειρον. Time and guilt are intertwined, because time is still perceived as a mythological power and not yet a feature of human rationality.[2] Thus, rationality reaches

---

[1] "ἀρχήν τῶν ὄντων τὸ ἄπειρον. ἐξ ὧν δὲ ἡ γένεσίς ἐστι τοῖς οὖσι, καὶ τὴν φθορὰν εἰς ταῦτα γίνεσθαι) κατὰ τὸ χρεών· διδόναι γὰρ αὐτὰ δίκην καὶ τίσιν ἀλλήλοις τῆς ἀδικίας κατὰ τὴν τοῦ χρόνου τάξιν" [Whence things have their origin, Thence also their destruction happens, As is the order of things; For they execute the sentence upon one another – In conformity with the ordinance of Time] (Diels and Kranz 1952, p. 15).

[2] This connection between time and guilt is also apparent in another mythology of this time – the Babylonian Enuma Eliš, in which human destiny is thought of as being trapped in the offering of sacrifices to compensate for the guilt of the gods.

as far as πέρας, and it is the central notion for understanding the κόσμος as being finite and ordered. The ἄπειρον is irrational, so it is not accessible and should be avoided. This view becomes clear in the philosophy of Parmenides (540–480 BC), in which his teaching of being, as the primary object of philosophy, is clearly identified with the πέρας.[3] This view is also still the case in Plato's philosophy. The ἄπειρον is neither desirable nor rationally accessible. However, Plato encourages us to extend the realm of rational accessibility, as he argues in his *Philebos*.[4] There he writes that the ἄπειρον is opposed to the ἕν and should be ignored only when all rational possibilities are realized.[5] This process is to be followed because all rationally accessible being is being in between the ἕν and the ἄπειρον, which are not rationally accessible, and it is a mixture of πέρας and ἄπειρον.[6]

It would be interesting to find out why and how the transformative process entered the Greek mind. This approach made it possible to think about the ἄπειρον existentially in a positive way, and it led toward the possibility of making it rationally accessible in science and mathematics.

1.3 From ἄπειρον to Potential Infinity: Aristotle

Aristotle was the first philosopher of ancient Greece to give a rational account of infinity.[7] In book III of his *Physics*,[8] which is about motion, he discussed whether or not the ἄπειρον exists at all. More importantly, he transformed it into a scientific concept, as opposed to the more or less mythological and religious notion of the ἄπειρον in Anaximander, to whom Aristotle alluded in his *Physics*, referring to it as something divine.[9] Because Aristotle could only think in his own philosophical categories, such as substance and accident, he claimed that the infinite could not exist in the same way as an infinite body does.[10] How then must the infinite be conceived of in a rational

---

[3] "Denn die machtvolle Notwendigkeit (ἀνάγκη) hält es (= das Seiende) in den Banden der Grenze (πέρας), die es rings umzirkt, weil das Seiende nicht ohne Abschluss (οὐκ ἀτελεύτητον τὸ ἐὸν) sein darf." *Parmenides* B 8, 30–32; see also B 8, 42–49.

[4] *Philebos* 16d–e; *Philebos* 23–25 (Plato 1972).

[5] "Des Unendlichen Begriff (τὴν δὲ τοῦ ἀπείρου ἰδέαν) soll man an die Menge nicht eher anlegen, bis einer die Zahl derselben ganz übersehen hat, die zwischen dem Unenlichen und dem Einen liegt (μεταξὺ τοῦ ἀπείρου τε καὶ τοῦ ἑνός), und dann erst jede Einheit von allem in die Unendlichkeit freilassen und verabschieden" (*Philebos*, 16d–e; Plato 1972). For an overview about the Greek ἄπειρον, see Turmakin (1943).

[6] *Philebos* 23b–26 (Plato 1972).

[7] This corresponds strikingly with the definition of time in Aristotle's philosophy. He avoids any allusion to the entanglement of time and guilt, as was the case with Anaximander. For Aristotle, time, as well as the ἄπειρον, can be subjected to rational measurement. His definition of time demonstrates this point: "τοῦτο γάρ ἐστιν ὁ χρόνος, ἀριθμὸς κινήσεως κατὰ τὸ πρότερον καὶ ὕστερον" (Aristotle 1993, *Physics* IV, 11, 219b2–3).

[8] "The study of Nature is concerned with extension, motion, and time; and since each one of these three must be either limited or unlimited (ἢ ἄπειρον ἢ πεπερασμένον) (. . .), it follows that the student of Nature must consider the question of the unlimited (ἄπειρον), with a view to determining whether it exists at all, and, if so, what is its nature" (Aristotle 1993, *Physics* III, 4, 202b30–36).

[9] "καὶ τοῦτ' (=ἄπειρον) εἶναι τὸ θεῖον· ἀθάνατον γὰρ καὶ ἀνώλεθρον, ὥς φησιν ὁ Ἀναξίμανδρος καὶ οἱ πλεῖστοι τῶν φυσιολόγων" (Aristotle 1993, *Physics* III, 203b14–15).

[10] "It is further manifest that infinity (ἄπειρον) cannot exist as an actualized entity (ἐνεργεία) and as substance (ἐνεργεία) or principle (ἀρχή)" (Aristotle 1993, *Physics* III, 5, 204a21); "From all these considerations it is evident that an unlimited body (σῶμα ἄπειρον) cannot exist in accomplished fact" (Aristotle 1993, *Physics* III, 5, 206a8–9; whole argument in 204a–206a9).

Table 1.1

	Actuality	Potentiality
Finiteness	———	nonexistent,
	finite line	self-contradictory
Infinity	nonexistent,	———
	self-contradictory	infinite line

way? That is, how can it be conceived in a way that transcends the religious feelings about the infinite in Anaximander's account?

Aristotle discussed infinity in the framework of his philosophical distinction between *potentiality* and *actuality*. Within this context, Aristotle said that the infinite cannot exist as an *actual infinity*. He said that the infinite is only a *potential infinity*, because it is the possibility of endless action.[11] Thus, the infinite exists in the mode of potentiality[12] as an endless approximation.[13] Aristotle associates the latter mainly with mathematical procedures, such as addition or division.[14] It can be demonstrated in a Table 1.1. Actual infinity and potential finiteness are nonexistent, because they are self-contradictory. Thus, Aristotle's major claim is that infinity only exists in a potential manner.

However, increasing a quantum in the process of change within potentiality and actuality also includes time and movement. Therefore, Aristotle argued at the end of book III of his *Physics* (about motion) that time and motion are infinite.[15] However, it is not the case with thinking. Although thinking relates also to the infinity of movement and time, thinking itself is not infinite.[16]

On the other hand, Aristotle denied the possibility of an infinite space. The reason is that Aristotle did not have a notion of space that would be void of matter. Instead of space, Aristotle had the notion of place, which is always associated with matter. Thus, in Aristotle's philosophy, we find a finite world with respect to space.

This finiteness with regard to the process of thinking has an important consequence. Given that God can be understood as the thinking of thinking, the νόησις νοήσεως (Aristotle 1997, *Metaphysics*, XII, 9, 1074b33–35), God also must be understood as finite.

Nearly simultaneously and in close agreement with Aristotle's philosophical treatment of the issue, the Greek mathematician Eudoxos of Knidos (410 [408]–355 [347] BC), a contemporary of Aristotle, invented the method of exhaustion (Frank 1955; Merlan 1960; Lasserre 1966). In this method, an infinitely repeatable mathematical process is used to measure surface area. This process corresponds strikingly with

---

[11] "The unlimited, then, is the open possibility of taking more, however much you have already taken" (Aristotle 1993, *Physics* III, 6, 207a6; elsewhere in Aristotle 2003, *Metaphysics* IX, 6, 1048b10–18).

[12] "It results that the unlimited potentiality exists" [λείπεται οὖν δυνάμει εἶναι τὸ ἄπειρον] (Aristotle 1993, *Physics* III, 6, 206a18).

[13] "The illimitable, then, exists only the way just described – as an endless potentiality of approximation by reduction of intervals" (Aristotle 1993, *Physics* III, 6, 206b15).

[14] "It never exists as a *thing*, as a determined quantum does. In this sense, then, there is also illimitable potentiality of addition" (Aristotle 1993, *Physics* III, 6, 206b16).

[15] "Time and movement are indeed unlimited, but only as processes, and we cannot even suppose their successive stretches to exist" [ὁ δὲ χρόνος καὶ ἡ κίνησις ἄπειρά ἐστι] (Aristotle 1993, *Physics* III, 8, 208a20).

[16] "ὁ δε χρόνος καὶ ἡ κίνησις ἄπειρά ἐστι, καὶ ἡ νόησις, οὐχ ὑπομένοντος τοῦ λαμβανομένου" (Aristotle 1993, *Physics* III, 8, 208a20).

Aristotle's understanding of the working of a potential infinity in a process. Eudoxos's method was taken up by Euclid (365–300 BC) into his elements (books V and XII). It was preserved as mathematical heritage, and it was refined by Archimedes (287–212 BC), who studied in Alexandria under the supervision of the successors of Euclid.[17]

In sum, Aristotle seems to be a thinker who stressed finiteness. It is not clear if he rejected an all-encompassing being, as Anaximander did with the ἄπειρον, which he quotes in his *Physics*. It may well be that Aristotle's account of potential infinity, as quoted in his *Physics*, and Anaximander's account of the ἄπειρον cannot be reconciled.

Aristotle rejected the ability of the human mind, the νοῦς, to think the actual infinite, and he contended that God is not infinite. However, he held that the human mind can potentially think in infinite processes, and that time and motion are infinite, even though space is finite. What makes Aristotle's account important is his distinction between potential infinity and actual infinity. This distinction proved to be of great influence in the subsequent intellectual struggle about infinity. However, Aristotle also created a major obstacle for scientific progress in thinking about infinity. This obstacle was his metaphysical framework. It stressed finiteness, at least in his thinking about God's finiteness. Christian theology broke through this limitation of Aristotle's metaphysics and paved the way to a new understanding of infinity. This view will be outlined in the following section. Aristotle's account of infinity did not enter into the theological considerations of Christian theology until Aristotle's work was received for theology by St. Thomas Aquinas in the thirteenth century.

1.4 From Potential Infinity to Actual Infinity

1.4.1 The Ecstatic Move toward Infinity – Plotinus

In late antiquity a slow process of transformation was about to emerge for understanding God as infinite. It was partly based on biblical writings, and partly based on syncretism. Thus, the Hellenistic Jew, Philo of Alexandria, referred to God along the lines of the Jewish tradition as absolutely transcendent. That meant that God is strictly to be contrasted with all mundane properties (Siegfried 1970, p. 199sq).[18] Thus, God must be thought of as not determined, which opposes the Greek philosophical tradition. The latter identified rational access and determinateness (πέρας). In this sense, one can argue that Philo is the founder of apophatic theology, which necessarily entails infinity as a property of God.[19] However, this argument is not compelling, as Philo never applies ἄπειρον to God in his writings (Mühlenberg 1966, pp. 58–64). Furthermore, the idea that God is not measurable with respect to space and time, and that God is ineffable, occurred during this late period of antiquity in the Mandean religion (Brandt 1889), as well as in the philosophy of Apuleius (130 BC) (Zeller 1963, bd. III, 23, 210),

---

[17] The method of exhaustion to calculate the surface area of a circle is described in detail in Wilson (1995, pp. 293–97).

[18] Philo assigns properties to God that are all in contrast to mundane qualities like: uncreated, name, space, but encompasses all space, cognizable, and irrational affections (Siegfried 1970, pp. 199sq, 203, 207).

[19] This argument was exposed by Guyot (1906, pp. 31–32).

and later at the end of the third century after Christ in the Hermetic writings. These ideas can be interpreted as precursors of the idea of God's infinity (Zeller 1963, bd. III, 23, 226). Thus, the old Greek idea of finitude and determinateness slowly vanished. However, these traces remained as exceptions, and they did not lead to an intellectual breakthrough.

Nevertheless, the stage was prepared for Plotinus (204/05–269/70), who was the first influential thinker to claim God's infinity from a nonbiblical, philosophical tradition.[20] Plotinus transformed both the idea of the ἄπειρον, as it was conceived by Anaximander, on his way from mythology to logos, and the concept of potential infinity, as it was introduced by Aristotle. Finally, Plotinus transformed the considerations of Plato concerning the the ἄπειρον and the ἕν from mutual exclusion to the mutual inclusion of the divine being of the ἕν and the ἄπειρον.[21] In addition, he overcame the Greek mutual exclusion of perfection and infinity, thus allowing perfection only for finite, well-circumscribed, and defined entities. This transformation could have only arisen from a deeply existential, religious experience, to which, in fact, Plotinus gave testimony.[22] Entering into the realm of the infinite ἕν requires the self-transcending philosophical ἔρως, which dissolves in an ecstatic movement toward the divine ἕν and its infinity as the form of perfect wholeness and infinite life.[23] What are the properties of this divine infinity? The infinity of the ἕν is to be thought of neither in terms of space[24] nor in terms of quantity, such as measure or number.[25] The One is also above form, being, and determination.[26] In contrast to these properties, infinity has to be thought of as power.[27] How can one understand that Plotinus ascribes infinite power to the ἕν? Plotinus, as a Neo-Platonist, advocates an ontological hierarchy of beings. This hierarchy is generated by matter and form, so that the degree of ontological power is contingent on its degree of unity. The more internally united a certain being is, the more ontological power it can exercise. The ἕν has the highest degree of unity, and it is at the top of the hierarchy of being. Therefore, its power – meaning, what it can produce – is infinite.[28]

Thus, the heritage of philosophical antiquity, formerly strictly opposed to the idea of actual infinity, culminated in the idea of an infinite divine being, as intuitively grasped by Plotinus.

This religious experience of the infinite One is the hermeneutical key to understanding the numerous contexts in which he talks about infinity (Sleeman and Pollet 1980, pp. 117–19).[29] For our purpose, it is sufficient to glance at his treatment of infinity in

---

[20] For literature on Plotinus's treatment of infinity, see Sweeney (1992c) and Armstrong (1954–55).

[21] Plotinus, *Ennead* V.5, 10, 19.

[22] We have no personal records of Plotinus's life, but his disciple Porphyrios has written in his biography about Plotinus that his teacher had twice had a deep religious experience that was the core of his philosophy (Porphyrios 1958, Vita 23, 7sqq.; Plotinus, *Ennead* VI.7, 35; VI.9, 10).

[23] Plotinus, *Ennead* III.7, 5, 26.

[24] Plotinus, *Ennead* V.5, 10, 19–20.

[25] Plotinus, *Ennead* V.5, 11, 1–3.

[26] Plotinus, *Ennead* V.1, 19–26 (Sweeney 1992b, p. 184).

[27] Plotinus, *Ennead* V.5, 10, 21.

[28] "Power is rooted in unity" (Sweeney 1992b, pp. 198, 216).

[29] For example, Plotinus discusses infinity in the context of evil and matter. For an excellent overview of all other contexts, see Sweeney (1992b, pp. 175–95).

his understanding of time, rationality, and number, inasmuch as they all are related to the divine infinity of the ἕν.

Infinity and time: In *Ennead* III.7, Plotinus gives a sophisticated account of his understanding of the relation between earthly (passing) time (χρόνος) and heavenly (everlasting) eternity (αἰών). This distinction stems from Plato, and Plotinus elaborated on it.[30] He defines eternity as a form of completed infinite life, which is whole, in the sense of being completed, self-sustaining, and thus not extended into past and future.[31] This eternal infinite life is grounded in the unchanging ἕν.[32] However, it is related to time in its form as χρόνος. Time, as χρόνος, is the life of the soul (ψυχή),[33] insofar as the soul changes and moves.[34] This movement and change of the soul in time are related to infinity in the form of an endless process,[35] and it is distinct from the completed infinity as eternity. Although being in the infinite process of time, change, and movement, the soul and its time nevertheless are capable of imitating the completed infinite eternity, because the soul is its image (αἰών).[36] The soul can do so if it is striving to participate in the one and completed infinite eternal being.[37]

To sum up, Plotinus has two different concepts of infinity related to time: the concept of a completed, whole infinity related to eternity (αἰών), which reminds one of the actual infinity of Cantor, and the concept of an infinity in process related to time (χρόνος). The latter concept is capable of mirroring the first one. This situation leads to the problem of how to relate these two concepts to one another. The structure of human rationality with its ability to calculate infinite processes is to be considered, as well as its partaking in the infinite one being (Leisegang 1913; Clark 1944; Guitton 1959; Jonas 1962; Beierwaltes 1967).

Infinity and rationality: Plotinus distinguishes two kinds of human rationality, the logos (λόγος) and the nous (νοῦς).[38] The logos represents the human ability for calculation, thinking in notions and logical sequences. The nous is more comprehensive and touches on the divine infinity. The logos is operating in the realm of diversity and multiplicity, whereas the nous is thinking in unity and totality. More specifically, one can say that the nous includes unity and diversity at the same time – seemingly a paradox.[39] Because the ἕν is the ultimate source of the nous,[40] it is striving to reunite with its origin.[41] Insofar as the nous reaches the ἕν, it is infinite, not in the sense of quantity, but rather in the sense of power.[42]

Infinity and number: How do these two forms of rationality apply to infinity in the realm of numbers? Plotinus devotes a long chapter to numbers and their relation

---

[30] Plato, *Timaeus* 37 d.
[31] Plotinus, *Ennead* III.7, 5, 25–30.
[32] Plotinus, *Ennead* III.7, 11, 3–4.
[33] Plotinus, *Ennead* II.7, 59.
[34] Plotinus, *Ennead* III.7, 11, 44–45.
[35] Plotinus, *Ennead* III.7, 11, 44–45; also Plotinus, *Ennead* III.7, 11, 54–55.
[36] Plotinus, *Ennead* III.7, 11, 47.
[37] Plotinus, *Ennead* III.7, 11, 56–59.
[38] Plotinus, *Ennead* VI.9, 5, 9.
[39] Plotinus, *Ennead* VI.7, 14, 11.
[40] Plotinus, *Ennead* VI.9, 9, 1.
[41] Plotinus, *Ennead* VI.8, 18, 7.
[42] Plotinus, *Ennead* VI.9, 6, 10–12.

to infinity.[43] He is not very clear about the ontological status of numbers, and he discusses this problem at length, offering various ways of understanding numbers.[44] He distinguishes numbers in the transcendent, ideal realm[45] from numbers as the essence of created things,[46] and from numbers as quantities[47] and numbers as counted quantities in the human mind.[48] In addition, he struggles with the ontological status of infinity: does it exist really, ideally, or only in the mind?[49] Finally, after asking how the numbers are related to these various forms of numbers,[50] Plotinus comes to the conclusion that the concept of infinity – obviously understood as actual infinity – is not applicable to the concept of number, no matter what their ontological status might be,[51] because numbers are limited.[52] However, he concedes that infinity – as potential infinity – can be thought of as an infinite process of counting.[53] Nevertheless, he does not exclude the possibility that there might be infinite numbers in the transcendent realm, insofar as they are not subject to quantity and measurement[54] – a rather strange idea in our day.

Finally, Plotinus holds that the human logos, as part of the soul, is able to comprehend infinity in the realm of numbers in the sensible world as the potential infinity of endless counting. On the other hand, the nous might touch the nonmeasurable infinity of numbers in the realm of ideas.

As we have seen, Plotinus has a rather complex understanding of infinity that appears in various contexts, of which the infinity of the divine ἕν is the most important. However, this infinity as an all-encompassing oneness, totality, and transcendent reality cannot be conceptualized by the human mind, by logos. It is transrational.[55] It can only be touched by the nous in rare moments of religious elevation, which happens when the soul and the nous turn in an inner movement toward the ἕν. Then, suddenly – ἐξαίφνης – the human nous leaves behind all its limitations with regard to number, time, and rationality in order to become transformed and enlightened in the light of the infinite ἕν.[56] Thus, Plotinus was the founder of thinking about infinity in the context of

---

43   Plotinus, *Ennead* VI.6, 1–18; *Ennead* VI.6, 2, 1–3.

44   Plotinus, *Ennead* VI.6, 4, 1–25; 5, 1–51.

45   Plotinus, *Ennead* VI.6, 15, 34–35.

46   Plotinus, *Ennead* VI.6, 16, 26.

47   Plotinus, *Ennead* VI.6, 16, 19, 21, 23f.

48   Plotinus, *Ennead* VI.6, 16, 7. Compare also *Ennead* VI.6, 9, 33–39; *Ennead* VI.14, 48–51; 6, 15, 34–44.

49   Plotinus, *Ennead* VI.6, 3, 1–2.

50   Plotinus, *Ennead* VI.6, 17, 1.

51   Plotinus, *Ennead* VI.6, 17, 3.

52   Plotinus, *Ennead* VI.6, 18, 1.

53   Plotinus, *Ennead* VI.6, 18, 3.

54   Plotinus, *Ennead* VI.6, 18, 6–8.

55   His disciple Proclus tried to think about the rational accessibility of the infinite, but Proclus ended up finding it paradoxical to think about infinity in a rational way, and that goes beyond Plotinus's metaphysical speculation and existential elevation. He gave the following example: Imagine a circle with two equal semicircles divided by one diameter, that is to say, one diameter creates two parts. If one continues to divide a circle in this way indefinitely, one gets infinite processes of partition, but twice as many parts. This doubled infinity is a paradox (see Becker 1975, p. 273). Thus, the neo-Platonic tradition ended in a dead end. It was with Christian theology that infinity became an indispensable feature of God, and it emerged slowly as a rational concept.

56   Plotinus, *Ennead* V.5, 7, 34; *Ennead* VI.7, 36, 18. This is similar to Plato's in ὁμοιοῦσθαι θεῷ in *Politeia* 613a8–b1, 621c5 and *Theatet* 176b1 sqq, to which Plotinus in *Ennead* V.3, 4, 10sqq; *Ennead* V.8, 7, 3sqq; and *Ennead* VI.9, 11, 48–51 alludes.

apophatic theology. In fact, he may have deeply influenced the Cappadocian Fathers: Basil, Gregory of Nyssa, and Gregory of Nazianzus. However, if God is infinite, how can one think about him in a rational way? Later Christian theology paved the way for struggling in a new way with this question (Armstrong 1954–55; Sweeney 1957).

1.4.2 Gregory of Nyssa, Dionysius the Areopagite, Nicholas of Cusa, and Georg Cantor

1.4.2.1 Gregory of Nyssa

The biblical writings do not refer directly to God as infinite, although some texts allude metaphorically to God's infinity.[57] The first well-known Christian theologian, Origen, was still within the conceptual framework of Aristotle. Surprisingly, Origen indirectly claimed in his *De Principiis* that God is finite.[58] St. Augustine held that thinking of God must include infinity, as a result of his divine omniscience and foreknowledge. The latter attributes cannot be thought of without presupposing that God himself is infinite. He substantiated his claim for God's infinity on biblical grounds in his *City of God*.[59] Augustine illustrated his argument for God's foreknowledge by the example of the infinite number of integers that is present in God's mind. Although he argued that God's infinite mind comprehends the infinite number of integers, he rejected the idea of an actual infinity ("quamvis infinitorum numerorum nullus sit numerus"). It is interesting that St. Augustine did not make his claim for God's infinity from a religious experience, as did his Neo-Platonic brother Plotinus, to whom he owes, in other respects, so much. Rather, St. Augustine made his claim from a rational argument that was based on biblical resources. Minucius Felix[60] and Clement of Alexandria[61] also thought about God in terms of infinity.

---

[57] For instance in Ps. 143:3 and Ps. 145:5, Ps. 90:2, Ps. 139:7–12, Ps. 147:5, Job 38–41, Is. 51:6, and Deut. 33:27, often the word "ōlām" is attributed to God, which can mean eternity or a far distant time.

[58] There is no direct quotation of Origen contending the finiteness of God. However, the overall schema of his thought points to this direction. One statement in his *De Principiis* ("Fecit autem (=God) omnia numero et mensura; nihil enim deo vel sine fine vel sine mensura est"; Origen, De principiis, IV, 4, 8) can be interpreted in this sense (Origenes 1992, pp. 808–9).

[59] "Absit itaque ut dubitemus quod ei notus sit omnis numerus cuius intelligentiae, sicut in psalmo canitur, non est numerus. Infinitas itaque numeri, quamvis infinitorum numerorum nullus sit numerus, non est tamen inconprehensibilis ei cuius intelligentiae non est numerus. Quapropter si quidquid scientia conprehenditur scientis conprehensione finitur, profecto et omnis infinitas quodam ineffabili modo Deo finita est quia scientiae ipsius inconprehensibilis non est" [Let us then not doubt that every number is known to him "of whose understanding" as the psalm goes, "there is no set number" (Ps. 147:5). Accordingly, the infinity of number, although there is no set number of infinite numbers, nevertheless is not incomprehensible to him "of whose understanding there is no set number." Wherefore, if whatever is comprehended by knowledge is limited by the comprehension of him who knows assuredly all infinity also is in some ineffable way finite to God because it is not incomprehensible for his knowledge] (Augustine 1988, IV, Book XII, 19). Among other sources St. Augustine discusses with Faustus about the infinity of God with relation to the problem of evil is Contra Faustum Manicheum, libri triginta tres 25, 1–2.

[60] "Hic (= Deus) non videri potest: visu clarior est; nec conprehendi potest nec aestimari: sensibus maior est, infinitus inmensus et soli sibi tanus, quantus est, notus, nobis vero ad intellectum pectus angustum est, et ideo sic cum digne aestimamus, dum inaestimabilem dicimus" (Felix 1977, 18, 8).

[61] "For the One is indivisible; wherefore also it is infinite, not considered with reference to inscrutability, but with reference to its being without dimensions, and not having a limit" (Clemens Alexandrinus 1968, V, 12).

However, it was not until Gregory of Nyssa that Christian theologians started to think about God in terms of infinity in a theologically elaborated way as actual infinity. As in the case of St. Augustine, the infinity of God in Gregory's theology was a matter of rational, theological argument, rather than the result of a transformative religious experience, as in Plotinus's philosophy. Gregory claimed the infinity of God[62] in the Christological and Trinitarian debates of the late Arian heresies with Eunomius (?–395) (Vaggione 2000). In the Christological debate, Gregory wanted to make sure that Christ is not a creature, but rather he partakes in God's essence, and so he is designated as homoousios (ὁμοούσιος). For this argument, he needed to presuppose God's infinity.[63]

In order to follow his train of thought, we need to understand the basic claims of his Arian opponent. As a Christian in the Greek metaphysical tradition, Eunomius's point of view is driven by two major religious concerns. First, his particular theological concern is to maintain the idea of the absolute unity and sovereignty of God. Thinking of this unity in the context of his Greek tradition, it meant that one conceived of God's οὐσία as his immortality (ἀγεννησία). This stress on God's unity and sovereignty also includes his self-sufficiency (ἀνεπίδεκτος) (Mühlenberg 1966, pp. 96–98). In contrast to the ἀγεννησία as the οὐσία of God, the οὐσία of the Son is γεννητός. Thus, the οὐσία of God and that of the Son are distinct from one another, in contradiction to the orthodox tradition. Given these presuppositions – the οὐσία of the Son is γεννητός – it necessarily entails that the Son must be subordinate to the Father, given that it must have an origin, and thus the Son cannot be divine. As a result, the Son is a creature, and there was a time when the Son had not existed.

Secondly, as an Arian Christian dissociating from the Greek tradition, he wants to make sure that the creation is finite and limited in space and time. Arguing along the lines of Aristotle (inferring from the result to the cause), he claims that Christ as the mediating λόγος of creation must be finite in power (ἐνέργαια), because the creation is finite. This inference also means that the Son cannot be grounded in the οὐσία of the Father. Hence, the Son is a finite creature. To sum up, one can say that both theological concerns, the unity and sovereignty of God and the finitude of the creation in connection with his Greek habit of reasoning, necessarily led Eunomius to the Arian heresy and the destruction of the Trinity. To defend the doctrine of Trinity and the traditional Christological teaching of Christ's essential unity with the Father, Gregory had to find conclusive arguments. He found them in the idea of God's infinity. How did he argue?

Gregory's major concern was to secure the divinity of the Son, based on the compatibility of the ὁμοούσιος and the γεννητός of the Son. This compatibility was impossible with regard to the theological axioms of Eunomius, because they were logically consistent in themselves. Therefore, Gregory had to change the underlying

---

[62] The texts of the controversy between Gregory of Nyssa and Eunomius are available in the Werner Jaeger edition. The theological arguments for the infinity of God are elaborated in Gregory of Nyssa (1960a, 1960b). The text about the infinite spiritual ascension to God in Gregor's De vita Moysis is available in Gregory of Nyssa (1964).

[63] For relevant secondary literature on the issue of infinity in Gregory of Nyssa, see Hennessy (1963), Mühlenberg (1966), Balás (1966), Ferguson (1973), Brightman (1973), Duclow (1974), Wilson-Kästner (1978), De Smet (1980), Sweeney (1992b), and Karfíková (2001).

theological suppositions. This was exactly the way Gregory pursued. He did so in a twofold way:

1. He claimed that the ἀγεννησία of God is not a trait of the οὐσία, but it is only a property. Given this different interpretation of God, it follows that the whole argument of Eunomius collapses.[64]
2. Gregory went further and designed a new concept of God. On the one hand, this new concept made it impossible to set up the logical inferences of Eunomius, and on the other hand, it allowed a logical deduction as to the Son's divinity based on the compatibility of the ὁμοούσιος and the γεννητός. This feat was accomplished by introducing the idea of infinity (ἄπειρον and ἀόριστον) as a decisive trait of God. In order to avoid the counterargument of a pure ad hoc hypothesis, he had to make sure that the idea of God's infinity could be substantiated by reasonable arguments.

The second part of the argument will be the focus here, and Gregory's argument will be followed in some detail. In the first part of the argument, Gregory wanted to make sure that the human mind cannot grasp the οὐσία of God at all. Given this point, Eunomius's logical inference can no longer be upheld. Gregory introduced the notion of infinity – as actual infinity (ἄπειρον and ἀόριστον) – into God's οὐσία. As we shall see later, infinity is a notion to which the human rationality cannot be applied. Gregory had to show that the relating of this notion to the essence of God was not just an argument based on an ad hoc hypothesis. First of all, he did this by substantiating it with biblical quotations (Ps. 144: 3b + 5a) (Mühlenberg 1966, p. 102 [sect. 103, 38, 21sq]; p. 160 [4, 9sq]). Secondly, and more convincingly, he used the common ground of the God of Greek metaphysics and its properties, on which Eunomius also agreed. By a logical inference, Gregory showed that the two metaphysical properties of God in the Greek tradition, simplicity and inalterability (ἀναλλοίωτος), necessarily entail God's infinity (Aristotle 2003, 1073a11). According to Ekkehard Mühlenberg (1966, pp. 121–22), the argument based on God's unchangeability runs as follows:[65]

1. Logical Presupposition: With regard to mercy, power, wisdom, and so on, a limitation can only be obtained by the respective opposite.
2. Metaphysical Presupposition: The divine nature is unchangeable (ἀναλλοίωτος).

Logical conclusions are as follows:

1. Because God is unchangeable, there is no opposite in his essence.
2. If God in this way is superior to the realm of opposites, where nonbeing and evil are also located, then he must be perfect.
3. Because, according to presupposition (i), God cannot be limited in his goodness, he is unlimited.
4. Being without limits is identical with infinity.

As the next logical step, in agreement with Greek metaphysical heritage, Gregory claimed that infinity is not accessible by reason (Gregory of Nyssa 1960b,

---

[64] For further elaboration of this argument see Mühlenberg (1966, p. 101) and Gregory of Nyssa (1960b, p. 101, 13–14).

[65] For the argument based on God's simplicity, see Mühlenberg (1966, p. 125).

sect. 103, p. 38, 17–21). Thinking in notions always presupposes limitations in thinking (Mühlenberg 1966, pp. 102–4). Having proven the infinity of God on the bases of commonplace metaphysical assumptions of the Greek tradition, Gregory can claim that the whole logical deduction of Eunomius can no longer be maintained.

However, this whole argument would have remained unsatisfactory if Gregory had not been able to show directly that the divinity of the Son could be proven directly by applying the infinity of God in an argument. In fact, he used infinity to prove the compatibility of the ὁμοοῦσιος and the γεννητός of Christ by relating it to the concept of time and eternity.

First of all, he argued that time is created and belongs for this reason to the mundane realm, as Plato already stated.[66] As created time, it can be measured, and its limits can be determined in terms of beginning and ending. Thus, there is a mundane order of time. All this is not the case with eternity. It cannot be measured, and it cannot have a beginning and end; therefore, eternity has no order (Gregory of Nyssa 1960b, sect. 363, p. 134, 13–17). Eunomius made the mistake in his arguing – a category mistake in modern terminology – of holding that the γεννητός of the Son must have a beginning in time, with the consequence that the Son would be created (Gregory of Nyssa 1960b, p. 226, 7–8). That process applies mundane time to God's eternity. As a consequence, Gregory claims that the γεννητός of the Son cannot be understood in terms of a mundane time-causal relationship. On the contrary, as a form of infinity that is conceived as having no extension in time, eternity leads toward an interpretation of the γεννητός as a timeless eternal act that has no extension in time (Gregory of Nyssa 1960b, sect. 32, p. 226, 14–18). Therefore, the Son can be thought of as not being created and can uphold his divinity, combining the ὁμοοῦσιος and the γεννητός.

To sum up the results of Gregory's considerations, one can say that he argued for God's actual infinity from the standpoint of traditional Greek metaphysical claims about the nature of the divine. Furthermore, he used this argument to substantiate the divine nature of Christ. In addition, actual infinity cannot be grasped by human rationality. Hence, infinity was firmly introduced into theology as one of the major traits of God, but it was at the expense of its complete rational accessibility.

Gregory went a step further and asked in what way infinity can be conceived (Gregory of Nyssa 1960b, sect. 103, p. 39, 2sq; Mühlenberg 1966, p. 103). He concluded that the human mind (νοῦς) is not able to comprehend the infinity of God (Gregory of Nyssa 1960b, sect. 103, p. 38, 17–21; Mühlenberg 1966, p. 102).

However, he did not mean that there is no relation between the infinity of God and human beings. The ψυχή of humans is striving for God's infinity, not the νοῦς. In this sense, Gregory stresses the infinite way of the ascending soul to the infinite God.[67] Furthermore, this way is combined with the struggle for virtue (ἀρετή).[68]

To sum up, Gregory opened a number of new horizons (theological, spiritual, ethical, and intellectual) in his theology of infinity:

---

[66] Plato, *Timaeus* 52a–53a.

[67] "ὁδεύει πρὸς τὸ ἀόριστον" (Mühlenberg 1966, p. 159; Gregory of Nyssa 1960b, p. 247, 7–18).

[68] "The only destination of virtue is infinity" [τῆς δὲ ἀρετῆς εἷς ὅρος ἐστὶ τὸ ἀόριστον] (Mühlenberg 1966, p. 160; Karfíková 2001, pp. 47–81).

1. He firmly conceived God as infinite, and thus he overcame the view of the finite God of classical Greek metaphysics.
2. He raised the question as to how to relate to this infinite God.
3. He claimed that it is not the νοῦς but the ascending and purifying ψυχή and ἀρετή on the way to perfection that are the ways to approach the infinite God without ever reaching God.

Gregory's point of view gradually entered into the theological scholarship of the Eastern Church. His claim was taken up by John of Damascus (676–749) in his work *De Fide Orthodoxa*, and in this way it was transmitted to the Middle Ages.[69]

1.4.2.1.1. Ways to Relate to the Infinity of God. How is it possible to relate to the infinite God? In my opinion, it is possible to relate to the infinity of God in at least three ways:

1. Existential-religious by self-transcendence (Gregory of Nyssa, mystic ascension)
2. Symbolical-allegorical (Nicholas of Cusa)
3. Intellectual (Nicholas of Cusa, Georg Cantor)

These three modes of relating to infinity follow a logical sequence. It demonstrates how the relation to infinity emerged in the course of the historical development. In Gregory's writings we have already seen that his relation to infinity was both rational and also existential. He claimed that one comes nearer to God's infinity, but without attaining it, through the infinite process of purifying the ψυχή and the ἀρετή. His claim was different from Plotinus's account of reaching the "infinite one," the ἕν. Gregory had no concept of unification (ἕνωσις) with the infinite one in a movement of ecstasy (ἔκστασις), as did Plotinus, but rather he confined his claim to an infinite movement toward God's infinity (Böhm 1996, pp. 104–6).

However, is there a way from this existential movement toward infinity to an intellectual understanding, once an actual infinity was established by Gregory? The leap from existential commitment to intellectual understanding is not easy. This difficulty is proven by Dionysius the Areopagite, who was the most influential theologian of the early church. He was inspired by Gregory's theology of infinity (Völker 1955, pp. 200ff, 215), and he claimed that the infinity of God cannot be understood by thinking.

1.4.2.2 Actual Infinity and Apophatic Theology: Dionysius the Areopagite

There is neither a notion nor an intellectual concept that can cover all features of God's infinity; thus, Dionysius created apophatic theology, one feature of which is that all attempts to reach God's infinity are in vain. God is always ontologically (ὑπερουσίως) and epistemologically beyond the way humans think about God.

---

[69] In his work *De Fide Orthodoxa* John of Damascus refers to God as "unlimited," "infinite," and "unconceivable" (Damascenus 1955, I, 2; I, 4). Further, "infinite and inconceivable is the divine essence" (Damascenus 1955, I, 4).

In the locus classicus in his *Mystica Theologia V*, the Areopagite claimed that it is impossible to understand God in terms of human rationality.[70] This approach of the Areopagite seems to end up in an intellectual dead end: God's actual infinity cannot be discerned by the human intellect. On the contrary, one has to leave behind the limitations of thinking. One has to get rid of thinking itself in order to get just a glimpse of God's actual infinity. This claim of apophatic theology could have been the end of all intellectual attempts to understand God's infinity. How did this theological endeavor go on?

In the subsequent historical development, there was no further step in Eastern theology that went beyond the Areopagite. In Western theology, actual infinity, as a property of God, was developed in various contexts and by various theologians. One way was by arguing with the Aristotelian concept of potential infinity, once Aristotle was received in Western theology. This line was pursued by theologians such as Duns Scotus and St. Thomas Aquinas (Sweeney 1992a, pp. 413–37). Duns had a distinct concept of God's infinity, arguing against Aristotle that the infinity of God cannot be conceived of in a potential manner, because that would presuppose thinking about infinity in terms of quantity, which he thought to be inappropriate for God. God cannot be described in a quantitative way. Besides, he thought that the concept of infinity is superior to that of a first cause (Pannenberg 1988, p. 379; Dettloff 2002; Dreyer and Ingham 2003; Honnefelder 2005). Thus, the nominalistic-voluntaristic tradition of thinking about God's infinity emerged, which was further pursued by William of Occam and John Buridan.

St. Thomas Aquinas contends the infinity of God (Aquinas 1934, I, 7, 1–4), arguing against the negative attitude of ancient Greek philosophy and toward and for an actual infinity. He combines the concept of infinity of the early Church with Greek hylomorphism, and he asks the question as to whether God can be perfect if he is infinite. In this way he takes up Greek thinking that perfection can only be thought of as being related to finite entities. His solution is that God's infinity can be reconciled with his perfection only if one thinks of God as being pure form, which is possible. Therefore, one can argue that Thomas transformed the Greek's mutual exclusion of perfection and infinity on the basis of his understanding of God as pure form. In this sense St. Thomas had a weak understanding, so to speak, of God's actual infinity.[71] It is an understanding of infinity that stresses infinity as a quality, or more precisely as a quality belonging to

---

[70] "Far more ascending we proclaim now that he, the first principle is neither soul nor spirit. He has no power of imagination nor opinion nor reason nor recognition (νόησις). God cannot be expressed in words nor can he be understood by thinking. He is neither number, nor order (τάξις), neither magnitude nor minimum, neither equality nor non-equality, neither similarity (ὁμοιότης) nor non-similarity. (. . .) He is not being (οὐσία), not eternity, not time. He cannot be understood by thinking, he is not knowledge (ἐπιστήμη), not truth, not dominion, not wisdom, not one (ἕν), not unity (ἑνότης), not god-likeness (θεότης), not mercy, not spirit (πνεῦμα) as we understand it. (. . .) There is no word (λόγος), no name, no knowledge (γνῶσις) about him" (Dionysios Areopagita 1956, V, 1045D–1048B, p. 171; Ritter and Heil 1991).

[71] "Unde infinitum secundum quod se tenet ex parte formae non determinatae per materiam, habet rationem perfecti. Illud autem, quod est maxime formale omnium, est ipsum esse, ut ex superioribus patet. Cum igitur esse divinim non sit esse receptum in aliquot, sed ipse sit suum esse subsistens, ut supra ostensum est, manifestum est quod ipse Deus est infinitus et perfectus" (Aquinas 1934, I, 7, 1).

God's essence. In this sense, Thomas rejects an interpretation of infinity as a quantity, as it was understood by ancient philosophers (Aquinas 1934, I, 7, 1). He does not refer to the theological heritage of Gregory and the Areopagite, but to John of Damascus (Johannes Damascenus 1955, I, 2, 4).[72]

Turning to the question as to whether or not infinity is possible in the world or in mathematics, Thomas changes from a qualitative notion of infinity to a quantitative notion of infinity. In this context, his distinction between essence and existence is operative. In God, essence and existence coincide; therefore, infinity can be attributed to God.[73] However, this coincidence of essence and existence is not applicable to his creatures. Therefore, in the final analysis infinities cannot be infinities in creation.[74] He thus rejects not only the existence of infinite magnitudes in the finite creation[75] but also their existence as infinite magnitudes in geometry (Aquinas 1934, I, 7, 3)[76] and actual infinity in mathematics.[77] Thus, he did not contribute to science by his understanding of infinity.

Basically, it was the simultaneous application of the distinction of esse and essentia both to the infinite God and to the creatures that led St. Thomas Aquinas to the conviction that neither in the world nor in mathematics can an actual infinity be possible. Then, of course, the whole picture changes as soon as the God-world relation is no longer seen in the categories of essence and esse. In fact, this is the case with Nicholas of Cusa.

Besides the huge Aristotelian influence on Western rational theology, there has also been a tremendous influence by the Areopagite on Western theology. The next decisive step toward an intellectual clarification of actual infinity was made by a German adherent and admirer of the Areopagite. This step is found in the contribution by Nicholas of Cusa, whose thought will now be examined.

---

[72] "Sed contra est quod Damascenus dicit (De fide orthodoxa, lib. 1, cap. 4): 'Deus est infinitus et aeternus et incircumscriptibilis'" (Aquinas 1934, I, 7, 1).

[73] "Ad primum sic proceditur. Videtur quod aliquid aliud quam Deus posit esse infinitum per essentiam. Virtus enim rei proportionatur essentiae ejus. Si igitur essential Dei est infinita, oportet quod ejus virtus sit infinita. Ergo potest producere effectum infinitum, cum quantitas virtutis per effectum cognoscatur" (Aquinas 1934, I, 7, 2).

[74] "Ad primum ergo dicendum quod hoc est contra rationem facti, quod essential rei sit ipsum esse ejus, quia esse subsistens non est esse creatum: unde contra rationem facti est, quod sit simpliciter infinitum. Sicut igitur Deus, licet habeat potentiam infinitam, non tamen potest facere aliquid non factum, hoc enim esset contradictoria esse simul; ita non potest facere aliquid infinitum simpliciter" (Aquinas 1934, I, 7, 2 ad1).

[75] Again he relates infinity to the hylomorphism, arguing that matter in no way can be conceived of as infinite, whereas only form might have an infinite magnitude. Real infinity thus can be attributed only to God: "dicendum quod aliquid praeter Deum potest esse infinitum secundum quid, sed non simpliciter" (Aquinas 1934, I, 7, 2 resp.).

[76] St. Thomas Aquinas refers to geometrical objects when he claims that in mathematics there are no infinite magnitudes: "Unde, cum forma quanti, inquantum hujusmodi, sit figura, oportebit, quod habeat aliquam figuram. Et sic erit finitum: est enim figura, quae termino vel terminis comprehenditur" (Aquinas 1934, I, 7, 3).

[77] "Ad tertium dicendum quod, licet quibusdam positis, alia poni non sit eis oppositum: tamen infinita poni opponitur cuilibet speciei multitudinis. Unde non est possible esse aliquam multitudinem actu infinitam" (Aquinas 1934, I, 7, 4 ad tertium).

1.4.2.3 Actual Infinity and the Watershed of Its Rational Understanding:
Nicholas of Cusa

Nicholas of Cusa was deeply influenced by Pseudo-Dionysius. He was the author most often quoted in Nicholas of Cusa's writings.[78] He was especially influenced by the Areopagite's approach to apophatic theology. First of all, Nicholas of Cusa agreed with the Areopagite that God is infinite.[79,80] Secondly, he also equated God's infinity with God's unity, which means that God's infinity and unity are identical (Schulze 1978, p. 95).[81,82]

Now the question is in what way he differed from the tradition of apophatic theology. That is, how did he differ with respect to the constraints that the limitations of *language*, the *categories*, and *logic* put on the human ability to have access to infinity?

Regarding *language,* Nicholas of Cusa agreed with Pseudo-Dionysius, who said in his early writing about the "docta ignorantia" that God cannot be reached by means of language.[79]

With regard to the applicability of the Aristotelian *categories* to infinity, he raised the question as to how to relate the category of quantity to infinity. He argued that infinity defies any characterization in terms of the category of quantity.

Furthermore, there is a question as to how traditional logic, or other axioms of reasoning in the finite realm, such as "the whole is larger than the parts," can be applied to infinity (Knobloch 2002, p. 227). Are the rules of logic valid in an infinite realm, which at the same time – as we have seen – coincides with unity? For instance, how can the rule of contradiction be used in an infinite unity, given that contradiction requires at least two propositions, which cannot be found in a nonquantitative infinite unity?[80]

Despite these obstacles to a rational account of infinity, however, Nicholas of Cusa went beyond the apophatic tradition, insofar as he attempted by means of symbolic expressions to find an intellectual understanding of God's infinity. This development must be explained in more detail, because this innovative intellectual and theological endeavor exercised an enormous impact on the subsequent understanding of infinity, as well as on natural science and mathematics.

1.4.2.3.1 Nicholas of Cusa's New Approach to Infinity. First of all, Nicholas of Cusa held that God is infinite, just as Gregory and the Areopagite did.[81] However, he introduced a new and quite innovative thought about this infinity, which is not found in the writings of either Gregory or Pseudo-Dionysius. He himself described this thought as "never heard before" ("*prius inaudita*"; Nikolaus von Kues 1964, book II, chap. 11,

---

[78]   For example, in "docta ignorantia" (Nikolaus von Kues 1964, book I, chaps. 18, 24, 26).

[79]   "Docuit nos sacra ignorantia Deum ineffalbilem; et hoc, quia maior est per infinitum omnibus, quae nominari possunt; et hoc quidem quia verissimum, verius per remotionem et negationem de ipso loquimur, sicuti et maximus Dionysius, qui eum nec veritatem nec intellectum nec lucem nec quidquam eorum, quae dici possunt, esse voluit" (Nikolaus von Kues 1964, book I, chap. 26, p. 292).

[80]   "Infinitas est ipsa simplicitas omnium, quae dicuntur, contradiction sine alteratione non est. Alteritas autem in simplicitate sine alteratione est, quia ipsa simplicitas" (Nikolaus von Kues 1967, chap. 13, p. 148).

[81]   For literature on Cusanus's concept of infinity, see Enders (2002).

[82]   "Debet autem in his profundis omnis nostril humani ingenii conatus esse, ut ad illam se elevet simplicitatem, ubi contradictoria coincident" (Flasch 1998, p. 46).

p. 388; cf. Flasch 1998, p. 47). He qualified the infinity of God as the "coincidentia oppositorum," the falling together of contradictions or opposites.[82] In different writings he approached infinity from different angles, such as language, the category of quantity, and even logic. In his "docta ignorantia," he related general notions of language, such as "maximum" and "minimum," to the "coincidentia oppositorum."[83] Elsewhere, he associated infinity with the category of quantity, claiming that infinity is not a matter of quantification.[84]

There is an inherent limitation to understanding infinity with the category of quantity. It is a limitation because this category cannot be applied ontologically to infinity, and also because human rationality operates epistemologically within the category of quantity. For example, notions such as "bigger," "smaller," and "unequal" cannot be related to infinity.[85] Infinity cannot be quantified, as it is beyond the category of quantity.[86] This understanding also puts constraints on mathematics in its reach for infinity, because all mathematical operations are based on this very category of quantity (Nikolaus von Kues 1967, pp. 362, 364; cf. Knobloch 2002, p. 225). There is no way from quantity to infinity;[87] therefore, infinity, the maximum, and the minimum are all transcendent terms.[88] Ontologically, there is no relation of finiteness – finite quantity – to infinity. Therefore, he objected to Aristotle's understanding of infinity as potential infinity, because that is based on an infinite progression of quantities (Knobloch 2002, p. 228).[89]

Epistemologically, infinity has no relation to our rationality ("intellectus"), which works by measuring quantities and finding proportions.[90] Infinity cannot be measured. However, it is the measure of everything else.[91] Furthermore, in considering the relation of infinity to logic, especially to the principle of contradiction, Nicholas of Cusa claimed that the infinite unity defies any logical treatment, because logical procedure requires at least a duality of entities, which is not the case in a unity.[92] Much more could be

---

[83] "Maximum itaque absolutum unum est, quo de omnia; in quo omnia, quia maximum. Et quoniam nihil sibi opponitur, secum simul coincidit minimum" (Nikolaus von Kues 1964, book I, chap. 2, p. 198).

[84] "Quia infinitum non habet partes, in quo maximum coincidit cum minimo" (Nikolaus von Kues 1964, book I, chap. 17, p. 246).

[85] "Infinitum ergo nec est dato quocumque aut maius aut minus aut inaequale. Nec propter hoc est aequale finito, quia est supra omne finitum, hoc est per se ipsum, tunc infinitum est absolutum penitus et incontrahibile" (Nikolaus von Kues 1967, chap. 13, p. 152).

[86] "Infinita quantitas non est quantitas, sed infinitas" (Nikolaus von Kues 1967, chap. 13, p. 152).

[87] "Hoc ex regula doctae ignorantiae constat, quae habet, quod in recipientibus magis et minus non est devenire ad maximum et minimum simpliciter" (Nikolaus von Kues 1967, p. 326).

[88] "Maximum autem et minimum, (...), transcendentes absolute significationis termini existunt, ut supra omnem contractionem ad quantitatem molis aut virtutis in sua simplicitate absoluta omnis complecantur" (Nikolaus von Kues 1964, book I, chap. 4, p. 206).

[89] "Finiti ad infinitum nulla est proportio" (Nikolaus von Kues 1967, chap. 23, p. 200).

[90] "Quomodo potest intellectus te capere, qui es infinitas? Scit se intellectus ignorantem et te capi non posse, quia infinitas es. Intelligere enim infinitatem est comprehendere incomprehensibile" (Nikolaus von Kues 1967, chap. 13, p. 148).

[91] "Infinitum non est mensurabile, quia infinitum est interminum. Non igitur potest claudi terminis cuiuscumque mensurare, sed ipsum est mensura omnium" (Nikolaus von Kues 1967, chap. 11, p. 686).

[92] "Infinitas est ipsa simplicitas omnium, quae dicuntur, contradiction sine alteratione non est. Alteritas autem in simplicitate sine alteratione est, quia ipsa simplicitas" (Nikolaus von Kues 1967, chap. 13, p. 148).

said about the appropriate interpretation of this thought. I skip it here and proceed to Nicholas of Cusa's own attempt at illustrating what he meant. He claimed that this "coincidentia oppositorum," as a feature of infinity, by far transcends the power and capability of human reason.[93] However, although human reason could not attain the "coincidentia oppositorum" in the infinity of God, there are other means of finding intellectual access to this realm. It is the way of symbolic illustration.

1.4.2.3.2 Symbolic Mathematical Illustration of Infinity: The First New Epistemological Approach. Symbolic illustration has to be explained in some detail. First of all, Nicholas of Cusa claimed that mathematics is most helpful in understanding God's infinity.[94] He justified this approach by an allusion to the works of Boethius and St. Augustine.[95] Nicholas elaborated this approach in his early writing *De Docta Ignorantia*, which was written about 1438 to 1439. However, the question remains as to how mathematics, which deals with finite quantities, can symbolize the infinity of God.[96] Nicholas of Cusa answered that it is necessary, first of all, to understand the mathematical properties of finite mathematical objects, and then to analyze the way in which these properties change if one increases these objects to infinity.[97]

As examples, he took a triangle and a circle. He argued that the infinite line, the circle, and the triangle coincide in infinity. The example of the circle is easy to understand. Nicholas of Cusa just took a segment of a circle and extended the diameter of the circle. This simple operation makes the circle infinite in diameter and makes the circumference a straight line.

By this illustration Nicholas of Cusa tried to show that opposites, in this case of a straight and a curved line, coincide and fall together in infinity.[98] He concluded after he made these illustrations.[99] In his early writing of the *De Docta Ignorantia*, he used this kind of mathematical symbolism to illustrate God's infinity and the "coincidentia oppositorum." However, in his later writings, which were more devoted to purely mathematical considerations, he tried to apply his thought of the "coincidenita

---

[93] "Hoc autem omnem nostrum intellectum transcendit, qui nequit contradictoria in suo principio combinare via rationis, quoniam per ea, quae nobis a natura manifesta fiunt, ambulamus; quae longe ab hac infinita virtute cadens ipsa contradictoria per infinitum distantia connectere simul nequit" (Nikolaus von Kues 1964, book I, chap. 4, p. 206).

[94] Nikolaus von Kues wrote eleven papers on mathematics within a period of 14 years (1445–59) (Flasch 1998, p. 171).

[95] "Ita ut Boethius, ille Romanorum litteratissimus, asseret neminam divinorum scientiam, qui penitus in mathematicis exercition careret, attingere posse" (Nikolaus von Kues 1964, book I, chap. 11, pp. 228–30).

[96] The three important works on the history of mathematics in Nikolaus von Kues's writings are Ubinger (1895, 1896, 1897), Nikolaus von Kues (1979), and Knobloch (2002).

[97] "Si finitis uti pro exemplo voluerimus ad maximum simpliciter ascendendi, primo necesse est figures mathematicas finitas considerare cum suis passionibus et rationibus, et ipsas rationes correspondenter ad infinitas tales figures transferre, post haec tertio adhuc altius ipsas rationes infinitarum figurarum transsumere ad infinitum simplex absolutissimum etiam ab omni figura" (Nikolaus von Kues 1964, book I, chap. 12, p. 232).

[98] "Unde hic videtur magna speculation, quae de maximo ex isto trahi potest; quomodo ipsum est tale, quod minimum est in ipso maximum, ita quod penitus omnem oppositionem per infinitum supergreditur" (Nikolaus von Kues 1964, book I, chap. 16, p. 242).

[99] "Adhuc circa idem: Linea finitia est divisibilis et infinita indivisibilis, quia infinitum non habet partes, in quo maximum coincidit cum minimo" (Nikolaus von Kues 1964, book I, chap. 17, p. 246).

1	2	3	4	5	6	7	8	9	10	11	12
1	4	9	16	25	36	49	64	81	100	121	144

Figure 1.1

oppositorum" to purely mathematical problems.[100] It can be seen in Nicholas of Cusa's work *De Mathematica Perfectione* (Nikolaus von Kues 1979, pp. 160–77). In this work he used the concept of "coincidentia oppositorum" as a tool for creating the new mathematical procedure of infinite approximation.

1.4.2.3.3 Approximate Mathematical Illustration of Infinity, Second New Epistemological Approach.

In his mathematical work *De Mathematica Perfectione*, Nicholas of Cusa gave a more elaborate account of his metaphysical principle concerning infinity. In fact, he was quite ambitious in wanting to bring mathematics to its perfection and completion by the application of the "coincidentia oppositorum."[101] He accomplished it by an approximate process of straightening a segment of a circle into a straight line.

By means of this new methodological approach of infinite approximation, he tried to calculate the circumference of a circle (Cantor 1892, pp. 176ff). In this way, infinity had become for him a methodological tool in mathematics. This elaboration of approximate processes in mathematics corresponded completely with his understanding of epistemology as an approximate process toward truth. Nicholas of Cusa substituted an infinite process for St. Thomas Aquinas's famous definition of truth as "veritas est adaequatio rei et intellectus." The epistemological consequences can hardly be overestimated, because, in the final analysis, he liberated rationality in theology from the trap of apophatic theology. Apophatic theology claimed that the infinite God is not accessible rationally. Therefore, rationality was not esteemed very highly in apophatic theology. Nicholas of Cusa, however, preserved the place of rationality in theology by transforming the ontological infinity of God into an infinity of epistemological processes, in mathematics, physical sciences, as well as philosophy. Thus, he created the epistemological prerequisites of modern natural science.

1.4.2.3.4 Infinity and the Real Numbers.

In dealing with the question of the applicability of the category of quantity to infinity, Nicholas of Cusa argued – as we have seen – that infinity defies being measured in terms of "bigger" or "smaller," or, to put it differently, he argued for the uniqueness of infinity (Knobloch 2002, p. 231). In this respect, he preceded the insight of Galileo Galilei – who might have been familiar with Nicholas of Cusa's writings – that the integers are as many as the numbers of the power two, although the numbers of the power two are obviously less than the integers (Rucker 1989, p. 19; Knobloch 2002, pp. 231ff).

Galileo Galilei could not solve this conceptual difficulty of comparing two infinite sets by using the quantitative comparison of "bigger" and "smaller" ($>$, $<$), other than

---

[100] In "de coniecturis" Nikolaus von Kues distinguishes between intellectus and ratio, which goes hand in hand with the distinction of philosophy and mathematics. However, he nevertheless uses his "coincidentia oppositorum" as a methodological tool in his later mathematical writings (Flasch 1998, p. 171).

[101] "My ambition is to bring mathematics to its completion by 'coincidentia oppositorum'" (Cohn 1960, p. 92; Nikolaus von Kues 1979, p. 161).

arguing that infinity is a paradoxical concept, because both sets are obviously equal, but at the same time the integers are more than the numbers of the power two.[102] This view opened up a completely new conceptual space, into which such different figures as Galileo Galilei, Bernard Bolzano, Georg Cantor, Kurt Gödel, and Gerhard Gentzen (transfinite induction) entered.

1.4.2.3.5 "Coincidentia Oppositorum" and the Relativity of Motion. Nicholas of Cusa's concept of the "coincidentia oppositorum" in infinity had an impact not only on mathematics but also on the physical sciences. This impact of his "coincidentia oppositorum" in infinity on the physical universe is due to a major shift in his teaching. He no longer held that only God is infinite, but because the world is the mirror of God,[103] therefore the world must also be conceived as infinite, especially with respect to space and motion. Nicholas of Cusa argued that the world cannot have a center, because in an infinite world the center coincides with the circumference.[104] A consequence is that the world must be regarded as infinite with respect to space. It also means, of course, that the earth can no longer be the center of the world – as taught by Aristotle – and it must have some kind of motion.[105] Thus, in the final analysis, Nicholas of Cusa's concept of infinity led to the concept of the relativity of motion, because a central point of reference in the world is denied. The principle of the relativity of motion was invented by purely philosophical considerations of infinity about 800 years before it was formulated by Albert Einstein, who saw it as a principle of the physical sciences.

To sum up, it can be argued that Nicholas of Cusa overcame the intellectual dead end of apophatic theology, by trying to illustrate infinity by means of mathematical symbols, and by applying the new thought of "coincidentia oppositorum" to God's infinity. However, this statement is only half of the truth. By also applying this new concept of "coincidenita oppositorum" to scientific problems, he paved the way to a kind of secularization of infinity. In different ways it became a scientific concept. The examples are (1) infinity of space, (2) the relativity of motion, (3) the approximate processes in mathematics and epistemology, and (4) the conceptual difficulties of dealing with infinity in terms of quantity. Thus, he paved the way to the modern sciences by thinking about infinity in many different ways.

1.4.2.4 The Mathematics of Actual Infinity: Georg Cantor

The chapter will conclude with some remarks about Georg Cantor and his achievements in giving a rational, mathematical account of infinity. Until Georg Cantor in the

---

[102] Later Cantor would argue that, in infinite sets, other arithmetic rules must be observed.

[103] "Consensere omnes sapientissime nostri et divinissimi doctores visibilia veraciter invisibilium imagines esse atque creatorem ita cognoscibiliter a creaturis videri posse quasi in speculo et in aenigmate" (Nikolaus von Kues 1964, book I, chap. 11, p. 228; Cassirer 1991, p. 24); "Quis melius sensum Pauli quam Paulus exprimeret? Invisibilia alibi ait aeterna esse. Temporalia imagines sunt aeternorum. Ideo si ea, quae facta sunt, intelliguntur invisibilia Dei conspiciuntur uti sunt sempiternitas virtus eius et divinitas. Ita a creatura mundi fit Dei manifestatio" (Nikolaus von Kues 1966, p. 270; Cohn 1960, p. 88).

[104] "Centrum igitur mundi coincidit cum circumferentia" (Nikolaus von Kues 1964, book II, chap. 11, p. 390).

[105] "Terra igitur, quae centrum esse nequit, motu omni career non potest" (Nikolaus von Kues 1964, book II, chap. 11, p. 390).

nineteenth century, the best minds in philosophy, natural sciences, and mathematics did not accept the concept of actual infinity; only the idea of potential infinity (in the sense of Aristotle) was accepted. The towering minds of people such as Galileo Galilei, Gottfried Wilhelm Leibniz, Baruch Spinoza, Isaac Newton, and even Carl Friedrich Gauss rejected actual infinity. It was rejected because they thought that this concept included antinomies, such as the one Galileo Galilei discovered. However, Bernard Bolzano (1955) showed that most of the antinomies of actual infinity could be reduced to paradoxes, and they could be resolved in a logical sense. In fact, antinomies would have been disastrous for mathematics, because they entailed the possibility of proving all mathematical conjectures, including those that are obviously, false.

Georg Cantor revived the tradition of Nicholas of Cusa, to whom he even alluded in the endnotes of his *Grundlagen*. Cantor is in fact the first person who claimed – despite the Kantian philosophy that prevailed then – that actual infinity could be an object of mathematical research. He said that the human *ratio* could create conceptual tools in order to discern its internal structure. Georg Cantor actually proceeded to do it. Here are a few important examples to show how Georg Cantor made infinity a subject of rational research. These examples fall into three different categories: (1) rational discernment of infinity, (2) real antinomies (logical contradictions), and (3) resolving the antinomies.

1.4.2.4.1 Rational Discernment of Infinity. Let me start with the *rational discernment* of the internal structure of infinity. Georg Cantor created a new kind of number. He defined the infinite number of the integers N as a new number that he called $\aleph_0$, or the first transfinite set, or the first cardinal number. This new number $\aleph_0 = N = \{1, 2, 3, \ldots, n\}$ serves as a kind of mathematical measurement device for the internal structure of infinity. This means that he substituted the "bigger" and "smaller" relation ($<, >$), and that was used by Galileo to compare the integers with the numbers of the power two, and that comparison created the paradox that was seen earlier. Thus, an infinite set itself became a mathematical measuring device for other infinities.

By relating $\aleph_0 = N = \{1, 2, 3, \ldots, n\}$ to the rational numbers Q and the real numbers R, he could show that these sets N and Q, on the one hand, and R, on the other hand, had a different cardinal number ("Mächtigkeit"). N and Q are countable infinities, whereas R is a noncountable infinity.[106] He called R the next higher cardinal number ("Mächtigkeit") $\aleph^*$ and showed that $\aleph^* > \aleph_0$. He also raised the question as to whether, between $\aleph^*$ and $\aleph_0$, there is another set of numbers with a specific cardinal number ("Mächtigkeit"), known as the continuum hypothesis. This hypothetical set he called $\aleph_c$. He showed that this hypothesis could be formulated by the mysterious relation $\aleph_c = 2^{\aleph_0}$. He also created an arithmetic for transfinite numbers, in which the commutative law was not valid. Then, he created higher orders of the $\aleph_s$, such as $\aleph_1, \aleph_2, \aleph_3, \ldots, \aleph_n$, and even $\aleph_n$.

With the help of this new concept of cardinal number ("Mächtigkeit"), he dealt with the paradox already noticed by Galileo Galilei, that a subset of N, like the integers of the

---

[106] This distinction within infinity was partly anticipated by Bernard Bolzano. He spoke about different "orders" of infinity. However, he did not arrive at Cantor's clear-cut notion of "Mächtigkeit" (cf. Bolzano 1955, §§ 21, 29, 33).

power 2, is equivalent to N itself. It is no longer a paradox, if one uses the N as a mathematical device to measure infinite sets. With this new device for mathematical measurement, he showed that the relations "bigger," "smaller," and "equivalent" $\{>, <, =\}$ could still be used in infinite sets without producing paradoxes. To put it differently, the arithmetic of transfinite sets obeys the following laws:

$$\aleph_0 + \aleph_0 = \aleph_0 \quad \text{and} \quad \aleph_0 \times \aleph_0 = \aleph_0.$$

It is a contradiction in finite sets. Another logical discernment of infinity was Georg Cantor's discovery that the power set of a cardinal number always had a higher cardinal number than the set itself:

$$\text{Card}\,(P(A)) > \text{Card}\,(A)$$

1.4.2.4.2 Real Antinomies. Georg Cantor discovered two different antinomies in his system, but he was not very concerned about them, because he thought he could resolve them. One of them was related to the idea of the totality of all cardinal numbers. How can one conceive of the set of all transfinite sets? He called this set ⊓, or absolute infinity, which means $⊓ = \aleph_1, \aleph_2, \aleph_3, \ldots, \aleph_n$. He showed that the set of all cardinal numbers results in a contradiction, which violates the logical consistency of all his mathematics, and which brings disastrous effects on the logical foundation of mathematics. To explain it in more detail, we can use the insight of Georg Cantor, mentioned earlier, that the cardinality of the set of all subsets of a given set is bigger than the cardinality of the set itself. As soon as one applies this insight to the absolute infinity ⊓, looking for the cardinality of the sets of all subsets of ⊓ becomes a logical contradiction, and an antinomy occurs. Let's look at that in more detail.

Let ⊓ be the set of all sets. According to Georg Cantor's law of the cardinality of subsets [Card (P (A)) > Card (A)], it follows that

$$\text{Card}\,(p(⊓)) > \text{Card}\,(⊓)$$

On the other hand, as the set of all sets, ⊓ includes the power set (Potenzmenge) P (⊓). A consequence for the cardinality is that

$$\text{Card}\,(⊓) > \text{Card}\,(P(⊓))$$

Obviously, we have now created an antinomy in the very heart of Georg Cantor's notation of infinity. Two contradicting assertions about infinity occurred by applying the Cantorian theorem regarding the cardinality of sets and subsets to infinity, which are

$$\text{Card}\,(⊓) > \text{Card}\,(P\,(⊓))$$

and

$$\text{Card}\,(P(⊓)) > \text{Card}\,(⊓)$$

It must have been a disaster to have such a logical contradiction in the heart of the theory. Although Georg Cantor knew about this contradiction as early as 1895, he did not care about it very much. The reason he did not care very much sheds light on his understanding of infinity and on the way he resolved this contradiction.

Table 1.2

Levels of Infinity	Epistemology	Ontology
Absolute Infinity (Actual Infinity) Georg Cantor: ה Nicholas of Cusa: "coincidentia oppositorum" The Areopagite, Gregory of Nyssa: apophatic theology	Intuition (νόησις)	Being
The Transfinite (Actual Infinity: $\aleph_0, \ldots,$ $\aleph_1, \ldots, \aleph_w$): Georg Cantor	Discursive quantitative rationality (διάνοια)	Becoming
Potential Infinity: Aristotle	Discursive quantitative rationality (διάνοια)	Becoming
Finiteness (πέρας), to be avoided due to lack of form: ἄπειρον	Sensual experience (αἴσθησις, δόξα)	Phenomenal world of senses

1.4.2.4.3 Resolving the Antinomy. Georg Cantor resolved this contradiction by claiming that ה, the absolute infinity, cannot be an object of *quantitative, discursive, rational* operation. It cannot be understood by logical discernment, but only by intuitive insight.[107] Furthermore, it cannot be recognized, for it can only be accepted without any further discursive rational activity and logical discernment.[108] "Logical discernment" in this context intended to show that he did not conceive of ה = {$\aleph_0, \ldots, \aleph_1, \ldots,$ $\aleph_n$} as a set. Instead, he called it an "inconsistent plurality," to which his theorem of the cardinality of sets could not be applied without creating logical inconsistencies.[109] Hence, he avoided the logical contradictions of ה by excluding it from being part of sets. Instead, he created another type of sets, which he called the "inconsistent plurality." However, this logical differentiation between *sets* and *inconsistent pluralities* was more than just a formal logical operation.

This ה, Georg Cantor claimed, is God, the creative source of all quantities existing in the world, and an intuitive insight of God is possible. It was the transformative experience of this ה that helped Cantor, according to his own words, to find the transfinite numbers with all their strange mathematical properties.[110]

The following Table 1.2 summarizes all the epistemological, ontological, and mathematical levels of infinity and the way they are related to each other:

---

[107] Georg Cantor made an allusion to this kind of *intuitive insight* in a letter to Philip Jourdain from 1903: "I have 20 years ago *intuitively* realized (when I discovered the Alefs themselves) the undoubtedly correct theorem, that except the Alefs there are no transfinite cardinal numbers" (Bandmann 1992, p. 282).

[108] "The absolute infinity can only be accepted, but not be recognized, not even nearly recognized" (Bandmann 1992, p. 285).

[109] More explicitly, Georg Cantor worked with two theorems to avoid the contradiction. Theorem A: The system ה of all $\aleph$'s is in its extension similar to Ω and is for this reason also an inconsistent plurality. Theorem B: The system ה of all alephs is noting else as the system of all transfinite cardinal numbers (Bandmann 1992, p. 281). In a letter from 1897 to David Hilbert, Georg Cantor introduced this distinction between normal sets and inconsistent pluralities. He wrote, "The totality of all Alefs is a totality which cannot be conceived as a distinct well-defined set. If this was the case, it would entail another distinct Alef following this totality, which would at the same time belong to this totality and not belong to it. This would be a contradiction. Totalities, which cannot be conceived from our perspective as sets (...) I called already years ago absolute infinite totalities and have distinguished them very clearly from the transfinite sets" (Bandmann 1992, p. 287).

[110] For the purposes of this chapter, the important publications, in which Cantor develops his thoughts about infinity, are Cantor (1883, 1932).

This diagram reveals to us that thinking and experiencing infinity also provide a story of liberation. The step from finiteness (πέρας) to potential infinity and the transfinite is associated with the liberation from a purely sensual encounter of the world in favor of a rational relation. The step from the transfinite to absolute infinity is the liberation from purely rational quantitative thinking to the intuitive insight of the unity and infinity of the all-encompassing infinity of God. One can get a vague intuitive glance of God, but not a rational account.

1.5 Infinity in Modern Theology and the Evolution of the Religious Self

Ever since the concept of infinity was introduced in theology as a property of God, both from the apophatic tradition and from the received Aristotelian tradition, theologians have refused to think about the infinity of God in terms of quantity, such as it is obviously done in mathematics. If theologians think about the infinity of God, they do so in stressing that it has to be understood in a qualitative manner, but they do so without clearly defining what that means (Pannenberg 1988, p. 430).[111] Thus, the concept of infinity was not elaborated as a property of God, and it never attained great prominence. This situation may be due to the conceptual difficulties of giving a rational account of infinity that is free of any relation to quantity.[112] This difficulty may also explain why infinity as a trait of God is not mentioned at all in the theology of Karl Barth. As we have seen, however, it proved to be very fruitful as it migrated to other fields.

However, the concept of infinity became a stimulating idea in modern theology dating back to the Enlightenment. Since the Enlightenment, a new thread of theological reasoning has emerged. Starting with Friedrich Daniel Ernst Schleiermacher (1768–1834), infinity again became a sphere of religious experience, and so it was made a feature of religious anthropology, rather than a property of God.[113] One can speak of an "anthropological turn" concerning the theological teachings about infinity. It will be argued here that, since Schleiermacher, the idea of infinity has helped to establish the idea of the "religious Self." One can distinguish three phases. The first phase can be called the phase of "discovering the religious Self," and it is associated with the work of Schleiermacher. The second phase can be called "constituting the religious Self," and it is associated with Søren Kierkegaard. The third phase can be called the phase

---

[111] This may be due to a remark of the philosopher Georg Wilhelm Friedrich Hegel, who regarded infinity as a quantity in the realm of finite entities, which can be understood by human reason, as "bad infinity." As opposed to this kind of bad infinity, real infinity is a qualitaitve property of God: "Nach dieser Betrachtung sond wohl zu unterscheiden die zwei Unendlichkeiten: die wahre und die bloß schlechte des Verstandes. So ist denn das Endliche Moment des göttlichen Lebens" (Hegel 2000, p. 192).

[112] An exception in this regard is the work of Bernard Bolzano. He thinks about the properties of God in terms of quantity and rejects Hegel's objections about a quantitative notion of infinity. He attributes to God properties such as power, knowledge, volition, and truth in terms of quantity (see Bolzano 1955, sects. 11, 25).

[113] In both editions of the Glaubenslehre, Schleiermacher regards infinity as an arguable property of God and treats this question only in an appendix of his teaching about the properties of God.

of "transcending the religious Self," and it can be associated with the work of Paul Tillich.[114]

1.5.1 Friedrich D. E. Schleiermacher and the Discovery of the Religious Self

This discovery can be traced back as early as Schleiermacher's *Reden* from 1799 and in his *Monologen* from 1800,[115] both being inspired by his early study of Spinoza,[116] as well as by his early links to German romanticism. In the context of his new understanding of religion as a"taste for infinity,"[117] based on experience, and as "sentiment,"[118] he identified infinity as a major trait of this religious anthropology. There is a special focus on infinity in the second of his five *Reden*, in which he clearly associates infinity with intuition (Anschauung) and sentiment (Gefühl). He complains that his adversaries are lacking this higher form of religious life.[119] This second "Rede" culminates in the famous last sentence, in which he intertwines infinity with eternity and

---

[114] The history of the religious Self has already been studied by Mark C. Taylor; however, he confines his study to the relation between Kierkegaard and Hegel and omits the importance of the concept of infinity for this process (see Taylor 1980).

[115] Schleiermacher's understanding of religion is a matter of a long discussion and controversy. In our context it is not possible or necessary to enter into this controversy.

[116] In the early working period of his youth, between 1787 and 1796, Schleiermacher elaborated two treatises on Spinoza, in which the concept of infinity as a property of God plays an important role. The first has the title "Spinozismus" and dates from probably 1793–94 (Schleiermacher 1984, pp. 515–58). The second one has the title "Kurze Darstellung des Spinozistischen Systems," probably also written in 1793–94 (Schleiermacher 1984, pp. 559–82). His thoughts on divine infinity are based on Spinoza's proposition XLVI, "Cognitio aeternae, et infinitae essentiae Dei, quam unaquaeque idea involvit, est adaequata, et perfecta," and proposition XLVII, "Mens humana adaequatum habet cognitionem aeternae, et infinitae essentiae Dei," in his *Tractatus de Intellectus Emendatione, Ethica*. Especially interesting is that Spinoza claims in proposition XLVII that the human mind has access to the infinite essence of God, an idea that will occur later in Schleiermacher's *Reden* in 1799 and *Monologen* in 1800. In particular, he evokes Spinoza in his second "Rede": "Opfert mit mir ehrerbietig eine Locke den Manen des heiligen verstoßenen Spinoza! Ihn durchdrang der hohe Weltgeist, das Unendliche war sein Anfang und Ende" [Sacrifice with me reverently a curl to the Mans of the holy and outcasted Spinoza! He was permeated by the high Weltgeist [world spirit, universal spirit], the infinite was his beginning and end] (Schleiermacher 1967, p. 52).

[117] Schleiermacher identifies religion as one form of innate human activities: "Praxis ist Kunst, Spekulation ist Wissenschaft, Religion ist Sinn und Geschmack fürs Unendliche" [Practise is artistry, speculation is science, religion is sense and taste for infinity] (Schleiermacher 1967, p. 51).

[118] "Ihr Wesen (=Religion) ist weder Denken noch Handeln, sondern Anschauung und Gefühl" [Its essence (=religion) is neither thinking nor acting, but intuition and sentiment] (Schleiermacher 1967, p. 49). The German word "Anschuung" is important in the writings of Immanuel Kant in his Critique of Pure Reason, to which Schleiermacher obviously alludes here in this context and can only be translated with some loss of meaning. The translation of "Anschauung" with "intuition" is in accord with the translation of the Kantian term.

[119] "Weil es an Religion gebrach, weil das Gefühl des Unendlichen sie nicht beseelte [because the sentiment of infinity did not ensoul them], und die Sehnsucht nach ihm und die Ehrfurcht vor ihm ihre feinen luftigen Gedanken nicht nötigte, eine festere Konsistenz anzunehmen, um sich gegen diesen gewaltigen Druck zu erhalten. Vom Anschauen muss alles ausgehen [everything has to start with intuition], und wem die Begierde fehlt, das Unendliche anzuschauen [to intuit infinity], der hat keinen Prüfstein und braucht freilich auch keinen, um zu wissen, ob er etwas Ordentliches darüber gedacht hat" (Schleiermacher 1967, pp. 51–52). Intuition and sentiment are also associated with infinity (Schleiermacher 1967, pp. 58, 60, 119).

immortality.[120] It is clear that this deep emotional involvement with the infinite defies, at first glance, any possibility of rational conceptualization. It is an expression of a genuine, original, and transformative religious experience, which transcends the anti-nomies that derive from a rational account of infinity, as it is described in the *Critique of Pure Reason* by Immanuel Kant. In this sense, infinity is for Schleiermacher a qualita-tive religious entity, and it transcends the quantitative limitations of space and time, to which the antinomies of infinity in the Kantian sense were due. Thus, with Anschauung and Gefühl he identifies aspects of the religious Self. However, it is interesting to note that Schleiermacher tried to conceptualize this new kind of religious experience with the idea of a mutual entanglement of the finite and the infinite. In this way he was attempting to overcome what he thought of as a false dichotomization, although this attempt did not lead him to a clear-cut concept.[121]

One year after his *Reden*, in 1800, the religious evaluation of infinity became a new twist in his *Monologen*. Having realized the indwelling of infinity in finiteness already in the *Reden* as a new idea, he offers – perhaps even unaware – in the *Monologen* a solution that proved to be very important for his whole scientific life as a theolo-gian. The *Monologen* expose the transformation of infinity as a matter of intuition (Anschauung) and sentiment (Gefühl) to a matter of action and self-education or self-cultivation (Bildung[122]), an iterative, open process of self-organization.[123] Therefore, infinite processes in all contexts, such as hermeneutics, theory of science, development of the religious consciousness, and even mathematics, became increasingly important for Schleiermacher's scientific thinking.[124] There is evidence of Schleiermacher's early

---

[120] "Die Unsterblichkeit darf kein Wunsch sein, wenn sie nicht erst eine Aufgabe gewesen ist, die Ihr gelöst habt. Mitten in der Endlichkeit eins werden mit dem Unendlichen und ewig sein in einem Augenblick, das ist die Unsterblichkeit der Religion" [Immortality must not be a desire, if it has not been a task, which you have solved. In the middle of finiteness to become one with infinity and being eternal in one moment, that is immortality of religion] (Schleiermacher 1967, p. 99).

[121] Schleiermacher thinks of finite entities as mirroring infinity, as a "picture of infinity": "Alles was ist, ist für sie (=die Religion) notwendig, und alles was sein kann, ist ihr ein wahres, unentbehrliches Bild des Unendlichen; wer nur den Punkt findet, woraus seine Beziehung auf dasselbe sich entdecken läßt" [Everything, that is, is necessary for it (=for religion), and everything, which can be, is for it a real and indispensable picture of infinity; whoever finds the point, from which his relation to infinity can be discovered] (Schleiermacher 1967, pp. 58–59). "Freilich ist es eine Täuschung, das Unendliche grade außerhalb des Endlichen, das Entgegengesetzte außerhalb dessen zu suchen, dem es entgegengesetzt wird" [Sure enough it is a deception, to seek the infinite outside finiteness, the opposition (=Entgegensetzung) beyond that to which it is opposed] (Schleiermacher 1967, p. 107).

[122] The German notion "Bildung" can hardly be translated. It is much more than education. It includes elements such as emotion, esprit, culture, self-control, discipline, and knowledge.

[123] "Unendlich ist was ich erkennen und besitzen will, und nur in einer unendlichen Reihe des Handelns kann ich mich selbst ganz bestimmen. Von mir soll nie weichen der Geist, der den Menschen vorwärts treibt, und das Verlangen, das nie gesättigt von dem, was gewesen ist, immer Neuem entgegen geht. Das ist dem Menschen Ruhm, zu wissen, dass unendlich sein Ziel ist, und doch nie still zu stehen im Lauf" [Infinite is what I want to recognize and hold, and only in an infinite succession of action can I totally determine myself. The spirit should not give way from me, who pushes human being forward, and the longing, that never can be satisfied by that, which has been, and always moves forward to novelties. This is the glory of man to know that infinite is his destination and never to stop in his course] (Schleiermacher 1978, p. 89).

[124] This has been shown in important works about the relation of Schleiermacher's theology to science. The first one is Dittmer (2001, pp. 23–24 [mathematics], 142–44 [metaphysics], 476–78 [hermeneutics]). The author shows that infinite processes are the key of Schleiermacher's concept of science (Wissenschaftslehre).

interest in mathematics (Mädler 1997, pp. 226–27), especially about his interest in the mathematical concept of functions, which are needed to calculate infinite processes. However, Schleiermacher's relation to mathematics as a theologian was much more than just accidental. Later in his dialectics he designed a way of thinking that tried to incorporate variety and unity on a high level of abstraction to secure recognition as the understanding of structure. In this context, he invented a structure-oriented way of thinking, which proved to be the starting point for his student Hermann Grassmann (1809–77). The latter eventually became a mathematician, and he confessed that he owed the basic tenets of his "Ausdehnungslehre" – a precursor of modern linear algebra as the science of mathematical structures – to the lecture Schleiermacher gave on Dialectics. He is also known as being the inventor of the vector and tensor calculus.[125] As a result, we see a gradual development of Schleiermacher's treatment of infinity, starting with a genuine experience, leading to the discovery of the religious Self, moving forward to the considerations about the relation of infinity and finiteness, focusing on infinite processes in various contexts, including mathematics, and ending up giving birth to a new mathematical concept that lies at the very core of modern mathematical reasoning. Again, we see the transformative power of infinity, in both science and religion.

1.5.2 Kierkegaard and the Constitution of the Religious Self

Later in the nineteenth century this thread toward the religious Self was made even stronger when pursued by Kierkegaard and associated with the religious experience of infinity. Especially in his book *The Sickness unto Death*, Kierkegaard turns nearly to a pure anthropological understanding of infinity insofar as he puts the development of the Self into the focus of his attention (e.g., Hannay 1993, pp. 31–40). Becoming a Self is the paramount task of every human being. This task can only be resolved if desperation, as a basic feature of every human being, is overcome by being connected to God. In particular, he interprets the striving for infinity as an expression of despair and alienation from God.[126] Thus, this striving for infinity has to be overcome in order to constitute oneself as a real religious Self.[127] The constitution of the religious Self is threatened by the mutual tension of four factors (Kierkegaard 2002, pp. 27–40).

In order to constitute the religious Self, this tension has to be eased and the mutually exclusive factors have to become part of a synthesis in the Self.[128] This synthesis,

---

He elaborates in a more extensive way a research result of Inken Mädler (1997), who had shown that Schleiermacher was very much influenced by the contemporary development of mathematics, especially the mathematics of infinite processes, such as limiting value problems.

[125] The fact that Schleiermacher triggered, by his theological reasoning, mathematical inventions is hardly known in either theological or mathematical circles. It was uncovered by the American historian of mathematics Albert C. Lewis in several publications (Lewis 1975, 1977, 1979; see also Mädler 1997, pp. 264, 266, 271, 292, 294).

[126] "However while one form of despair moves ferociously into infinity and looses itself, another form of despair let take away its Self by others" (Kierkegaard 2002, p. 29).

[127] "Thus, every human existence, which is alleged infinite or wants to be infinite, even every moment, in which a human existence is infinite or only wants to be it, is desperation" (Kierkegaard 2002, p. 29).

[128] "The Self is constituted out of infinity and finitude" (Kierkegaard 2002, p. 29).

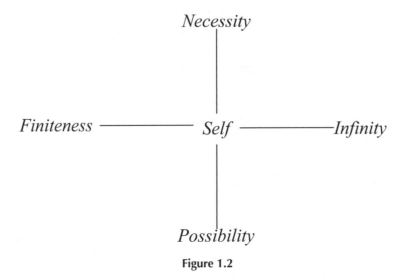

Necessity

Finiteness ——————— *Self* ——————*Infinity*

Possibility

Figure 1.2

which constitutes the religious Self, is an ongoing process that is only possible if a relation to God is established.[129]

1.5.3 Paul Tillich and the Transcending of the Religious Self

Whereas in Kierkegaard's theology infinity plays a role in constituting the religious Self, in the twentieth-century theology of Paul Tillich the religious Self is already presupposed and infinity is mainly related to human religious Self experience in the form of infinite Self-transcendence. In fact, Tillich defines religion as Self-transcendence.[130] Although he maintains the infinity of God,[131] infinity defies the possibility of rational conceptualization and is only accessible to symbolic understanding and ecstatic elevation.[132] Nevertheless, his focus is on the basic religious act of infinite Self-transcendence.[133]

1.6 Summary and Prospect

In this historical survey, the argument has been made for the creative, and even transformative, power of the concept of infinity, especially with regard to the interaction

---

[129] "The Self is the conscious synthesis of finitude and infinity, which relates to itself. Its task is, to become itself, which can be realized only by the relation to God" (Kierkegaard 2002, p. 28).

[130] "We have religion defined as self-transcendence into the dimension of the spirit" (Tillich 1984, bd. III, p. 118).

[131] "God is infinite, insofar as he is the creative ground of finitude and creates in eternity the finite possibilities in himself" (Tillich 1984, bd. II, p. 101).

[132] "When we say: 'God is the infinite or the absoluteness or being-itself' we speak at the same time rationally and ecstatically. This describes exactly the borderline, where symbolic and non-symbolic talk coincide" (Tillich 1984, bd. II, p. 16).

[133] "Infinity is: finitude with infinite Self-transcendence" (Tillich 1984, bd. I, p. 223).

between theology on the one hand and mathematics on the other. As a result of a long historical process, we can identify certain current traits of the concept of infinity in mathematics and theology:

1. Even though the infinity of God was introduced into Christian theology in order to overcome the limitations of ancient Greek metaphysics, and even though the progress in mathematics was stimulated by the theological idea of infinity, identifying infinity as an important aspect of God is nearly abandoned in contemporary theology. If it still exists, it is only used more or less in a metaphorical way.

2. Since the Enlightenment, however, starting with Schleiermacher, infinity in the realm of religion and theology has been serving a different function. As we have seen, the idea of infinity has played the role of discovering, constituting, and transcending the religious Self. The concept of infinity has migrated from a property of God to a factor in religious anthropology.

3. This loss of the concept of infinity in theology as a trait of God is unacceptable, although there are good reasons not simply to use Cantor's ideas for a theological concept of God's infinity, because that entails thinking about it in terms of quantification. On the other hand, a purely qualitative, metaphorical, or symbolic way of talking about God's infinity, as is the case in many contemporary theologies, is also unsatisfactory. For my understanding, the theological task today is not so much to understand infinity as a property of God, as well as how to describe this property, but rather to understand God as the creator in his infinite creativity. Any robust theology of creation has to come up with an exploration of God's infinite creative possibilities.

4. In the realm of mathematics, Cantor's work on infinity in set theory is still of paramount importance, with all its implications on logic and the continuum. However, it would be far more interesting and forward-looking to follow the line of thought introduced by the Hilbert program. It is commonly argued that Hilbert's program failed as a result of Gödel's arguments. He argued against the simultaneous validity of completeness and consistency of arithmetic structures with a certain degree of complexity. Nevertheless, only a few years later Gerhard Gentzen, a scholar of the Hilbert school, has shown – going beyond Gödel – that with stronger proof methods, such as "transfinite induction," the Hilbert program can be carried out further, going beyond the limitations for which Gödel had argued (Gentzen 1936a, 1936b). It seems to me that following the way that Gentzen has proposed could lead to new insights in infinite levels of complexity in arithmetic – analogs to Cantor's levels of infinity in set theory.

References

Aquinas, T. 1934. *Summa Theologica*, bd. 1, Gottes Dasein und Wesen. Salzburg, Leipzig, and Regensburg: Verlag Anton Pustet.

Aristotle. 1993. *Physics*, books I–IV, Loeb Classical Library, G. P. Goold (ed.). Translated by P. H. Wicksteed and F. M. Cornford. Cambridge: Harvard University Press.

Aristotle. 1997. *Metaphysics*, books X–XIV, Loeb Classical Library, G. P. Goold (ed.). Translated by C. G. Armstrong. Cambridge: Harvard University Press.

Aristotle. 2003. *Metaphysics*, books I–IX, Loeb Classical Library, J. Henderson (ed.). Translated by H. Tredennick. Cambridge: Harvard University Press.

Armstrong, A. H. 1954–1955. Plotinus' doctrine of the infinite and its significance for Christian thought. *Downside Review* 73: 47–58.

Augustine. 1988. *City of God*, vol. 4, books XII–XV, Loeb Classical Library, G. P. Goold (ed.). Translated by P. Levine. Cambridge: Harvard University Press.

Balás, D. L. 1966. *Metousia Theou: Man's Participation in God's Perfections According to Saint Gregory of Nyssa*. Rome: Pontificium Institutum S. Anselmi, Studia Anselmiana 55.

Bandmann, H. 1992. *Die Unendlichkeit des Seins. Cantor's Transfinite Mengenlehre und ihre metaphysischen Wurzeln*. Frankfurt: Verlag Peter Lang.

Becker, O. 1975. *Grundlagen der Mathematik in geschichtlicher Entwicklung*. Frankfurt: Suhrkamp Verlag.

Beierwaltes, W. 1967. *Plotin. Über Ewigkeit und Zeit (Enneade III 7)*. Frankfurt: Klostermann Verlag.

Böhm, T. 1996. *Theoria, Unendlichkeit, Aufstieg. Philosophische Implikationen zu De Vita Moysis von Gregor von Nyssa*. Leiden and New York, Köln: E. J. Brill.

Bolzano, B. 1955. *Paradoxien des Unendlichen*. Hamburg: Felix Meiner Verlag.

Brandt, A. J. H. W. 1889. *Die mandäische Religion. Ihre Entwicklung und geschichtliche Bedeutung. Erforscht, dargestellt und beleuchtet*. Leipzig: J. C. Hinrich'sche Buchhandlung (reprint: Amsterdam, 1973).

Brightman, R. S. 1973. Apophatic theology and divine infinity in St. Gregory of Nyssa. *Greek Orthodox Theological Review* 18: 97–114.

Cantor, G. 1883. *Grundlagen einer allgemeinen Mannichfaltigkeitslehre. Ein mathematisch-philosophischer Versuch in der Lehre des Unendlichen*. Leipzig: Teubner Verlag.

Cantor, G. 1932. *Gesammelte Abhandlungen mathematischen und philosophischen Inhalts*. Berlin: Verlag Julius Springer (reprint: 1980).

Cantor, M. 1892. *Vorlesungen über die Geschichte der Mathematik*, bd. II. Leipzig: Teubner Verlag.

Cassirer, E. 1991. *Das Erkenntnisproblem*, bd. I. Darmstadt: Wissenschaftliche Buchgesellschaft (reprint).

Clark, G. H. 1944. The theory of time in Plotinus. *Philosophical Review* 53: 337–58.

Clemens Alexandrinus. 1968. *Stromata*, O. Staehlin (ed.). Lichtenstein and Nendeln: Kraus Reprint.

Cohn, J. 1960. *Geschichte des Unendlichkeitsproblems im abendländischen Denken bis Kant*. Hildesheim: Georg Olms Verlagsbuchhandlung.

De Smet, R. V. 1980. Stretching forth to infinity: The mystical doctrine of Gregory of Nyssa. In *Prayer and Contemplation*, C. M.Vadakkekara (ed.), pp. 331–48. Bangalore: Asian Trading Corporation.

Dettloff, W. 2002. Johannes Duns Scotus. Die Unverfügbarkeit Gottes. In *Theologen des Mittelalters. Eine Einführung*, U. Köpf (ed.). Darmstadt: Wissenschaftliche Buchgesellschaft.

Diels, H., and Kranz, W. 1952. *Die Fragmente der Vorsokratiker*. Berlin: Weidmann Verlag.

Dionysios Areopagita. 1956. *Mystische Theologie*, J. Gebser (ed.) Schriftenreihe "Weisheitsbücher der Menschheit." Translated by W. Tritsch. München-Planegg: O. W. Barth Verlag.

Dittmer, J. M. 2001. *Schleiermacher's Wissenschaftslehre als Entwurf einer prozessualen Metaphysik in semiotischer Perspektive. Triadizität im Werden*. Berlin and New York: de Gruyter.

Dreyer, M., and Ingham, M. B. 2003. *Johannes Duns Scotus zur Einführung*. Hamburg: Junius Verlag.

Duclow, D. F. 1974. Gregory of Nyssa and Nicholas of Cusa: Infinity, anthropology and the via negativa. *Downside Review* 92: 102–8.

Enders, M. 2002. Unendlichkeit und All-Einheit. Zum Unendlichkeitsgedanken in der philosophischen Theologie des Cusanus. In *Nikolaus Cusanus zwischden Deutschland und Italien*, M. Thurner (ed.), pp. 383–441. Berlin: Akademieverlag.

Felix, M. 1977. *Minucius, Octavius, Lateinisch/Deutsch*, B. Kytzler (ed.). Stuttgart: Reclam Verlag.

Ferguson, E. 1973. God's infinity and man's mutability: Perpetual progress according to Gregory of Nyssa. *Greek Orthodox Theological Review* 18: 59–78.

Flasch, K. 1998. *Nikolaus von Kues, Geschichte einer Entwicklung*. Frankfurt: Klostermann Verlag.

Frank, E. 1955. Die Begründung der mathematischen Naturwissenschaft durch Eudoxos. In *Knowledge, Will, and Belief: Collected Essays*, E. Frank (ed.), pp. 134–57. Chicago and Zürich: Artemis Verlag.

Gentzen, G. 1936a. Die Widerspruchsfreiheit der reinen Zahlentheorie. *Mathematische Annalen* 112: 493–565.

Gentzen, G. 1936b. Die Widerspruchsfreiheit der Stufenlogik. *Mathematischde Zeitschrift* 41: 357–66.

Gregory of Nyssa. 1960a. *Gregorii Nysseni Opera*, vol. I, Contra Eunomius Libros I–II, W. Jaeger (ed.). Leiden: E. J. Brill.

Gregory of Nyssa. 1960b. *Gregorii Nysseni Opera*, vol. II, Contra Eunomius Liberum III, W. Jaeger (ed.). Leiden: E. J. Brill.

Gregory of Nyssa. 1964. *Gregorii Nysseni Opera*, vol. VII, pars I, Gregorii Nysseni De Vita Moysis, W. Jaeger, H. Langenbeck, and H. Musurillo (eds.). Leiden: E. J. Brill.

Guitton, J. M. P. 1959. *Le Temps et l'Eternité chez Plotin et Saint Augustin*. Paris: Vrin.

Guyot, H. 1906. *L'Infinité Divine: Depuis Philon Le Juif Jusqu'a Plotin*. Paris: Alcan.

Hannay, A. 1993. *Kierkegaard*. London: Routledge.

Hegel, G. W. F. 2000. *Vorlesung über die Philosophie der Religion*. Frankfurt: Suhrkamp Verlag.

Hennessy, J. E. 1963. "The Background, Sources and Meaning of Divine Infinity in St. Gregory of Nyssa." Ph.D. thesis, Fordham University, ETD Collection for Fordham University. Paper AAI6305594.

Hilbert, D. 1925. Über das Unendliche. *Mathematische Annalen* 95: 161–90. Reprint in *From Frege to Gödel, A Source Book in Mathematical Logic, 1879–1931*, J. van Heijenoort (ed.), pp. 367–92. Lincoln, NE: iUniverse.com, Inc. (1967).

Honnefelder, L. 2005. *Johannes Duns Scotus*. München: Verlag C. H. Beck.

Johannes Damascenus. 1955. De fide orthodoxa, Saint John Damascene. In *Versions of Burgundio and Cerbanus*, E. M. Buytaert (ed.). St. Bonaventure, NY: The Franciscan Institute.

Jonas, H. 1962. Plotin über Zeit und Eweigkeit. In *Politische Ordnung und menschliche Existenz*, A. Dempf et al. (eds.). Festgabe für Eric Voegelin, pp. 295–319. München: Verlag C. H. Beck.

Karfíková, L. 2001. Die Unendlichkeit Gottes und der unendliche Weg des Menschen nach Gregor von Nyssa. *Sacris Erudiri (Brugge)* 40: 47–81.

Kierkegaard, S. 2002. *Die Krankheit zum Tode*. Hamburg: Europäische Verlagsanstalt.

Knobloch, E. 2002. Unendlichkeit und Mathematik bei Nikolaus von Kues. In *Chemie-Kultur-Geschichte, Festschrift für Hans-Werner Schütt*, A. Schürmann and B. Weiss (eds.). Berlin: Verlag für Geschichte der Naturwissenschaften und der Technik (GNT-Verlag).

Lasserre, F. 1966. *Die Fragmente des Eudoxos von Knidos (Texte und Kommentare 4)*. Berlin: de Gruyter.

Leisegang, H. 1913. *Die Begriffe Zeit und Ewigkeit im späteren Platonismus, Beiträge zur Geschichte der Philosophie des Mittelalters XIII 4*. Münster: Aschendorff.

Lewis, A. C. 1975. "An Historical Analysis of Grassmann's Ausdehnungslehre of 1844." Ph.D. thesis, University of Texas.

Lewis, A. C. 1977. H. Grassmann's Ausdehnungslehre and Schleiermacher's Dialektik. *Annals of Science* 34: 103–62.

Lewis, A. C. 1981. Friedrich Schleiermacher's influence on H. Grassmann's Mathematics. In *Social History of Nineteenth Century Mathematics. Papers from a Workshop Held in Berlin, July 5–9, 1979*, H. Mehrtens, H. J. Bos, and I. Schneinder (eds.), pp. 246–54. Basel and Boston: Birkhäuser.

Mädler, I. 1997. *Kirche und bildende Kunst der Moderne. Ein an F.D.E. Schleiermacher orientierter Beitrag zur theologischen Urteilsbildung*. Tübingen: Mohr Siebeck.

Merlan, P. 1960. The life of Eudoxos. In *Studies in Epicurus and Aristotle, Klassisch philologische Studien 22*, pp. 98–104. Wiesbaden: Harrassowitz Verlag.

Mühlenberg, E. 1966. *Die Unendlichkeit Gottes bei Gregor von Nyssa: Gregors Kritik am Gottesbegriff der klassischen Metaphysik*. FKDG 16; Göttingen: Vandenhoeck & Ruprecht.

Nikolaus von Kues. 1964. *Philosophisch-Theologische Schriften*, vol. I, Lateinisch-Deutsch,. L. Gabriel (ed.). Wien: Herder Verlag.

Nikolaus von Kues. 1966. *Philosophisch-Theologische Schriften*, vol. II, Lateinisch-Deutsch, L. Gabriel (ed.). Wien: Herder Verlag.

Nikolaus von Kues. 1967. *Philosophisch-Theologische Schriften*, vol. III, Lateinisch-Deutsch, L. Gabriel (ed.). Wien: Herder Verlag.

Nikolaus von Kues. 1979. *Die mathmatischen Schriften, übersetzt von Josepha Hofmann und mit einer Einführung versehen von Josef Ehrenfried Hofmann*. Hamburg: Meiner Verlag.

Origenes. 1992. *Vier Bücher von den Prinzipien (Origenis De Principiis Libri IV)*, H. Görgemanns and H. Karpp (eds.). Darmstadt: Wissenschaftliche Buchgesellschaft.

Pannenberg, W. 1988. *Systematische Theologie*, bd. 1. Göttingen: Vandenhoeck & Ruprecht.

Pascal, B. 1978. *Pensées. Über die Religion und über einige andere Gegenstände*. Heidelberg: Verlag Lambert Schneider.

Plato. 1970. *Theatet*, werke VI, G. Eigler (ed.). Darmstadt: Wissenschaftliche Buchgesellschaft.

Plato. 1971. *Politeia*, werke IV, G. Eigler (ed.). Darmstadt: Wissenschaftliche Buchgesellschaft.

Plato. 1972. *Philebos*, werke VII, G. Eigler (ed.). Darmstadt: Wissenschaftliche Buchgesellschaft.

Plato. 1983. *Parmenides*, werke V, G. Eigler (ed.). Darmstadt: Wissenschaftliche Buchgesellschaft.

Plotin. 1956–71. *Plotins Schriften*, bd. I–VI, R. Harder and R. Beutler (eds. and trans.). Hamburg: Felix Meiner Verlag.

Porphyrios. 1958. *Über Plotins Leben und die Ordnung seiner Schriften*. Hamburg: Felix Meiner Verlag.

Ritter, A. M., and Heil, G. 1991. *Corpus Dionysiacum II*. Berlin and New York: de Gruyter.

Rucker, R. 1989. *Die Ufer der Unendlichkeit*. Frankfurt: Wolfgang Krüger Verlag.

Schleiermacher, F. D. E. 1967. *Über die Religion. Reden an die Gebildeten unter ihren Verächtern*. Göttingen: Vandenhoeck & Ruprecht.

Schleiermacher, F. D. E. 1978. Monologen. In *Neujahrspredigt von 1792. Über den Wert des Lebens*, F. M. Schiele (ed.). Hamburg: Felix Meiner Verlag.

Schleiermacher, F. D. E. 1984. Spinozismus, Spinoza betreffend aus Jacobi. In *Kritische Gesamtausgabe*, bd. I, Jugendschriften 1787–96, G. Meckenstock (ed.), pp. 515–58. Berlin and New York: de Gruyter.

Schulze, W. 1978. *Zahl, Proportion, Analogie. Eine Untersuchung zur Metaphysik und Wissenschaftshaltung des Nikolaus von Kues, Buchreihe der Cusanusgesellschaft*, bd. VII. Münster: Aschendorff Verlag.

Siegfried, C. 1970. *Philo von Alexandrien als Ausleger des A.T.* Aalen: Scientia-Verlag (reprint).

Simplicius. 1954. *Aristotelis Physicorum libros quattuor posteriors commentaria (Commentaria in Aristotelum Graeca 10)*, H. Diels (ed.). Berlin: de Gruyter (reprint).

Sleeman, J. H., and Pollet, G. 1980. *Lexicon Plotinianum, Ancient and Medieval Philosophy Series*. Leiden, Leuven: E. J. Brill.

Spinoza, B. 1989. *Tractatus de Intellectus Emendatione. Ethica*, bd. II, K. Blumenstock (ed.). Darmstadt: Wissenschaftliche Buchgesellschaft.

Sweeney, L. 1957. Infinity in Plotinus. *Gregorianum* 38: 521–35, 713–32.

Sweeney, L. 1992a. Bonaventure and Aquinas on the divine being as infinite. In *Divine Infinity and Medieval Thought*, L. Sweeney (ed.), pp. 413–37. Frankfurt and New York: Peter Lang Verlag.

Sweeney, L. (ed.).  *ne Infinity in Greek and Medieval Thought*. Frankfurt and New York: Peter Lang Verla~~

Sweeney, L. 1992c. ~~tinus. In *Divine Infinity in Greek and Medieval Thought*, L. Sweeney (ed.), pp. 167–222~~ ~~d New York: Peter Lang Verlag.

Taylor, M. C. 1980. *J~~lfhood, Hegel & Kierkegaard*. Berkeley: University of California Press.

Tillich, P. 1984. *Systen~~logie*. Frankfurt: Evangelisches Verlagswerk.

Übinger, J. 1895. Die m~~n Schriften des Nikolaus Cusanus. In *Philosophisches Jahrbuch der Görres Gesellsch~~ ~~301–17, 403–22.

Übinger, J. 1896. Die ma~~Schriften des Nikolaus Cusanus. In *Philosophisches Jahrbuch der Görres Gesellschaj~~ ~~–66, 391–410.

Übinger, J. 1897. Die math~~hriften des Nikolaus Cusanus. In *Philosophisches Jahrbuch der Görres Gesellschaft*,~~ ~~4–59.

Vaggione, R. P. 2000. *Eun~~us and the Nicene Revolution*. Oxford: Oxford University Press.

Völker, W. 1955. *Gregor von~~tiker*. Wiesbaden: Franz Steiner Verlag.

Wilson, A. M. 1995. *The Infin~~*. Oxford: Oxford University Press.

Wilson-Kästner, P. 1978. God~~is relationship to creation in the theologies of Gregory of Nyssa and Jonathan Edwa~~~~s 21: 305–21.

Zeller, E. 1963. Die Philosophie~~n ihrer geschichtlichen Entwicklung: Drei Teile, jeder in zwei Abteilungen. Hildesh~~s Verlag.

PART TWO

Perspectives on Infinity from Mathematics

CHAPTER 2

The Mathematical Infinity

Enrico Bombieri

2.1 Early History

In my youth I read a popular book, written by the famous physicist Gamow (1947), aimed at guiding the reader to a glimpse of modern science, with special emphasis on the microcosm of the atom, the macrocosm of the galaxies all the way back to the Big Bang, and Einstein's theory of relativity along the way. It was a fascinating book indeed. The title was *One, Two, Three... Infinity*, in reference to how counting may have started in primitive tribes as "One, two, three... many."[1]

We may smile, thinking proudly of how far ahead we have gone in our understanding of counting, but in a certain sense we have not made much progress beyond this. Studies have shown that the average person, when shown a multitude of objects, is not able to recollect more than seven of them with accuracy, if not even less. So our inner way of counting may still be today, "One, two, three,... seven... many." On the other hand, there are ways of understanding the very large and the incredibly small, and mathematics provides the tools to do so.

What is infinity? Is it the inaccessible, the uncountable, the unmeasurable? Or should we consider infinity as the ultimate, complete, perfect entity? Can mathematics, the science of measuring, deal with infinity? Is infinity a number, or can it be treated as such?

The concept of infinity plays a fundamental, positive role in today's mathematics, but it was not always so positive in antiquity. The Greek mathematicians and philosophers took, at times, a negative view of infinity.

For Pythagoras, the Eleatic school, and the philosophers Parmenides and Plato, infinity was accepted as a negative concept: it could not be reached or described in finite terms; it was the irrational; it was formless because it could not be increased or decreased. Simply put, it was inaccessible. In arithmetic and geometry, it meant that unending mathematical constructions were not allowed.

Zeno based his proofs of the impossibility of movement on such ideas (Achilles will never reach the tortoise, because in the time he reaches the distance that separates him

---

[1] This is not an exaggeration. The Pirahã Amazonian tribe actually counts "One, two, many."

from the tortoise, the tortoise will have moved forward a little more). Zeno's paradox is not a joke. It leads to very meaningful questions, among them whether we accept infinite divisibility and the concept of a continuum, or not. (In Zeno's case, we deal with space and time, which I will leave to the physicist, but in mathematics it is again a fundamental question.) Every mathematical concept had to be described in finite, precise terms. A straight line was a good geometrical concept because of its constancy of direction, and so were circles, squares, triangles, and polygons built of triangles. Ellipses, parabolas, and hyperbolas could be considered part of geometry, because they were conic sections (the intersection of a plane and a circular cone in three-dimensional space), but they were not part of the Euclidean geometry of circles and lines.

The first limitations of such an ideal view of geometry appeared with the discovery of the irrational. How was it possible that the diagonal of a square, obviously a good geometric object, could not be commensurable with the side of the square? The proof of such a statement is very simple. Suppose that the diagonal and side of a square were in a rational ratio m/n in whole numbers. Then Pythagoras's theorem would give $m^2 = 2n^2$. Hence, m would be even and we could write $m = 2p$, with p an integer. Therefore, $4p^2 = 2n^2$ and $n^2 = 2p^2$. This process could be repeated without end, leading to an impossible "infinite descent" of positive integers. Fermat's technique of "infinite descent" is today an important tool of number theory.

2.2 Three Famous Problems of Antiquity

Although this paradox could be resolved by accepting the fact that geometric constructions belong to a mathematical world ampler than pure arithmetic, new difficulties arose. The problem of constructing a segment equal to the side of a cube with volume twice the volume of a given cube could not be solved within the scope of an Euclidean geometry that allowed only lines and circles. The problem of trisecting an arbitrary angle suffered the same fate, although in both cases solutions could be found by means of three-dimensional geometry or by mechanical constructions. Archimedes himself found a simple mechanical way of trisecting the angle using only ruler and compass.

On the other hand, the quadrature of the circle, namely, the geometric construction of a square with the same area as a given circle, could not be solved in this way. The problem became famous, and even Dante Alighieri (1966–1967), mentions it in his *Commedia* when confronted with the impossibility of understanding the divine mystery of the Trinity:

Dante, Paradiso, Canto XXXIII:
Qual è 'l geomètra che tutto s'affige
per misurar lo cerchio, e non ritrova,
pensando, quel principio ond' elli
   indige,
tal era io a quella vista nova:
veder voleva come si convenne
l'imago al cerchio e come vi
   s'indova;
ma non eran da ciò le proprie penne:

The Longfellow translation:
As the geometrician, who endeavours
To square the circle, and discovers not,
By taking thought, the principle he wants,
Even such was I at that new apparition;
I wished to see how the image to the circle
Conformed itself, and how it there finds
   place;
But my own wings were not enough for
   this,

Dante was not off the mark, because a proof of the impossibility of a purely geometric solution to this problem was eventually found by Lindemann in 1882. This result represents a triumph of modern mathematics, in view of its profound historical meaning and philosophical implications.

These three problems of antiquity have been the hunting grounds of literally thousands and thousands of amateur mathematicians, all in search of satisfying their egos.

The reader will find an edifying account of the subject, up to the year 1872, in the book *A Budget of Paradoxes* by Augustus De Morgan, astronomer and logician De Morgan (1872). Notwithstanding the complete solution, in the negative, of all three problems, the flood of false solutions continues unabated. The truth has not yet reached our politicians. In the book *Mathematical Carnival* by Martin Gardner, a selection of his Mathematical Games columns in *Scientific American*, we find the following:

> ... On June 3, 1960, the Honorable Daniel K. Inouye, then a representative from Hawaii, later a senator and member of the Watergate investigation committee, read into the Congressional Record (Appendix, pages A4733-A4734) of the 86th Congress a long tribute to Maurice Kidjel, a Honolulu portrait artist who has not only trisected the angle but also squared the circle and duplicated the cube. Kidjel and Kenneth W.K. Young have written a book about it called 'The Two Hours that Shook the Mathematical World', and a booklet, 'Challenging and Solving the Three Impossibles'. Through a company called the Kidjel Ratio they sell this literature along with the Kidjel Ratio calipers with which one can apply the system. In 1959 the two men lectured on their work in a number of U.S. cities, and a San Francisco radio station, KPIX, produced a documentary about them called 'The Riddle of the Ages'. According to Inouye, 'The Kidjel solutions are being taught in hundreds of schools and colleges throughout Hawaii, the United States, and Canada.

Indeed, a very instructive story.

The purist view that permeates Euclid's geometry can also be found in the arithmetic of the *Elements*. The following example deals with prime numbers, a favorite subject of number theorists. In the ninth book of his *Elements* (IX.20) Euclid notes that "Given an arbitrary number of primes, there exists another prime different from them." He gives a startlingly simple proof of this statement:[2] Consider the product of the given primes and add 1. Then we get a number, larger than 1, that divided by anyone of these primes yields a remainder 1, so it is not divisible by any of them. However, this number can be factored as a product of prime numbers necessarily distinct from the given primes, concluding the proof.

Nowadays, we refer to this as Euclid's proof that the sequence of primes is infinite, but we must note that Euclid studiously avoids using the word "infinite" in the statement of the proposition.

With Archimedes, things are quite different. Archimedes often used limiting processes. His method for calculating 2π, the length of the circumference of a circle of radius 1, is exemplary. He started from the remark that this number is always larger than the perimeter of a regular polygon inscribed in the circle and smaller than the

---

[2] See Busard (2001, p. 227). Euclid's writing is always rather formal and lengthy, so we give his proof in modern terms.

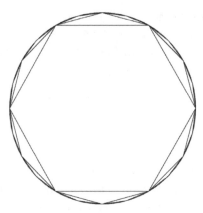

Figure 2.1. Archimedes's approximations to the circle.

length of the corresponding polygon circumscribed to the circle (Fig. 2.1). Starting with a given regular polygon with n sides, there is a simple geometrical construction for doubling the number of sides to $2n$. When n gets larger and larger, the two polygons become closer and closer to the circle, and their perimeters yield better and better approximations to π. Archimedes used this method, starting with the hexagon and doubling the sides four times to reach 96 edges, to prove rigorously that $3\frac{10}{71} < \pi < 3\frac{1}{7}$. In fact, $3\frac{1}{7}$ is an excellent simple approximation to π, with an error less than 0.002.

He used similar ideas to compute exactly the area between the chord and an arc of a parabola, by a procedure that may be considered a precursor of the differential and integral calculus of Leibniz and Newton. Justly famous is his result that the area of a sphere is two-thirds of the area of the circumscribed cylinder (including caps) of height equal to the diameter of the sphere. Such a result could not be proved by direct geometrical methods; the proof required a passage to the limit, precisely the kind of infinite process abhorred by the Pythagorean school. Archimedes himself considered it to be his greatest achievement.

Archimedes died at the end of the siege of Syracuse, in 212 BC. His tomb was identified by Cicero, as told in his *Tusculanae Disputationes*, when he was questor in Sicily in 75 BC, by recalling certain doggerel lines he had heard, that a sphere and a cylinder had been put over his grave; indeed, in the cemetery there was a column, hidden in a thicket of bushes, on which there were inscribed a sphere within a cylinder.

2.3 Archimedes and Aristotle

Archimedes' struggle with infinity also appears in two other remarkable works. In the so-called problem of the sandreckoner in his *Psammites* (or *Arenarius*), he invented a new number system that enabled him to count the number of grains of sand needed to fill the universe (as described by the Greek astronomer Aristarchus). In substance, he discovered the power of the exponential function and of iteration. There is no question in my mind that this was indeed a very daring idea at the time. This work contributed to the fame of Archimedes in antiquity, as attested from the line of the poet Silius

Italicus (Geymonat 2006, p. 110): "*Non illum mundi numerasse capaci harenas vana fides*," which may be approximately translated as "It was not an empty belief that he had counted all the grains of sands that the world contains."

A second problem,[3] attributed to Archimedes, appeared in a Greek epigram of twenty-four verses, published by Gotthold Ephraim Lessing in 1773, and is referred to as the cattle problem of Archimedes. The problem is to find the composition of the Herd of the Sun, namely, the number of the white, black, spotted, and brown bulls, and the number of cows of corresponding colors, with the numbers in question satisfying nine relations. There are eight unknowns, and seven relations are linear (as an example, it is required that the number of white bulls is five-sixths of the number of black bulls, plus the number of brown bulls). The remaining two relations are arithmetic in nature and ask that the white bulls, together with the black bulls, can be aligned to form a perfect square with the same number of bulls in each line. In a similar way, the spotted bulls, together with the brown bulls, can be corralled in a perfect triangle made of rows of lines, with the number of bulls increasing by one from a line to the next.

The problem reduces to finding the smallest solution, satisfying certain divisibility conditions, of a certain Pell equation[4] $Ax^2 + 1 = y^2$. The Lessing version leads to $A = 4,729,494$ and, in the end, to the truly gigantic smallest solution

$$x = 50,549,485,234,315,033,074,477,819,735,540,408,986,340$$

with a corresponding integer y that I will not write down. From this solution we can go back to the smallest solution of the original problem, and it turns out that the total number of cattle has $206,545$ digits, as proved by Amthor in 1880 (see Vardi [1998] for a modern treatment and a discussion of the history of the problem). Nygrén (2001) proposed a more elementary approach to the problem that, at least in principle, would give a procedure for finding the solution using an abacus.

Was this really the original Archimedes' problem? Probably not. Some scholars maintain that the epigram text is corrupted and proposed alternative versions leading to solutions of a few hundred digits that, in view of Archimedes's known feat with the sandreckoner, were within his range. (See Bartocci and Vipera [2004] for a discussion of alternative versions and solutions.) An amusing alternative explanation can be found in Archimedes' writings (see http://www.groups.dcs.st.ac.uk/~history/Mathematicians). In his treatise *On Spirals*, Archimedes tells us that he was in the habit of sending statements of his latest theorems to his friends in Alexandria, but without giving proofs. Since some of the Alexandrine mathematicians there had claimed his results as their own, Archimedes says that on the last occasion when he sent them theorems he included two that were false: "*. . . so that those who claim to discover everything, but produce no proofs of the same, may be confuted as having pretended to discover the impossible.*" Could it be that Archimedes formulated such a problem, apparently

---

[3] See the account of this problem in Dickson (1952, XII 5, p. 342). A precursor is in Homer's *The Odyssey*, book XII, lines 194–98.

[4] This classical equation was well known in antiquity. The attribution to Pell is a misnomer originating with Euler. See Dickson (1952, p. 354).

simple, but with a solution that no one, including Archimedes, could write down, for the purpose of unmasking his false competitors?

Archimedes' evaluation of the area of the sphere, his interest in extremely large numbers, and his use of passage to the limit all indicate his acceptance of infinity in mathematics. Instead, in Aristotle's view, infinity is only something potential, but cannot ever be actually reached. Thus, mathematicians do not need an actual infinity, rather only finite quantities, as well as constructions repeated as many times as the case may be. So, should we view Archimedes' evaluation of the area of the sphere as a theorem or only a truth that transcends the realm of mathematics, because achieved only through an infinite limiting process? Should we reject any result outright, on the ground that it is obtained by forbidden methods? Forbidden by whom, and why? Because of tradition? Because of the fear of challenging accepted wisdom?

2.4 Combinatorics and Infinity

Today, the matter remains the subject of hot debate among mathematicians. For example, are computer proofs acceptable? And if so, under which limitations? Rejecting infinity also leads to intrinsic difficulties. For example, as briefly discussed later, after the work of Paris and Harrington (1977) we know that there are simple, plausible statements in finite combinatorics that are provably true within the model of Zermelo-Fraenkel set theory, but are undecidable within the very natural, but much simpler, finitistic model P of Peano arithmetic.[5]

The combinatorial statements alluded to are a profound extension of the elementary (but not so easy to prove) "friends and strangers" theorem: Let n be given. Then in any party of sufficiently large size (determined by n) we can find either a subset of n people who know each other (i.e., a clique) or a subset of n people who never met before (i.e., an independent set). How big should the size of the party be? If we denote by $R(n, n)$ the minimum size allowable, it is very easy to show that $R(3, 3) = 6$, less easy that $R(4, 4) = 18$, and beyond this we know only some inequalities, for example, $43 \leq R(5, 5) \leq 49$ and $102 \leq R(6, 6) \leq 165$ (Fig. 2.2).

Even the simplest Ramsey numbers, the numbers $R(n, n)$, are extraordinarily difficult to determine, because the number of possible relations among n people grows incredibly fast with n. The mathematician Paul Erdős described the situation in this way: "*Imagine an alien force, vastly more powerful than us, landing on Earth and demanding the value of $R(5, 5)$, or they will destroy our planet. In that case, we should marshal all our computers and all our mathematicians, and attempt to find the value. But suppose, instead, that they asked for $R(6, 6)$, we should attempt to destroy the aliens.*"

---

[5] See Paris and Harrington (1977). From *Mathematical Reviews*, MR0491063 (58 #10343): "...D8. J. Paris and L. Harrington: A mathematical incompleteness in Peano arithmetic. This research paper is included here because of its importance. An arithmetically expressible true statement from finitary combinatorics is given which is not provable in Peano arithmetic P. The statement S in question is the strengthening of the finite Ramsey theorem by requiring the homogeneous set H to be 'relatively large', i.e., $|H| \geq \min H$. The statement S follows from the infinite Ramsey theorem and is equivalent in P to the uniform $\sum_1^0$-reflection principle for P. The corresponding Ramsey function exceeds almost everywhere all provably recursive functions of P."

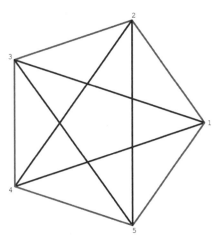

Figure 2.2. A proof that R(3, 3) ≥ 6. The vertices represent five people: those connected by an outer edge have met each other, and those connected by an inner edge never met before. This complete graph on five vertices has no triangle with edges of a same type.

The situation for the Paris-Harrington result is even worse. An apt description of this difficulty is that the problem is like the Hydra monster of Greek mythology, which grew back two heads when one was cut off (Fig. 2.3). Solving the Paris-Harrington riddle in Peano arithmetic is like killing the Hydra, because the examination of each case gives rise to several more subcases, and so on ad infinitum. A solution can be found only by an excursion all the way to infinity, where we can see what happens, and only then, bringing the knowledge so obtained back to the finite realm, can we get the answer.

More precisely, the Paris-Harrington problem arises as follows. Consider arbitrary colorings, with c colors, of all subsets of $\{1, 2, \ldots, r\}$ and ask for a subset E, of cardinality n, such that all subsets of E with a same cardinality have the same color. The fact that we can always do this, provided that r is sufficiently large as a function of n and c, is a theorem in P. The Paris-Harrington problem consists in adding the apparently mild restriction on E that its minimum element does not exceed n. The fact that we can always find such an E if r is sufficiently large remains true, but the corresponding function grows so incredibly fast that it is not computable in the finitistic model of P. The statement is, however, provable in the arithmetic of P after adding to it the transfinite induction up to the ordinal ϵ_0.

The first and simplest approach to the mathematical infinity is the counting without end. This naive definition is tautological because, as pointed out by Georg Cantor, infinity is then implicitly defined self-referentially as a process that goes on for infinite time. Mathematicians avoid such a paradox by axiomatizing the rules of mathematics in a small set of axioms, thus reducing any mathematical proof to a logical combination of such axioms. The pitfall of this approach is that, because of Gödel's incompleteness theorems, we cannot prove that the axioms themselves are not contradictory and never lead to statements that are true and false at the same time. At any rate, because no profound inner contradiction has been found in the body of all of mathematics done up to today, mathematicians tend to ignore the question and go on happily with their work, leaving the subtleties of foundational mathematics to historians, philosophers, and logicians. Here I will do the same and ignore foundational questions.

Figure 2.3. *83.AE.346, The J. Paul Getty Museum, Villa Collection, Malibu, California.* Attributed to Eagle Painter. Caeretan Hydria, Etruria about 525 BC. Terracotta. Object: H. and Diam., 17 9/16 × 13 in.; Object (rim): Diam., 9 in.

The big revolution in the role of the infinite in mathematics occurs with Leibniz and Newton, with the explicit introduction of the infinitesimal and the derivative, and of the inverse process of integration as a continuous sum, with the corresponding differential and integral calculus. (Some aspects of calculus were anticipated by Cavalieri and Fermat.) This led to two views, one being the acceptance of an actual notion of infinity in mathematics, the other regarding infinity only as a convenient abbreviation to indicate certain processes, thereby considering the whole process as a single object. Philosophers such as Berkeley and Hume took the opposite stance and wrote scathing rebuttals of the whole idea.

2.5 Euler and Infinity

We find the first point of view in Euler, the foremost mathematician of the eighteenth century and one of the greatest of all times. Here is how he proves that the sequence of primes is infinite. From the equation

$$\sum_{n=1}^{\infty} \frac{1}{n^s} = \prod_p \left(1 - \frac{1}{p^s}\right)^{-1},$$

where p runs over all prime numbers and $s > 1$, he goes boldly to the limit for $s \to 1$, writing the meaningless equation

$$\sum_{n=1}^{\infty} \frac{1}{n} = \prod_p \left(1 - \frac{1}{p}\right)^{-1}.$$

The left-hand side is the so-called harmonic series, which diverges to ∞. If there were only finitely many primes, the right-hand side of the equation would be a finite number, a contradiction that proves what we want. Euler goes even further, writing the formula

$$\sum_p \frac{1}{p} = \log \log \infty,$$

which is meaningless by itself. On the other hand, Euler's reasoning is simple, rooted on solid ground, and it is not difficult for the modern mathematician to reconstruct it, transforming the meaningless statement into an interesting result. It goes as follows.

Let us think of infinity as the infinity that arises from counting. Thus, $1 + 1 + 1 + \cdots = \infty$. A way to approach this is via the geometric series,

$$1 + x + x^2 + x^3 + \cdots = \frac{1}{1 - x},$$

then letting $x \to 1$. Thus, in a sense we can write, as Euler does, the formula

$$1 + 1 + 1 + \cdots = \frac{1}{0} = \infty.$$

If we take the logarithm of the geometric series, we have the series

$$x + \frac{x^2}{2} + \frac{x^3}{3} + \cdots = \log\left(\frac{1}{1 - x}\right),$$

showing that the harmonic series, namely, the sum of the reciprocal of the natural integers, diverges to infinity, but much more slowly than the sum $1 + 1 + 1 + \cdots$. We can express this fact by writing the meaningless equation

$$\sum_{n=1}^{\infty} \frac{1}{n} = \log \frac{1}{0} = \log \infty.$$

Finally, taking the logarithm of the (meaningless) expression relating the harmonic series to the product over primes, we easily complete the basic thoughts behind Euler's statement that

$$\sum_p \frac{1}{p} = \log \log \infty.$$

This kind of reckoning reminds us very much of Archimedes' *method*: a nonrigorous approach to problems, leading to a correct line of thought, that can be perfected to fully rigorous treatment at a later stage. For the example at hand, it is quite easy to transform it into a rigorous statement, working in Aristotelian fashion by using only finite quantities. The precise result can be stated as

$$\sum_{p \leq X} \frac{1}{p} = \log \log X + C + \varepsilon(X),$$

where C is a certain numerical constant, $X \geq 3$, and $\varepsilon(X) \to 0$ when X increases without limit. We see here an advantage of the finite method: whenever it works, it leads to quantitative statements, that is, information and knowledge that may be lost by going to the limit.

Euler was somewhat of an exception in his ability to handle the infinity correctly and with ease, even if not in a rigorous way. Others were not so successful, as witnessed by endless and at times acrimonious debates about the value of the infinite sum $1 - 1 + 1 - 1 + \cdots$ A century later, with the work of Cesàro laying the foundation of a theory giving precise meaning to such processes without a limit in the traditional sense (the theory of summability), it became possible to disentangle the heuristic methods of the past, separating meaningful ideas from paradoxical calculations.

2.6 Rigorous Proofs

Starting with Cauchy at the beginning of the nineteenth century, the finitistic view, avoiding the question of actually defining infinity or infinitesimals as precise mathematical entities, became generally accepted. The current way of teaching calculus, with the traditional epsilons and deltas indicating arbitrarily small numbers, is indeed the simplest example of this general trend. Thus, infinity took a precise meaning, especially in analysis, and specific rules and conditions to be met for handling infinite sums and integrals were formulated, thereby adding a new effective weapon in the armory of analysis.

Should we then say that the Aristotelian point of view, regarding the mathematical infinity as a convenient shorthand for potentially unlimited quantities or constructions, has won the day at last? Is infinity nothing other than a mathematical convention? Not so. Infinity has come back into mathematics in a far more powerful way, as well as in many different forms, to coexist with the finitistic approach of Aristotle. There is no need to coerce every proof into a finite argument.

In the view of many mathematicians, including myself, the intellectual contortions needed to remain within the realm of the finite (if possible at all) indicate that a wholesale rejection of infinity in mathematics is not a good thing. What really matters is the final understanding, coupled with a good foundation. Thus, the modern mathematician approaches the foundational question of the nature of infinity in a pragmatic way: what matters first is *knowledge*. How it is obtained is also very important, but that takes second stage.

This view leads right away to many questions. Is mathematics a science, hence needing validation through experimentation? Is verification of truth or falsity an absolute, or is it dependent on language, social factors, and time? Is validation *via* computer acceptable in mathematics? Here opinions differ and no consensus has been reached or is likely to be reached in the future. My opinion is that indeed mathematics, notwithstanding its abstractedness and appearance of absoluteness, has an experimental aspect:

The Experiment: Verification of truth or falsity.
The Input: The supposed proof of a mathematical statement.
The Hardware: Biological (i.e., a brain, firing signals along neurons).
The Software: Two-valued (i.e., True–False Boolean) logic.
The Output: True, False, Fail (i.e., an incomprehensible paper).

This approach has limitations. Already, a definition of truth in a language cannot be given within the given language, which must be extended in order to define truth. The value (True-False) of a proposition may depend on how we extend the language, so one must give away the notion of an absolute truth and replace it by a semantic definition. Finding a "good" model of mathematics (i.e., the language), compatible with the intuition we derive from the natural world and experience, is therefore of primary importance. The search here continues, but in any case it has become evident that the wholesale rejection of infinity is not a good thing. In this search, the mathematician is guided by principles that go beyond mathematics, namely, a search for simplicity, order, linearity. When confronted with two different routes for approaching a new concept or a problem, he will follow Ockham's choice, namely, the simple way. Only then will he not be mired forever in endless complications, and only then may he achieve the goals of his research.

2.7 Set Theory

The revolution that allowed infinity to enter mathematics in a very precise, meaningful way begins with the notion of set, introduced by Bernard Bolzano in 1847, one year before his death: a naive definition of a set is as *an aggregate of objects, irrespective of ordering*.

However, it was only with Georg Cantor that the mathematical foundation of set theory started. Before him, there was only one mathematical infinity, namely, the negation of the finite, the unreachable. In his paper of 1874, which was destined to change the course of mathematics, he noted that there were different orders of infinity. Cantor writes $\aleph_0$ for the infinity arising from the primitive counting as in Gamow's book, namely, $1, 2, 3 \ldots, \infty$. Then he shows that the continuum is a different type of infinity, because it cannot be counted. His argument to prove this is really remarkable, and I will repeat it here, albeit in a nonrigorous form.

Let us say that a real number between 0 and 1 is an infinite sequence such as $0.643546\ldots$, namely, 0. followed by an infinite sequence of decimal digits not all of them equal to 9 from some point onward. Suppose that we can count these real numbers. Then we may put them in a list such as

$$0.6435466754325346456001112\ldots$$
$$0.1000534536475455460438 60\ldots$$
$$0.00000000000010000453 4237\ldots$$
$$0.9999999996104567473 2017\ldots$$
$$0.2229556003330545645 01179\ldots$$

....

Now consider the sequence of digits $0.60095\ldots$, where the nth digit is precisely the nth digit of the nth element of the list, and construct a new number A by adding 1 to every digit from 0 to 7, but subtracting 1 if the digit is 8 or 9. In the preceding example, we get $A = 0.71186\ldots$ Then the new number A cannot be in the list, because, for every n, the nth digit of this number does not match the nth digit of the nth element of the list. Cantor then proceeds to define a kind of calculus on infinities. The continuum is written $2^{\aleph_0}$, the cardinality of the set of all subsets of the natural integers. (The notation 2^A for the set of subsets of A is consistent. The number of subsets of the set $\{1, 2, \ldots, n\}$ is 2^n.)

A famous question posed by Cantor was whether the first infinite cardinal larger than $\aleph_0$, denoted by $\aleph_1$, was indeed the continuum $2^{\aleph_0}$. Thus, we refer to $\aleph_1 = 2^{\aleph_0}$ as the continuum hypothesis CH. A common misconception is that CH is the statement $continuum = 2^{\aleph_0}$, which actually is the definition of the continuum. This mistake was perpetrated even by Gamow in his book, where he flatly identifies $\aleph_1$ with the number of points on a line (i.e., $2^{\aleph_0}$) and $\aleph_2$ with the numbers of all functions (i.e., $2^{2^{\aleph_0}}$, the next step of the generalized CH). It took ninety years to answer Cantor's question. The story of the problem, and its surprising solution, is instructive.

First of all, difficulties arise when we consider sets whose elements are sets. The first paradox (essentially equivalent to the Burali-Forti paradox) of set theory occurs when we try to give a meaning to a phrase such as "the set of all sets." Another problem, the famous Russell paradox that destroyed Frege's work in logic, arises when we accept the plausible idea that the totality of objects that have a given property forms a set. Russell's paradox consists in considering the set S of all sets that do not contain itself as an element. Now, if S is an element of S, then, by its very definition, S is not an element of S. On the other hand, if S is not an element of S, then S is an element of S. In either case, we have a contradiction.

Other paradoxes were found. In response, some mathematicians, including David Hilbert, took a constructivist view of mathematics, in which only finite, constructible objects were allowed, together with very precise rules of inference; going even further, the intuitionistic school of the Dutch mathematician Luitzen Brouwer asserted that we cannot have a mathematical object unless it comes with a way of constructing it. On the opposite side, other mathematicians refined the rules of set theory, trying to make them consistent. Eventually, through the work of Zermelo and Fraenkel, eight basic axioms were accepted as the foundation of set theory, in what is now called ZF theory. A ninth axiom, the axiom of choice AC, stating that from a collection of nonempty sets we can always choose a member from each set in the collection, was added in what is called the ZFC model of set theory. The sixth axiom of ZF is the existence of an infinite set, the acceptance of infinity.

The axiom of choice AC was shown, by Kurt Gödel and Paul J. Cohen, to be independent of ZF. Thus, if ZF is consistent, then not only ZFC is consistent, but also ZF together with the negation of AC is consistent. Most mathematicians accept AC, but some are not happy with it because it also yields results contrary to our intuition coming from the real world. A well-known example is the Banach-Tarski paradox: With AC, it is possible to decompose the closed unit ball in 3-space into five pieces and reassemble these pieces, without deforming them, to form two copies of the same ball. As strange as it may seem, there is no contradiction here. The pieces in question are not measurable and we cannot, for example, speak of their volume. In essence, what would be achieved by stretching is achieved here by allowing a decomposition in totally weird pieces.

What then is the status of CH? Gödel first proved that CH is consistent with ZFC, and finally Paul Cohen proved that its negation, ¬CH, is also consistent with ZFC. In other words, the CH, as well as its negation, can be added to ZFC as an axiom. However, the last word on the subject has not been written yet, and very interesting alternatives to ZFC have been put forward, as with Grothendieck's universes. In naive terms, a universe U is a set in which all the operations of set theory can be performed, without ever getting out of U. The existence of a universe U cannot be proved from ZF (it is equivalent to the existence of strongly inaccessible large cardinals). Universes provide models of ZF in which it makes sense to speak of "the set of all sets."

2.8 Geometry, Infinity, and the Peano Curve

In the process of this revision of analysis, intuitive concepts such as continuity went through a complete review. Pathological examples, showing that continuity could lead to unforeseen possibilities, were found. Famous examples are Weierstrass' construction of a function of one variable, continuous but nowhere differentiable, and Peano's example of a continuous map $\varphi : [0, 1] \to [0, 1] \times [0, 1]$ of the unit interval to the unit square, with range the whole unit square. These examples are counterintuitive to the notion of tangent of a curve, and to the very notion of curve. How was it possible to have a continuous curve completely filling a square? In a letter of May 20, 1893, to his friend Stieltjes (*Correspondance d'Hermite et de Stieltjes 1905*. T.2, letter 374), the great mathematician Charles Hermite wrote, referring to certain series that were not convergent and complimenting Stieltjes on his expansion in continued fraction of a complicated integral: "*Mais ces développements, si élegants, sont frappés de malédiction; leurs dérivees d'ordre $2m + 1$ et $2m + 2$ sont des séries qui n'ont aucun sens. L'Analyse retire d'une main ce qu'elle donne de l'autre. Je me détourne evec effroi et horreur de cette plaie lamentable des fonctions continues qui n'ont point de dérivées et je viens vous féliciter bien vivement de votre merveilleux développement en fraction continue de l'intégrale....*"

This purist view has changed today. Indeed, nowhere differentiable functions do appear quite often as solutions of very natural problems ranging from number theory to dynamical systems. Thus, they have lost their early negative connotation as useless pathological examples.

In geometry, one of the very first concepts that appear is the notion of dimension. A point has dimension 0, a curve has dimension 1, a surface has dimension 2, and so on. The dimension represents the degrees of freedom for motion on the geometric object. So, we live in three-dimensional space, four-dimensional if we take into account time (a different type of freedom there, because the motion in time is only forward). Algebraic geometers (they study geometric objects defined by systems of polynomial equations) routinely talked about curves having ∞^1 points, surfaces having ∞^2 points, and three-dimensional space having ∞^3 points. All this was put into discussion by Peano's example of a curve filling a square, which indicated that it was a "curve" of dimension 2. How could this happen? What was the correct notion of dimension?

It turns out that Peano's curve has self-intersections, and, hence, the corresponding map it gives from [0, 1] is not injective. In fact, any continuous surjective map like Peano's curve cannot be injective. The geometric notion of dimension, by now classic, is that the dimension of a space S is the smallest integer d such that any open cover of the space can be refined to a cover in which no point of S is covered more than $d + 1$ times. With this definition, the dimension turns out to be invariant by continuous injective mappings, thereby conforming to the naive intuition that, for example, a plane has "more points" than a line. This does not contradict the theorem of Cantor that there is a biunivocal map between the points of an interval and the points of a square; actually, it clarifies that any such map must be far from continuous. In retrospect, this should not be surprising, because continuity is a notion of ordering, while sets are unordered aggregates of elements. The peculiarity of the Peano curve is now explained.

Although until recently the Peano curve was considered by many to be a strange pathological object, it is of interest that such curves have found practical applications. For example, we may imagine data given by points in a square and an idea for sorting a set of particular points that consists in sorting them in the order by which they appear on the Peano curve. Sorting methods of this type are actually used in several sophisticated software applications for handling very large databases.

Another aspect of infinity appears in the notion of self-similarity. If one looks at a geographical map, one finds that the general shape of coastlines appears to be independent of the scale of the map. This phenomenon of self-similarity occurs quite often in mathematical models of dynamical systems (depending on parameters) in which the evolution of the system is determined by the repeated, in the limit infinite, iteration of simple rules. For certain values of the parameters the behavior of the dynamical system may be chaotic, and it becomes important to study the structure of the boundary between regular and chaotic behavior. It turns out that very often this boundary has a self-similarity property, namely, after a change of scale it looks essentially the same, and more and more so by iterating the change of scale ad infinitum, an echo of Voltaire's Micromégas universe.

The study of such sets, called fractals, has become of considerable importance for modeling chaotic phenomena. Today, fractals are no longer considered as "pathological" objects.

A fundamental concept of geometry is the notion of boundary. In antiquity, an argument given in favor of an infinite universe was the following: If the universe is finite, what lies beyond its boundary? If the earth is flat, what is its boundary and what could be beyond it? Against this argument stood the Zeno paradox, which could be used to explain that the boundary could never be reached, and the examples of the circle or the sphere, which are finite but without boundary.

Today, mathematical models can capture all such possibilities by means of simple examples. A disk with hyperbolic metric has its circumference as its boundary, but any curve reaching the boundary must have infinite length, making the boundary "unreachable." This is an example of a manifold with a complete metric. If instead we adopt the usual Euclidean metric, we can reach the boundary following a path of finite length, so the metric here is not complete. Geometers have studied ways by which one may enlarge a space with a metric that is not complete, so, to extend the space and the metric to a new space with a complete metric. This leads to new objects that capture all of the previous space but also have much better geometric properties, because paths will never stop abruptly at the boundary and can always be extended to infinite length. Also, geometry in infinite dimensions has become a basic tool in mathematics, with the theory of Hilbert spaces extending Euclidean geometry in the plane and space.

The quest for understanding the shape of our universe has gone today way beyond the simplistic models of the past, adding a time dimension to the usual three-dimensional space, adding curvature to space as in general relativity, and even adding an extra six or seven dimensions to unify microcosm and macrocosm as in the current models of the universe proposed by string theorists. Certainly, this challenge to unify the universe of the galaxies with the subatomic universe is one of the most exciting aspects of modern theoretical physics. If the universe is finite, as seems to be the consensus at the present moment, what will be its geometry?

2.9 A Success of Set Theory

It would be wrong to consider set theory as an esoteric small branch of mathematics, an end unto itself. It has produced a number of significant results in pure mathematics, all the way to elementary number-theoretic statements. Here is my favorite example of this.

A diophantine equation is, in its simplest version,

$$f(x_1, x_2, \ldots, x_n) = 0,$$

where f is a polynomial with integer coefficients, to be solved in integers $x_1, x_2, \ldots, x_n$. A related problem, namely, the solution in rational numbers, is a special case of the above when the polynomial f is homogeneous, that is, composed of monomials all with the same total degree d (excluding now the trivial zero solution).

Because zero is the only integer divisible by every natural integer, the first step in solving a diophantine equation is asking the solvability of a divisibility problem, namely, whether for any natural integer N we can find integers $x_1, \ldots, x_n$ (not all 0 if f is homogeneous) such that $f(x_1, \ldots, x_n)$ is divisible by N. It is quite easy to show that we need to prove such a thing just when N is a prime power p^n, for all primes p and all *n*. If we can do this for a given prime *p* and every *n*, we obtain what mathematicians call a *p*-adic solution to the problem.

It is not difficult to give examples, for any prime number *p*, of homogeneous polynomials *f* of degree *d*, in d^2 variables, for which the equation $f = 0$ has no *p*-adic solutions. On the basis of this and the evidence of several examples, Emil Artin conjectured that if we have more than d^2 variables, then the homogeneous problem of degree *d* always has a *p*-adic solution.

The first big surprise was the proof, by Ax and Kochen in 1965 (Ax and Kochen 1965), of Artin's conjecture with the additional condition that *p* is larger than some quantity dependent on *d* but independent of the polynomial *f*. Their proof relied in an essential way on ZFC and, initially, on the assumption of CH. The second surprise arrived with a construction, by G. Terjanian (Terjanian, G. 1966), of a homogeneous polynomial of degree 4 in eighteen variables but without 2-adic solutions. Thus, the additional constraint imposed by Ax and Kochen cannot be avoided. Although a constructive proof of the Ax-Kochen theorem was eventually found by Paul Cohen (1969), the work by Ax and Kochen showed the power of mathematical logic in tackling, with success, difficult questions in mainstream mathematics.

2.10 The Turing Machine and the Busy Beaver

The Turing machine is a famous thought experiment by Alan Turing that shows how logical calculation works. It is like a typewriter: a tape, infinite in both directions, and a tape head for read and write. It has only two symbols, 1 and 0, representing one bit. The machine, at any given moment, is in one of a possible *N* states; there is also an additional special state *H*, called the halting state. There is a table *I* of instructions (*Sb*), as follows: Given a state *S* and a bit *b*, the instruction gives a new state *Snew*, a new bit *bnew*, and a move *M* of the tape by one step, which can be only either *R* to the right or *L* to the left.

The Turing machine works as follows:

INITIALIZATION: a tape of symbols (the input) and a table *T* of instructions (the program)
RUN:
    READ the current symbol *b* on the tape
    DO with current state *S* and instruction *Sb*
        IF *Sb* = *H* then STOP
        ELSE
            MOVE to state *Snew*
            WRITE the symbol *bnew* on the tape
            MOVE the tape one step right or left according to instruction *Sb*
    CONTINUE
END:

The fundamental fact is that such a simple machine can emulate any computer and any finite program. As a very simple example of how such a machine works, consider a Turing machine with only one state *A* and the halting state *H*, with instruction table

$A0 \rightarrow (A1, R)$, $A1 \rightarrow (H)$, and an input tape. The machine starts in state A, printing 1 as long as it reads 0, and will keep moving to the right until it reads 1, when it will stop having printed a string of consecutive $1's$ to its left; it will not stop otherwise.

Is it possible to understand whether such a Turing machine will stop or not? The simplest way to approach the problem is to consider a blank tape of 0 as the initial tape and machines with N states only and two symbols 0 and 1. There are only finitely many possible tables of instructions. Consider now only the tables I for which the Turing machine stops. Is it possible to determine the maximum number of ones that may be printed on the tape, the maximum length needed for the tape, and the maximum number of steps before the Turing machine halts? These problems are all of a similar nature, and mathematicians call the first one the "busy beaver problem." The beaver (i.e., a Turing machine with N states, two symbols, and a table of instructions) starts working in a river (the blank tape) carrying sticks (the bits equal to 1), to build the biggest dam (the largest number of 1 when the machine halts). If the beaver never stops working, call it a "foolish beaver." Otherwise, there will be a "champion beaver" (possibly more than one) finishing with the biggest dam. Call it a "busy beaver" and let $\Sigma(N)$ be the number of sticks (i.e., 1's on the tape) at the end of the work of the busy beaver.

For small values of N, the value of $\Sigma(N)$ can be computed by a clever analysis of all cases, so it is known that $\Sigma(1) = 1$, $\Sigma(2) = 4$, $\Sigma(3) = 6$, $\Sigma(4) = 13$. However, the number of Turing machines to analyze grows very quickly, because there are $(4n + 4)^{2n}$ Turing machines with n states and two symbols. There are already $63, 403, 380, 965, 376$ Turing machines with five states and two symbols, so a case-by-case analysis is difficult, to say the least. However, by exploiting symmetries and finding conditions necessary for the Turing machine to stop, mathematicians have shown that for determining $\Sigma(5)$ there remain 164 Turing machines for which the halting problem has not been solved yet (there is no a priori reason why the halting problem should be decidable for all of them). Current lower bounds for the next values of the busy beaver function are $\Sigma(5) \geq 4098$ (by Marxen and Buntrock in 1989), $\Sigma(6) > 2.50 * 10^{881}$ (by Ligocki and Ligocki in 2007). If $S(n)$ is the maximum number of steps that an n-state, two-symbol Turing machine as shown above can perform before halting, it is known that $S(1) = 1$, $S(2) = 6$, $S(3) = 21$, $S(4) = 107$, $S(5) > 47, 176, 870$, and $S(6) > 8.929 * 10^{1762}$.

One should not be misled into thinking that the working of Turing machines with only a few possible states and symbols must be simple. Any description of a Turing machine (i.e., states, symbols, rules, initial tape) can be encoded in a tape of symbols. Turing showed that a universal machine U exists that emulates any Turing machine M. In his words, "It is possible to invent a single machine which can be used to compute any computable sequence. If this machine U is supplied with a tape on the beginning of which is written the S.D[6] of some computing machine M, then U will compute the same sequence as M" (see Davis 1965, p. 127).

The existence of U was, at the time, a great surprise, and it may be considered as the theoretical equivalent of the modern computer. Note, however, that, unlike the computer, the Turing machine has no fast access to its memory (the tape), so the Turing

---

[6] The S.D is the "standard description" given by Turing for a machine.

machine is not a good model of a computer with respect to execution time. At any rate, it turns out that one can actually produce relatively simple examples of universal Turing machines with very few states and symbols. Minsky gave an example of such a machine U with seven states and four symbols and showed that any Turing machine can be represented by a machine with only two symbols. It is not known what is the smallest number of states for a machine U with two symbols, although there are some candidates for it requiring only four states.[7]

THE THIRD BUSY BEAVER

$$S = A\,B\,C\,H$$

$$I = A0 \rightarrow 1RB,\ \ A1 \rightarrow 1LC,\ \ B0 \rightarrow 1LA,\ \ B1 \rightarrow 1RB,\ \ C0 \rightarrow 1LB,\ \ C1 \rightarrow H$$

The program runs as follows, where the highlighted box shows the current position of the tape head before the next reading step:

A ...	0	0	0	0	**0**	0	0	0	...
B ...	0	0	0	0	1	**0**	0	0	...
A ...	0	0	0	0	**1**	1	0	0	...
C ...	0	0	0	**0**	1	1	0	0	...
B ...	0	0	0	**1**	1	1	0	0	...
A ...	0	0	**0**	1	1	1	0	0	...
B ...	0	0	1	**1**	1	1	0	0	...
B ...	0	0	1	1	**1**	1	0	0	...
B ...	0	0	1	1	1	**1**	0	0	...
B ...	0	0	1	1	1	1	**0**	0	...
B ...	0	0	1	1	1	1	**0**	0	...
A ...	0	0	1	1	1	1	**1**	0	...
C ...	0	0	1	1	1	**1**	1	0	...
H	HALT								

giving a string of six 1's, taking thirteen steps to end, and using six cells of the tape (two to the right and three to the left). Note also that this busy beaver is not the slowest, because $S(3) = 21$; this number of steps is attained by another busy beaver, thus showing that there may be more than one champion beaver.

Notwithstanding these results, it was proved by Tibor Radó in 1962 (Radó 1962; Lin and Radó 1965) that the function $\Sigma(N)$ is not computable by a Turing machine, because it grows faster than any function computable by a Turing machine. In other words, there is no Turing machine that for all N computes $\Sigma(N)$ on input N. The existence of uncomputable functions (i.e., not representable by any computer program) gives us a hint of the unapproachability of infinity by finite means. The halting problem for Turing machines is undecidable, so there is no finite algorithm that, for every Turing machine, decides whether a Turing machine will halt or not.

---

[7] A recent claim (2007) for a machine U with two states and three symbols, for which a \$25,000 prize has been awarded, has given rise to controversy because it allows the use of special infinite input data, which is excluded in classical models of Turing machines. See the criticism in http://cs.nyu.edu/pipermail/fom/2007-October/012156.html and the response in http://forum.wolframscience.com/showthread.php?s=&theadid=1472.

Thus, even the study of finite algorithms immediately brings us to the threshold of infinity and to the undecidable.

2.11 The Infinite in the Finite: P versus NP

The advent of computers has brought back the Aristotelian, finitistic point of view closer to the real world, so a new notion of finite has taken hold. On the one hand, we have the finite reachable by a computing machine, exemplified by short programs with short running time; on the other hand, we have the impossibly large finite, exemplified by short programs with a running time greater than the age of the universe, exemplified by the busy beaver program in two symbols and six states, with a set of only twelve instructions and a running time in excess of 10^{865} steps.

The new central question posed by such a notion is best described by means of an example. Suppose we write a very large integer, say, of 500 digits, and ask to factor it. Suppose we also ask the help of the Oracle of Delphi to solve the problem, and that we do get an answer. Then we can quickly check the truth of the oracle by multiplying together the factors. Because multiplication of two numbers can be done in a few steps, numbering not more than the square of the number of bits needed to describe the two factors (think of the number of digits), we speak of multiplication, hence also of checking factorization, as having polynomial complexity.

On the other hand, factoring large numbers (say, of several hundred digits) remains extremely difficult and still represents a challenge well beyond the power of computers and mathematicians. For a theoretical model of quantum computing, Peter Shor (1994) has proved that this problem is in P. However, for classical computing it is so far only in a subexponential, but not polynomial, complexity class; opinions are divided on whether it may belong to P also in this case.

This motivates the following two definitions.

The class of problems soluble in polynomial time is denoted by P. Instead, the class NP of problems solvable in nondeterministic polynomial time is the class of problems solvable in polynomial time, but with the help of an oracle: the oracle gives us a solution, checkable in polynomial time. Of course, all this needs precise definitions of what we mean by a computer and by an oracle, but this is outside of the scope of this exposition. It is obvious that $P \subset NP$.

The big discovery, by Stephen Cook in 1971 (Cook 1971), was that there are problems that are NP-complete: A problem is NP-complete if

(i) it is in NP, and
(ii) it is NP-hard, that is, every problem in NP can be reduced to it, in polynomial time.

The conclusion is that if there is one NP-complete problem in the class P, then every problem in NP belongs to P; in other words, $NP = P$.

It is not obvious why NP-complete problems should exist at all. Why should every problem in a very large class such as NP be equivalent, in polynomial time, to a fixed single problem? If we could show that there is an NP-complete problem solvable deterministically in polynomial time, then it would mean that $NP = P$ and the presence of an oracle would be irrelevant for solving problems in P. If checking the validity of a proposed solution could be done in polynomial time, the same would be true for the apparently much harder problem of producing an actual solution. For example, the problem of finding a clique of size k in a graph (an arrangement of points, called vertices, some of which are joined by segments, called edges) is NP-complete. Obviously, it is in P: the oracle simply points out to the viewer the k vertices of a clique, and the viewer checks the presence of the $k(k-1)/2$ edges forming the clique. However, finding the clique without the help of the oracle is very hard, as our experience with the Ramsey number $R(6, 6)$ shows (at least, according to Erdős).

By now, thousands of classical problems in finite combinatorics have been shown to be NP-complete, and most mathematicians believe that NP is not equal to P. Notwithstanding very deep studies of the question, the problem remains open. Is it possible that the solution to this basic problem, with its finitistic formulation, will require a daring excursion into the realm of infinity? Only time will tell. Certainly, the computer has shown us, in a dramatic way, the distinction between the "finite" in real life and the "finite" beyond our grasp, as in George Gamow's *One, Two, Three . . . Infinity*.

Acknowledgments

The author ackowledges the following additional source, besides those quoted in the References, used for the compilation of this chapter: Wikipedia, the Free Encyclopedia, in http:/en.wikipedia.org/wiki, for technical information and the Erdős quote. Figures 2.1 and 2.2 were drawn using the Mathematica software program. The picture of the Hydra is by permission from the J. Paul Getty Museum, Villa Collection, Malibu, California.

References

Ax, J., and Kochen, S. 1965. Diophantine problems over local fields. I. *American Journal of Mathematics* 87: 605–30.

Bartocci, U., and Vipera, M.C. 2004. Variazioni sul problema dei buoi di Archimede, ovvero, alla ricerca di soluzioni "possibili" . . . http://www.cartesio-episteme.net/mat/cattle-engl.htm.

Busard, H. L. L. 2001. *Johannes de Tinemue's Redaction of Euclid's Elements, the So-called Adelard III version*. vol. 1. Stuttgart: Franz Steiner Verlag.

Cohen, P. J. 1969. Decision procedures for real and p-adic fields. *Communications on Pure and Applied Mathematics* 22: 131–51.

Cook, S. 1971. The complexity of theorem proving procedures. In *Proceedings, Third Annual ACM Symposium on the Theory of Computing*, ACM, New York, pp. 151–58.

Correspondence d'Hermite et de Stieltjes. 1905. T.II. Paris: Gauthier-Villars.

Dante Alighieri. 1966–1967. *La Commedia secondo l'antica vulgata a cura di Giorgio Petrocchi*. Edizione Nazionale a cura della Società Dantesca Italiana. Milano: Arnoldo Mondadori Editore.

Davis, M. (ed.) 1965. *The Undecidable*. Hewlett, NY: Raven Press. http://cs.nyu.edu/pipermail/fom/2007-October/012156.html. http://forum.wolframscience.com/showthread.php?s=&threadid=1472.

Dickson, L. E. 1952. *History of the Theory of Numbers*, vol. 2. Reprint New York: Chelsea.

De Morgan, A. 1872. *A Budget of Paradoxes*. London: Longmans, Green.

Gamow, G. 1947. *One, Two, Three...Infinity: Facts and Speculations of Science.* Illustrated by author. Rev. ed. 1961. London: Macmillan.

Gardner, M. 1975. *Mathematical Carnival.* New York: Borzoi Books, Alfred A. Knopf. p. 264. http://www-groups.dcs.st-and.ac.uk/~history/Mathematicians.

Geymonat, M. 2006. *Il grande Archimede.* 2nd ed. Roma: Sandro Teti Editore.

Netz, R., Saito, K., and Tchernetska, N. 2001. A new reading of Method Proposition 14: Preliminary evidence from the Archimedes palimpsest. I. *Source and Commentaries in Exact Sciences* 2: 9–29.

Nygrén, A. 2001. A simple solution to Archimedes' cattle problem. University of Oulu, Linnanmaa, Oulu, Finland. Acta Universitatis, Ouluensis, Scientiae Rerum Naturalium. http://herkules.oulu.fi/isbn9514259327/index.html?tulosta=yes.

Paris, J., and Harrington, L. 1977. A mathematical incompleteness in Peano arithmetic. In *Proof Theory and Constructive Mathematics*, with contributions by C. Smorynski, Helmut Schwichtenberg, Richard Statman, et al. *Handbook of Mathematical Logic*, part D, pp. 819–1142. *Studies in Logic and the Foundations of Math*, vol. 90. Amsterdam: North-Holland.

Radó, T. 1962. On non-computable functions. *Bell System Technical Journal* 41: 877–84.

Lin, S., and Radó, T. 1965. Computer studies of Turing machine problems. *Journal of the Association of Computing Machinery* 12: 196–212.

Shor, P. W. 1994. Algorithms for quantum computation: Discrete logarithms and factoring. In *35th Annual Symposium on Foundations of Computer Science, Santa Fe, New Mexico*, pp. 124–34. Los Alamitos, CA: IEEE Computer Society Press.

Terjanian, G. 1966. Un contre-example à une conjecture d'Artin. *Compte Rendus de L'Académie des Sciences Paris Séries A-B* 262: A612.

Vardi, I. 1998. Archimedes' cattle problem. *American Mathematical Monthly* 105: 305–19.

CHAPTER 3

Warning Signs of a Possible Collapse of Contemporary Mathematics

Edward Nelson

I rejoice that we live in a world of boundless, infinite possibilities, one in which, with Blake, we can see a world in a grain of sand and a heaven in a wildflower, hold infinity in the palm of our hand and eternity in an hour. I rejoice that the sacred scriptures of our faith portray a God who listens to prayer, who loves us and longs to lead us. I rejoice that my chosen line of work, mathematics, has enabled me to bring into being new things that did not exist before and to greet with wonder and awe many amazing inventions of my fellow workers. I rejoice that daily we live immersed in infinity, that we have the freedom not only to make choices but at times to be the agent, by will or by grace, to sing to the Lord a new song.

Is infinity real? For example, are there infinitely many numbers? Yes indeed. When my granddaughter was a preschooler, she asked for a problem to solve. I gave her two seventeen-digit numbers, chosen arbitrarily except that no carrying would be involved in finding the sum. When she summed the two numbers correctly, she was overjoyed to hear that she had solved a mathematical problem that no one had ever solved before.

The celebration of infinity is the celebration of life, of newness, of becoming, of the wonder of possibilities that cannot be listed in a finished, static rubric. The very etymology of the word infinite is "unfinished." As Aristotle observed, infinity is always potential and never actual or completed.

So what are we to make of the contrasting notion of a completed infinity? I confess at the outset to the strong emotions of loathing and feeling of oppression that the contemplation of an actual infinity arouses in me. It is the antithesis of life, of newness, of becoming – it is finished.

Consider a cosmology in which space is actually infinite. Assuming Euclidean geometry for simplicity of discussion, divide space into cubes of side 100^{100} light-years, called "local regions," and call two local regions "widely separated" in the case that they are at least $100^{100^{100}}$ light-years apart. If the world is deterministic, then it is a tale told by an idiot, full of sound and fury, signifying nothing. Otherwise, chance plays a role. There can be no causal influence between widely separated local regions, and it is a simple and unavoidable result of probability theory that if we have an actual infinity of independent trials of an event, then if the event is possible at all it is certain to occur, and certain to occur infinitely often. Thus, there will be infinitely many local regions with meetings like ours in San Marino where exactly the same words are spoken, and since it is *possible* for it to snow in August in San Marino, there will be infinitely many local regions with a blizzard at a meeting otherwise just like ours. A world in which space is actually infinite is a tale told by infinitely many idiots, full of sound and fury, signifying everything conceivable.

Turning from cosmology to theology, I want to emphasize that in this paragraph I am speaking personally. I have no wish to belittle thoughts that others may find helpful in attempting partially to understand the infinite mystery of God, or even to appear to do so. I shall just describe some thoughts that I personally find unhelpful. I am uneasy when abstract concepts, such as "omniscient" and "omnipotent," are posited as attributes of God. To my mind, these are formal linguistic constructions, unrelated to anything in our experience, with no clear meaning and prone to paradox. ("I can find a stone too heavy to lift, which is something God can't do.") I have the same reservations about the concept of actual infinity. To me these are all cold and debatable notions, and I find their application to God unscriptural and unhelpful to my understanding or worship.

It is widely believed that there is a clear and correct theory of actual infinity in mathematics. Certainly if there is not, then there cannot be a clear and correct use of actual infinity in cosmology or any other branch of science. I want to examine that belief. Let's turn to the concept of actual infinity in mathematics.

Actual infinity entered fully into mathematics with the work of Georg Cantor. My introduction to Cantor's theory, and a thrilling one it was, occurred at the age of sixteen when I picked up a copy of Bertrand Russell's *Introduction to Mathematical Philosophy* in Italian translation. A set is said to be of cardinality $\aleph_0$ in case it can be put in one-to-one correspondence with the natural numbers $0, 1, 2, \ldots$. To illustrate the axiom of choice, Russell invents the tale of the millionaire who bought $\aleph_0$ pairs of shoes and $\aleph_0$ pairs of socks. Then he has $\aleph_0$ shoes (as opposed to pairs of shoes), since we can make a one-to-one correspondence between the shoes and the natural numbers: mark with 0 the left shoe of the first pair, with 1 the right shoe of the first pair, with 2 the left shoe of the second pair, and so on. But how, Russell asks, can we do this with socks, where there is nothing to distinguish one sock in a pair from the other? This requires choosing a sock in each pair, and the axiom of choice permits this. This fable convinced me that the axiom of choice is true. Well, we are all Platonists in our youth.

The most impressive feature of Cantor's theory is that he showed that there are different sizes of infinity, by his famous diagonal argument. But Russell applied this argument to establish his paradox: the set of all sets that are not elements of themselves both is and is not an element of itself. Actually, Russell's paradox was in response to Frege's work, not Cantor's. Frege gave a clear and precise account of his work, making it possible for Russell to show that it was wrong, whereas Cantor's work was in parts so vague and imprecise that, as Pauli said of another theory, it was not even wrong.

This is the first warning sign of trouble in contemporary mathematics: the intuitive notion of infinite sets leads to a contradiction.

Important progress was made by Zermelo, who wrote axioms for set theory, one of which was unclear. His work was later extended by Fraenkel and now we have ZFC, Zermelo-Fraenkel set theory with the axiom of choice, which is commonly taken as

Figure 3.1. The tale of ω.

the foundational theory for contemporary mathematics. But how do we know that ZFC is a consistent theory, free of contradictions? The short answer is that we don't; it is a matter of faith (or of skepticism). I took an informal poll among some students of foundations, and by and large the going odds on the consistency of ZFC were only 100 to 1, a far cry from the certainty popularly attributed to mathematical knowledge.

But I don't want to say more about set theory. To my mind, the trouble – and I believe there *is* trouble – lies deeper. Let's consider arithmetic. The natural numbers, which henceforth I'll just call numbers, begin with 0, and each is followed by its successor, obtained by adding 1. The notion of successor is more basic than that of addition, so it is customary to introduce the symbol S for successor. Thus the numbers are pictured as in Figure 3.1.

The tale of ω goes on forever, so to endow it with physical reality we would need a cosmology in which space is actually infinite.

The principal tool for proving theorems in arithmetic is induction, which may be stated as follows: if a property of numbers holds for 0, and if it holds for the successor of every number for which it holds, then it holds for all numbers. This appears to be innocuous. Suppose that a property satisfies the two hypotheses (called, respectively, the basis and the induction step) and we want to prove that the property holds for some number, say SSS0. We know that it holds for 0 by the basis, so it holds for S0 by the induction step, and it holds for SS0 and then for SSS0 for the same reason, concluding the proof.

One problem is that the notion of "property" is vague. We could reify it as a collection of numbers, but this leads us back into set theory. The other way – and this is what today is meant by Peano Arithmetic (P) – is to introduce a formal language and replace the vague notion of property by the precise, syntactical, and concrete notion of a formula of the language. Then the axioms of P are as follows:

AXIOM 1. Not $Sx = 0$.
AXIOM 2. If $Sx = Sy$, then $x = y$.
AXIOM 3. $x + 0 = x$.
AXIOM 4. $x + Sy = S(x + y)$.
AXIOM 5. $x \cdot 0 = 0$.
AXIOM 6. $x \cdot Sy = (x \cdot y) + x$.
AXIOM 7. If $\varphi(0)$, and if for all x, $\varphi(x)$ implies $\varphi(Sx)$, then for all x, $\varphi(x)$.

In Axiom 7, the induction axioms, φ is any formula of the language of arithmetic. This formulation of P is still not satisfactory because of the use of "Not," "if . . . then," "implies," and "for all." Let's rewrite the axioms using the usual symbolism of mathematical logic:

AXIOM 1. $\neg\, Sx = 0$.
AXIOM 2. $Sx = Sy \rightarrow x = y$.

AXIOM 3. $x + 0 = x$.

AXIOM 4. $x + Sy = S(x + y)$.

AXIOM 5. $x \cdot 0 = 0$.

AXIOM 6. $x \cdot Sy = (x \cdot y) + x$.

AXIOM 7. $\varphi(0) \,\&\, \forall x[\varphi(x) \to \varphi(Sx)] \to \forall x[\varphi(x)]$.

This is not a wanton move, made to keep the sacred mysteries within the priesthood, but an absolutely vital step. In the crowning achievement of his extraordinary career, David Hilbert created proof theory, which for the first time enabled one to discuss mathematical theories with the same precision and clarity that previously had been reserved for mathematical objects, such as groups and topological spaces. The key idea was the formalization of logic, introducing formal symbols, such as ¬, →, and ∨ (which as a useful mnemonic may be read as "not," "implies," and "for all") that are not assigned any meaning; they are simply combined according to certain explicit rules. This separation of syntax from semantics enables one to treat mathematical reasoning itself by the methods of mathematics. Surprisingly, the way to a deep understanding of mathematical reasoning lay in stripping it of meaning. All future progress in mathematical logic – by Gödel, Gentzen, Cohen, and others – depended on this fundamental insight. Having emphasized this point, from now on I'll largely use ordinary language for greater readability, but it must be taken as a surrogate for formal logical symbolism.

Formulas and proofs are combinations of symbols formed according to certain explicit rules. Details can be found in any book that treats the predicate calculus. This formalization of proof makes precise the notion of rigorous argument actually used in mathematical practice, and when properly presented it is far closer to actual practice than mathematicians usually think possible.

Notice that no concept of actual infinity occurs in P, or even in ZFC for that matter. The infinite Figure 3.1 is no part of the theory. This is not surprising, since actual infinity plays no part in mathematical practice (because mathematics is a human activity carried out in the world of daily life). Mathematicians of widely divergent views on the foundations of mathematics can and do agree as to whether a purported proof is indeed a proof; it is just a matter of checking. But each deep open problem in mathematics poses a challenge to confront potential infinity: can one find, among the infinite possibilities of correct reasoning, a proof?

How do we know that P is a consistent theory, free from contradiction? That is, how do we know that we cannot prove both a formula and its negation? The usual answer is this: the actual infinity of all numbers, as in Figure 3.1, provides a model of the theory; Axioms 1 through 7 are true in the model, and the syntactical rules of the theory are truth-preserving: if the premises of a rule of inference are true, so is the conclusion of the rule. Hence every theorem of the theory is true, and since a formula and its negation cannot both be true, the theory is consistent.

To study this argument we must examine the notion of truth in arithmetic. I'll illustrate it by discussing the twin primes conjecture, one of the famous open problems of mathematics. Two primes are called twin primes if they differ by 2; for example, 11 and 13 are twin primes. The twin primes conjecture is that there are infinitely many twin primes. This can be expressed by a formula in the language of arithmetic: "for all n there exists p with p greater than n such that p and $p + 2$ are primes." How can we tell whether this is true or false by looking at Figure 3.1? We may have found a certain number n and searched fruitlessly after that for a very long time, acquiring the suspicion that the conjecture is false, but to be certain we would need to go on forever, completing an actually infinite search. Or contrariwise, we may keep on finding larger and larger twin primes and acquire the suspicion that the conjecture is true, but to be certain we would need to go on forever, completing an actually infinite search. This, of course, is not possible, and some students of foundations deny that, in the phrase beloved by many philosophers, there is a fact of the matter as to whether the conjecture is true or false. But, someone may say, God is omniscient, and in the divine mind there is a fact of the matter. I want to discuss this assertion, which I'll call divine omniscience in arithmetic, from two points of view.

A mathematician's reverie: Wouldn't it be great to have a god who would do actually infinite searches for me and tell me whether formulas of arithmetic are true or false? But lo! Here is a little brass one that will fit nicely on my desk. I'll question it: "Is the twin primes conjecture true?" "Yes." Well, that's good to know, in its way. "Is there a proof in Peano Arithmetic?" "Yes." OK, but what I really want is to see a proof. "In the standard encoding, is the first bit of the shortest proof 0?" "Yes." "Is the second bit 0?" "No." Aha! Then I know that the second bit is 1. (Continuing this extended game of twenty questions, the mathematician obtains a proof.) Hooray! I'll be famous! . . . But mathematics used to be fun, and this is more tedious than the income tax. I'll just mop up the remaining millennium questions and with my six million dollars I'll abandon mathematics and take up beekeeping.

This conceit says something about mathematicians. We are not much concerned with truth; what interests us is proof. And apart from the external trappings of fame and fortune, the driving motivation for doing mathematics is to have fun. I don't feel that this fact requires apology; just don't let the funding agencies know.

But the story tells us nothing about truth in arithmetic. Let's look at divine omniscience in arithmetic again. Consider a different question, "Did Hansel and Gretel drop an even number of bread crumbs or an odd number?" It must be one or the other; there is no third possibility. It is impossible to find a proof one way or another from the facts we know. Nevertheless, someone might say, there is a fact of the matter in the divine mind. But the fallacy here is obvious. The bread crumbs are just a product of human imagination; the story was simply made up. You can see where I am heading.

The notion of the actual infinity of all numbers is a product of human imagination; the story is simply made up. The tale of ω even has the structure of the traditional fairy tale: "Once upon a time there was a number called 0. It had a successor, which in turn had a successor, and all the successors had successors happily ever after."

Some mathematicians, the fundamentalists, believe in the literal inerrancy of the tale, whereas others, the formalists, do not. When mathematicians are doing mathematics, as opposed to talking about mathematics, it makes no difference: the theorems and proofs of the ones are indistinguishable from those of the others.

Let us examine the fundamentalist belief in the existence of the completed infinity ω in the light of monotheistic faith. It is part of monotheistic faith, as I understand it, that everything in creation is contingent; I AM WHO I AM is not constrained by necessity. Are we to believe that ω is contingent, that the truths of arithmetic might have been different had it pleased God to make them so? Or are we to believe that ω is uncreated – existing in its infinite magnitude by necessity, as it was in the beginning, is now, and ever shall be? But these are unreal questions, like "can $\aleph_2$ angels dance on the head of a pin?"

Hilbert, spurred on by the incisive criticisms of classical mathematics by his antagonist Brouwer, realized that the usual argument for the consistency of P is unsatisfactory. A semantic proof of consistency, by appeal to a model, simply replaces the question of the consistency of a simpler theory, such as P, by the question of the consistency of the far more complicated set theory in which the notion of a model can be expressed. He proposed to establish the consistency of classical mathematics, beginning with arithmetic, by concrete syntactical means. It is well known that his program was overly ambitious and that it was shown to be impossible of realization by Gödel. But I want to focus on another aspect of the Hilbert program here. Hilbert is revered and reviled as the founder of formalism, but his formalism was only a tactic in his struggle against Brouwer to preserve classical mathematics: in his deepest beliefs he was a Platonist (what I have called, as a polemical ploy, a fundamentalist). Witness his saying, "No one shall expel us from the Paradise that Cantor has created." Hilbert's mistake, which a radical formalist would not have made, was to pose the problem of *proving* by finitary means the consistency of arithmetic, rather than to pose the problem of *investigating* by finitary means whether arithmetic is consistent.

Gödel's second incompleteness theorem is that P cannot be proved consistent by means expressible in P, provided that P is consistent. This important proviso is often omitted. This theorem I take to be the second warning sign of trouble in contemporary mathematics. Its straightforward significance is this: perhaps P is inconsistent. But this is not how his profound result was received, as a result of the a priori conviction of just about everyone that P *must* be consistent.

How can one doubt the consistency of P? Because of the untamed power of induction, which goes far beyond the original intuitive justification for it. Peano Arithmetic allows for the introduction of primitive recursive functions, as follows. To introduce a new primitive recursive function f, say, of two variables, one first posits a value for $f(x, 0)$ in terms of previously defined functions and then posits a formula for $f(x, Sy)$ in terms of $f(x, y)$ and previously defined functions. The coherence of this scheme depends on the assumption that all of the values introduced in this way denote numbers that can be expressed as numerals, meaning terms of the form SSS . . . 0, as in Figure 3.1. This can be proved in P using induction, but in doing so we extend the meaning of induction, originally justified by intuition on numerals, to new kinds of numbers – exponential numbers, superexponential numbers, and so forth – that are themselves created by induction. The reasoning that primitive recursion as a computational scheme always terminates – that is, that primitive recursive functions are total – is circular, for the number of steps required for the computation to terminate can only be expressed in terms of these new primitive recursive functions themselves. This is illustrated by a fable. A student went to a teacher and a dialogue ensued:

S: Will you teach me arithmetic?

T: Gladly, my boy. 0 is a numeral; if x is a numeral, so is Sx. Numerals are used to count things.

S: I understand.

T.   *Addition* is introduced as follows:

$$x + 0 = x,$$
$$x + Sy = S(x + y).$$

Try some addition.

S: I have encountered no difficulty in reducing sums of numerals to numerals.

T: *Multiplication* is introduced as follows:

$$x \cdot 0 = 0,$$
$$x \cdot Sy = x + (x \cdot y).$$

S: I've tried a few examples and reduced the products to numerals. But what about

$$SS0 \cdot \ldots \cdot SS0$$

with y occurrences of SS0, where y is a long numeral?

T: It reduces to the numeral denoted by SS0 $\uparrow$ y.

S: What is $\uparrow$?

T: It is *exponentiation*, introduced by

$$x \uparrow 0 = S0,$$
$$x \uparrow Sy = x \cdot (x \uparrow y).$$

S: I've tried a few very small examples, but what about

$$SS0 \uparrow \cdots \uparrow SS0$$

with y occurrences of SS0?[1]

T: It reduces to the numeral that SS0 $\Uparrow$ y denotes, where $\Uparrow$ is *superexponentiation*, introduced by

$$x \Uparrow 0 = S0,$$
$$x \Uparrow Sy = x \uparrow (x \Uparrow y).$$

S: I've looked at a couple of nontrivial examples but can't seem to make them work. I'll accept it on your authority. But what about

$$SS0 \Uparrow \cdots \Uparrow SS0$$

with y occurrences of SS0?

T: Introduce *supersuperexponentiation* . . .

S: Excuse me, teacher, but I am having trouble following you. I raised the question as to whether an arbitrarily long product of copies of SS0 can be reduced to a numeral. You responded in terms of $\uparrow$, and to the same question about $\uparrow$ in terms of $\Uparrow$, and so forth. You seem to be presenting us with a skyhook. We want to hang a heavy weight in midair, so we suspend it from a hook, which is itself suspended from a hook, which in turn is suspended from a hook, and so on forever.

---

[1] Infix symbols are to be associated from right to left.

T: Very good, my boy! That is an excellent metaphor – a skyhook is just what the higher arithmetic gives us. And it accommodates not only the primitive recursive functions we have been discussing but much more. Would you like to learn about the Ackermann function? General recursive functions?

S: Thank you, ma'am. Another day, perhaps.

Finitism is usually regarded as the most conservative of all positions on the foundations of mathematics, but finitism accepts all primitive recursive functions as being total on the grounds that the computations involved are finite. But I have argued that they are simply postulated to be finite, by a circular reasoning. The untamed use of induction in P is justifiable only by appeal to ω as a completed infinity. Finitism is the last refuge of the Platonist.

It is one thing to criticize a position but something else to show that it is problematic, so now I turn to a closer examination of finitism. Let us extend P by adjoining a new symbol, which in accord with the decision to use ordinary language I write as "is a counting number." We adjoin two new axioms:

AXIOM 8. 0 is a counting number.

AXIOM 9. If x is a counting number, then Sx is a counting number.

Call the extended theory P'. It is easy to see that if P is consistent then so is P'. This is because we have not defined the new notion of a counting number, and we could, if we wished, define "x is a counting number" to mean "$x = x$," so that all numbers would be counting numbers and Axioms 8 and 9 would hold trivially. But we don't do this; we leave it as an undefined notion.

It follows from Axioms 8 and 9 that 0, S0, SS0, and so forth, are counting numbers. In fact, our intuitive understanding of "number" is the same as that of "counting number." But we cannot prove that all numbers are counting numbers. One might be tempted to try to do so by induction, but the induction axioms of arithmetic were postulated for formulas φ of the specified language of arithmetic and "is a counting number" is not in that language. Moreover, there is a simple semantic argument using Gödel's completeness theorem, which he proved shortly before his more famous incompleteness theorems, showing that one cannot prove that all numbers are counting numbers, or even that if x and y are counting numbers then so is $x + y$.

But we can do something almost as good, called a relativization scheme. We define a new notion, that of *additionable number*, and show that not only are additionable numbers counting numbers but that the sum of two additionable numbers is again an additionable number.

10. DEFINITION. x is an additionable number in case for all counting numbers y, the sum $y + x$ is a counting number.

Then we have:

11. THEOREM. If x is an additionable number, then x is a counting number.

> **PROOF**   Let x be an additionable number. By Axiom 8, 0 is a counting number. By Axiom 10 applied to $y = 0$, $0 + x$ is a counting number, but $0 + x = x$.   □

12. THEOREM. 0 is an additionable number.

   PROOF.   Let y be a counting number. By Axiom 10, we need to show that $y + 0$ is a counting number. But $y + 0 = y$.   □

13. THEOREM. If x is an additionable number, so is Sx.

   PROOF   Let x be an additionable number and let y be a counting number; by Axiom 10, we need to show that $y + Sx$ is a counting number. But $y + Sx = S(y + x)$. By Axiom 10, $y + x$ is a counting number, so by Axiom 9, $S(y + x)$ is indeed a counting number.   □

14. THEOREM. If x_1 and x_2 are additionable numbers, so is $x_1 + x_2$.

   PROOF.   Let x_1 and x_2 be additionable numbers and let y be a counting number. By Axiom 10, we need to show that $y + (x_1 + x_2)$ is a counting number. But $y + (x_1 + x_2) = (y + x_1) + x_2$ (the associative law for addition), and $y + x_1$ is a counting number by Axiom 10, so $(y + x_1) + x_2$ is indeed a counting number, also by Axiom 10.   □

We can define an even stronger notion, *multiplicable number*, and show not only that multiplicable numbers are counting numbers but that the sum and product of two multiplicable numbers is again a multiplicable number.

15. DEFINITION. x is a multiplicable number in case for all additionable numbers y, the product $y \cdot x$ is an additionable number.

By the same kind of elementary reasoning, one can prove the following theorems:

16. THEOREM. If x is a multiplicable number, then x is an additionable number.
17. THEOREM. If x is a multiplicable number, then x is a counting number.
18. THEOREM. 0 is a multiplicable number.
19. THEOREM. If x is a multiplicable number, so is Sx.
20. THEOREM. If x_1 and x_2 are multiplicable numbers, so is $x_1 + x_2$.
21. THEOREM. If x_1 and x_2 are multiplicable numbers, so is $x_1 \cdot x_2$.

The proof of the last theorem uses the associativity of multiplication. The significance of all this is that addition and multiplication are unproblematic. We have defined a new notion, that of a multiplicable number, that is stronger than the notion of counting number, and proved not only that multiplicable numbers have successors that are multiplicable numbers, and hence counting numbers, but that the same is true for sums and products of multiplicable numbers. For any specific numeral SSS . . . 0 we can quickly prove that it is a multiplicable number.

But now we come to a halt. If we attempt to define "exponentiable number" in the same spirit, we are unable to prove that if x_1 and x_2 are exponentiable numbers then so is $x_1 \uparrow x_2$. There is a radical difference between addition and multiplication, on the one hand, and exponentiation, superexponentiation, and so forth, on the other hand. The obstacle is that exponentiation is not associative; for example, $(2 \uparrow 2) \uparrow 3 = 4 \uparrow 3 = 64$, whereas $2 \uparrow (2 \uparrow 3) = 2 \uparrow 8 = 256$. For any specific numeral SSS . . . 0 we can indeed prove that it is an exponentiable number, but we cannot prove that the world of exponentiable numbers is closed under exponentiation. And superexponentiation leads us entirely away from the world of counting numbers.

The belief that exponentiation, superexponentiation, and so forth, applied to numerals yield numerals is just that – a belief. Here we have the third, and most serious, warning sign of trouble in contemporary mathematics.

PART THREE

Technical Perspectives on Infinity from Advanced Mathematics

CHAPTER 4

The Realm of the Infinite

W. Hugh Woodin

4.1 Introduction

The twentieth century witnessed the development and refinement of the mathematical notion of infinity. Here, of course, I am referring primarily to the development of set theory, which is that area of modern mathematics devoted to the study of infinity. This development raises an obvious question: is there a nonphysical realm of infinity?

As is customary in modern set theory, V denotes the universe of sets. The purpose of this notation is to facilitate the (mathematical) discussion of set theory – it does not presuppose any meaning to the concept of the universe of sets.

The basic properties of V are specified by the ZFC axioms. These axioms allow one to infer the existence of a rich collection of sets, a collection that is complex enough to support all of modern mathematics (and this, according to some, is the only point of the conception of the universe of sets).

I shall assume familiarity with elementary aspects of set theory. The *ordinals* calibrate V through the definition of the cumulative hierarchy of sets (Zermelo 1930). The relevant definition is given as follows:

Definition 1. Define for each ordinal α a set V_α by induction on α.

(1) $V_0 = \emptyset$.

(2) $V_{\alpha+1} = \mathcal{P}(V_\alpha) = \{X \mid X \subseteq V_\alpha\}$.

(3) If β is a limit ordinal, then $V_\alpha = \cup\{V_\beta \mid \beta < \alpha\}$.

There is a much more specific version of the question raised concerning the existence of a nonphysical realm of infinity: is the universe of sets a nonphysical realm? It is this latter question that I shall focus on.

There are a number of serious challenges to the claim that the answer is yes. But where do these issues arise? More precisely, for which ordinals α is the conception of V_α meaningful?

The first point that I wish to make is that for a rather specific *finite* value of n, the claim that V_n exists is a falsifiable claim; moreover, this "possibility" is consistent with our collective (formal) experience in mathematics to date. The details are the subject of the next section, and this account is a variation of that given in Woodin (1998). I will continue the narrative by bringing in the basic arguments of Woodin (2004, 2009), ultimately defining a position on mathematical truth that is the collective conclusion of these three papers.

4.2 The Realm of the Finite

By Gödel's Second Incompleteness Theorem, any system of axioms of *reasonable* expressive power is subject to the possibility of inconsistency, including, of course, the axioms for number theory. A natural question is how profound an effect could an inconsistency have on our view of mathematics or indeed on our view of physics.

For each finite integer n, $|V_{n+1}| = 2^{|V_n|}$ and so even for relatively small values of n, V_n is quite large. Is the conception of V_{1000} meaningful? What about the conception of V_n, where $n = |V_{1000}|$?

By a routine Gödel sentence construction, I produce a formula in the language of set theory that implicitly defines a property for finite sequences of length at most 10^{24}. For a given sequence this property is easily decided; if s is a sequence with this property, then s is a sequence of nonnegative integers each less than 10^{24} and the verification can be completed (with appropriate inputs) in significantly fewer than 10^{48} steps.

If there exists a sequence with this property, then the conception of V_n is meaningless where $n = |V_{1000}|$. I use the bound 10^{24} in part because the verification that a candidate sequence of length at most 10^{24} has the indicated property is arguably physically feasible. The question of whether or not there is a sequence of length 10^{24} concerns only the realm of all sequences of length 10^{24} and so is certainly a meaningful question for this realm. The existence of such a sequence has implications for the nonexistence of V_n, where $n = |V_{1000}|$, which is a vastly larger realm. This is entirely analogous to the well-known and often-discussed situation of the number theoretic statement,

"ZFC is formally inconsistent."

This statement concerns only V_ω (i.e., the realm of number theory), and yet its truth has implications for the nonexistence of the universe of sets, again a vastly larger realm.

The philosophical consequences of the existence of a sequence of length 10^{24} as described earlier are clearly profound, for it would demonstrate the necessity of the finiteness of the universe. Clearly such a sequence does not exist. However, this property has the feature that if arbitrarily large sets *do* exist, then there is no proof of *length* less than 10^{24} that no such sequence of length at most 10^{24} can have this property. I shall make these claims more precise.

Is the existence of such a sequence a meaningful question for our actual physical universe? A consequence of quantum mechanics (as opposed to classical mechanics) is that one could really build (on Earth, today) a device with a *nonzero* (although ridiculously small) chance of finding such a sequence if such a sequence exists, which is the other reason for the explicit bound of 10^{24}. So, the claim that no such sequence exists is a prediction about our world.

Now the claim that there is no such sequence is analogous to the claim that there is no formal contradiction in set theory or in set theory together with large cardinal axioms. I do not see any credible argument at present for the former claim other than the claim that the conception of V_n is meaningful where $n = |V_{1000}|$ (although in $2^{10^{26}}$ years there will be such a credible argument). But then what can possibly provide the basis for the latter claim other than some version of the belief that the conception of the universe of sets is also meaningful?

4.2.1 Preliminaries

I shall assume familiarity with set theory at a naive level and below list informally the axioms. I do this because I will need a variation of this system of axioms, and this variation is not a standard one.

AXIOM 0. There exists a set.

AXIOM 1 (Extensionality). Two sets A and B are equal if and only if they have the same elements.

AXIOM 2 (Pairing). If A and B are sets, then there exists a set $C = \{A, B\}$ whose only elements are A and B.

AXIOM 3 (Union). If A is a set, then there exists a set C whose elements are the elements of A.

AXIOM 4 (Powerset). If A is a set, then there exists a set C whose elements are the subsets of A.

AXIOM 5 (Regularity or Foundation). If A is a set, then either A is empty (i.e., A has no elements) or there exists an element C of A that is disjoint from A.

AXIOM 6 (Comprehension). If A is a set, and $P(x)$ formalizes a property of sets, then there exists a set C whose elements are the elements of A with this property.

AXIOM 7 (Axiom of Choice). If A is a set whose elements are pairwise disjoint and each nonempty, then there exists a set C that contains exactly one element from each element of A.

AXIOM 8 (Replacement). If A is a set, and $P(x)$ formalizes a property that defines a function of sets, then there exists a set C that contains as elements all the values of this function acting on the elements of A.

AXIOM 9 (Infinity). There exists a set W that is nonempty and such that for each element A of W there exists an element B of W such that A is an element of B.

Axiom 6 and **Axiom 8** are really infinite lists or schemata corresponding to the possibilities of the *acceptable properties*. These axioms are vague in that it may not be clear what an acceptable property is. Intuitively these properties are those that can be expressed using only the fundamental relationships of equality and set membership and are made mathematically precise through the use of formal mathematical logic.

Axioms 0 through **8** are (essentially) a reformulation of the axioms of number theory. It is the **Axiom of Infinity** that takes one from number theory to set theory. An exact reformulation of the number theory is given by **Axioms 0** through **8** together with the negation of **Axiom 9**. Mathematical constructions specify objects in the universe of sets; this is the informal point of view I shall adopt. For example, by using a property that cannot be true for any set, $x \neq x$, one can easily show using **Axiom 0** and **Axiom 4**

that there exists a set with no elements. By **Axiom 1** this set is unique; it is the *empty set* and is denoted by ∅.

4.2.2 Finite Set Theory

The formal versions of the axioms in the preceding section are the ZFC axioms, which are a specific (infinite) theory in the formal first-order language for set theory – a specific list is given in Woodin (1998) for the formal language, $\mathcal{L}(\hat{=}, \hat{\in})$, of set theory. This theory is too strong for my purposes. The following axioms describe the universe of sets under the assumption that for some finite ordinal α, $V = V_{\alpha+1}$.

> **AXIOM 0.** There exists a set.
>
> **AXIOM 1 (Extensionality).** Two sets A and B are equal if and only if they have the same elements.
>
> **AXIOM 2 (Bounding).** There exists a set C such that every set is a subset of C.
>
> **AXIOM 3 (Union).** If A is a set, then there exists a set C whose elements are the elements of A.
>
> **AXIOM (4a) (Powerset).** For all sets A either there exists a set B whose elements are all the subsets of A, or there exists a set C such that every set is a subset of C and such that A is not an element of C.
>
> **AXIOM (4b) (Powerset).** For all sets A, either every set is a subset of A, or there exists a set B such that B is an element of A and such that A does not contain all the subsets of B, or there is a set C whose elements are all the subsets of A.
>
> **AXIOM 5 (Regularity).** If A is a set, then either A is empty or there exists an element C of A that is disjoint from A.
>
> **AXIOM 6 (Comprehension).** If A is a set and $P(x)$ formalizes a property of sets, then there exists a set C whose elements are the elements of A with this property.
>
> **AXIOM 7 (Axiom of Finiteness).** If A is a nonempty set, then there is an element B of A such that for all sets C, if C is an element of A, then B is not an element of C.

The two forms of the **Powerset Axiom** are needed to compensate for the lack of the **Pairing Axiom**, and the **Bounding Axiom** eliminates the need for the **Axiom of Replacement**. Note that the set specified by the **Bounding Axiom** must be unique (by the **Axiom of Extensionality**). **Axioms 1** through **6** imply that for some ordinal α, $V = V_{\alpha+1}$. By **Axiom 7**, this ordinal is finite, so these axioms actually do imply the assertion, "*For some finite ordinal α, $V = V_{\alpha+1}$.*" As a consequence, one can show that these axioms also imply the **Axiom of Choice**.

The formal versions of these axioms define the theory with which I shall be working; it is our base theory, and I denote it by ZFC_0.

4.2.3 The Formula

I first discuss the standard example of a Gödel sentence modified to our context (in the language $\mathcal{L}(\hat{=}, \hat{\in})$ and relative to the theory ZFC_0). This is the sentence, Ξ_0, which asserts that its negation, $(\neg\Xi_0)$, can be proved from the theory ZFC_0.

By the usual arguments, it follows (within our universe of sets) that the theory ZFC_0 does not prove Ξ_0 and ZFC_0 does not prove $(\neg\Xi_0)$; i.e., the sentence Ξ_0 is *independent* of the theory ZFC_0. I give the argument.

For trivial reasons, ZFC_0 cannot prove Ξ_0. This is because

$$(\{\emptyset\}, \in) \vDash ZFC_0$$

and clearly

$$(\{\emptyset\}, \in) \vDash (\neg\Xi_0)$$

because $\emptyset$ is not a formal proof. Therefore, I have only to show that

$$ZFC_0 \nvdash (\neg\Xi_0).$$

Assume toward a contradiction that

$$ZFC_0 \vdash (\neg\Xi_0).$$

Then for all sufficiently large finite ordinals, n,

$$(V_n, \in) \vDash \text{``} ZFC_0 \vdash (\neg\Xi_0)\text{,''}$$

and so $(V_n, \in) \vDash \Xi_0$. But for all finite ordinals $n > 0$, $(V_n, \in) \vDash ZFC_0$, and so

$$(V_n, \in) \vDash (\neg\Xi_0),$$

which is a contradiction.

The sentence Ξ_0 is too pathological even for my purposes; a proof of $(\neg\Xi_0)$ cannot belong to a model of ZFC_0 with any extent beyond the proof itself. The sentence I seek is obtained by a simple modification of Ξ_0 that yields the sentence Ξ.

Informally, the sentence Ξ asserts that there is a proof from ZFC_0 of $(\neg\Xi)$ of length less than 10^{24} and further that V_n exists where $n = |V_{1000}|$. As I have already indicated, the choice of 10^{24} is only for practical reasons. There is no corresponding reason for my particular choice of n; one could quite easily modify the definition by requiring that the choice of n be larger.

The formal specification of Ξ is a completely standard (although tedious) exercise using the modern theory of formal mathematical logic; this involves the formal notion of proof defined so that proofs are finite sequences of natural numbers, and so forth (Woodin 1998).

In our universe of sets $(\neg\Xi)$ is true, so there *is* a proof of $(\neg\Xi)$ from ZFC_0. It is not clear just how short such a proof can be. This is a very interesting question. The witness for Armageddon (although with the end of time comfortably distant in the future) is a proof of $(\neg\Xi)$ from ZFC_0 of length less than 10^{24}.

It is important to emphasize that while ZFC_0 is a very weak theory, in attempting to prove $(\neg\Xi)$ from ZFC_0, one is free to augment ZFC_0 with the axiom that V_n exists where $n = |V_{1000}|$. This theory is not weak, particularly as far as the structure of binary sequences of length 10^{24} or even of length $10^{10^{10}}$ is concerned.

The sum total of human experience in mathematics to date (i.e., the number of manuscript pages written to date) is certainly less than 10^{12} pages. The shortest proof from ZFC_0 that no such sequence exists must have length greater than 10^{24}. This is arguably beyond the reach of our current experience, but there is an important issue that concerns the *compression* achieved by the informal style in which mathematical arguments are actually written. This is explored a little bit further in Woodin (1998).

With proper inputs and global determination, one could verify with current technology that a given sequence of length at most 10^{24} is a proof of $(\neg\Xi)$ from ZFC_0. However, we obviously do not expect to be able to find a sequence of length less than 10^{24} that is a proof of $(\neg\Xi)$ from ZFC_0. This actually gives a prediction about the physical universe because one can code any candidate for such a sequence by a binary sequence of length at most 10^{26}. The point is that, assuming the validity of the quantum view of the world, it is possible to build an actual physical device that must have a nonzero chance of finding such a sequence if such a sequence can exist. The device simply contains (a suitably large number of independent) modules, each of which performs an independent series of measurements that in effect *flips a quantum coin*. The point, of course, is that by quantum law *any outcome is possible*. The prediction is simply that any such device must fail to find a sequence of length less than 10^{24} that is a proof of $(\neg\Xi)$ from ZFC_0. One may object that the belief that any binary sequence of length 10^{26} is *really* a possible outcome of such a device requires an extraordinary faith in quantum law; but any attempt to build a quantum computer that is useful (for factoring) requires the analogous claim where 10^{26} is replaced by numbers at least as large as 10^5.

This, of course, requires something like quantum theory. In the universe as described by Newtonian laws, the argument just described does not apply because truly random processes would not exist. One could imagine proving that for a large class of chaotic (but deterministic) processes ("mechanical coin flippers"), no binary sequence of length 10^{24} that actually codes a formal proof can possibly be generated. In other words, for the nonquantum world, the prediction that no such sequence (as just presented) can be generated may *not* require that the conception of V_n is meaningful where $n = |V_{1000}|$.

Granting quantum law, and based only on our collective experience in mathematics to date, how can one account for the prediction (that one *cannot* find a sequence of length less than 10^{24} that is a proof of $(\neg\Xi)$ from ZFC_0) unless one believes that the conception of V_n is meaningful where $n = |V_{1000}|$?

Arguably (given current physical theory) this is *already* a conception of a nonphysical realm.

4.3 Beyond the Finite Realm

In this section we briefly summarize the basic argument of Woodin (2009), although here our use of this argument is for a different purpose.

> **Skeptic's Attack:** The mathematical conception of infinity is meaningless and without consequence because the entire conception of the universe of sets is a complete fiction. Further, all the theorems of set theory are merely finitistic truths, a reflection of the mathematician and not of any genuine mathematical "reality."

Throughout this section, the "Skeptic" simply refers to the metamathematical position that denies any genuine meaning to a conception of uncountable sets. The counterview is that of the "Set Theorist."

Set Theorist's Response: The development of set theory, after Cohen, has led to the realization that there is a robust hierarchy of strong axioms of infinity.

Elaborating further, it has been discovered that, in many cases, very different lines of investigation have led to problems whose degree of unsolvability is exactly calibrated by a notion of infinity. Thus, the hierarchy of large cardinal axioms emerges as an intrinsic, fundamental conception within set theory. To illustrate this, I discuss an example from modern set theory that concerns infinite games.

Suppose $A \subset \mathcal{P}(\mathbb{N})$, where $\mathcal{P}(\mathbb{N})$ denotes the set of all sets $\sigma \subseteq \mathbb{N}$ and $\mathbb{N}$ is the set of all natural numbers: $\mathbb{N} = \{1, 2, \ldots, k, \ldots\}$.

Associated to the set A is an infinite game involving two players, Player I and Player II. The players alternate declaring at stage k whether $k \in \sigma$ or $k \notin \sigma$:

Stage 1: Player I declares $1 \in \sigma$ or declares $1 \notin \sigma$;
Stage 2: Player II declares $2 \in \sigma$ or declares $2 \notin \sigma$;
Stage 3: Player I declares $3 \in \sigma$ or declares $3 \notin \sigma$; . . .

After infinitely many stages, a set $\sigma \subseteq \mathbb{N}$ is specified. Player I wins this run of the game if $\sigma \in A$; otherwise, Player II wins. (Note: Player I has control of which odd numbers are in σ, and Player II has control of which even numbers are in σ.)

A *strategy* is simply a function that provides moves for the players given just the current state of the game. More formally, a strategy is a function

$$\tau : [\mathbb{N}]^{<\omega} \times \mathbb{N} \to \{0, 1\},$$

where $[\mathbb{N}]^{<\omega}$ denotes the set of all finite subsets of $\mathbb{N}$. At each stage k of the game the relevant player can choose to follow τ by declaring "$k \in \sigma$" if

$$\tau(a, k) = 1$$

and declaring "$k \notin \sigma$" if $\tau(a, k) = 0$, where

$$a = \{i < k \mid \text{"}i \in \sigma\text{" was declared at stage } i\}.$$

The strategy τ is a *winning strategy* for Player I if, by following the strategy at each stage k where it is Player I's turn to play (i.e., for all odd k), Player I wins the game *no matter how Player II plays*. Similarly, τ is a *winning strategy* for Player II if, by following the strategy at each stage k where it is Player II's turn to play (i.e., for all even k), Player II wins the game *no matter how Player I plays*.

The game is *determined* if there is a *winning strategy* for one of the players. Clearly it is impossible for there to be winning strategies for *both* players.

It is easy to specify sets $A \subseteq \mathcal{P}(\mathbb{N})$ for which the corresponding game is determined; however, the problem of specifying a set $A \subseteq \mathcal{P}(\mathbb{N})$ for which the corresponding game is *not* determined turns out to be quite a bit more difficult. The *Axiom of Determinacy* (AD) is the axiom that asserts that for all sets $A \subseteq \mathcal{P}(\mathbb{N})$, the game given by A, as described previously, is determined. This axiom was first proposed by Mycielski and Steinhaus (1962) and contradicts the Axiom of Choice, so the problem here is whether the Axiom of Choice is necessary to construct a set

$$A \subseteq \mathcal{P}(\mathbb{N})$$

for which the corresponding game is not determined. Clearly, if the Axiom of Choice is necessary, then the existence of such set A is quite a subtle fact.

The unsolvability of this problem is exactly calibrated by large cardinal axioms. The relevant large cardinal notion is that of a *Woodin cardinal*, which I shall not define (Kanamori 1994). The ZF axioms are the ZFC axioms but without the Axiom of Choice. The issue of whether the Axiom of Choice is needed to construct a counterexample to AD is exactly the question of whether the theory ZF + AD is formally consistent.

Theorem 2.   *The two theories,*

(1) ZF + AD,

(2) ZFC + *"There exist infinitely many Woodin cardinals"*

are equiconsistent.

4.3.1 A Prediction and a Challenge for the Skeptic

Is the theory ZF + AD really formally consistent? The claim that it is consistent is a prediction that can be refuted by finite evidence (a formal contradiction). Taking an admittedly extreme position, I claim in Woodin (2009) the following:

> *It is only through the calibration by a large cardinal axiom in* **conjunction** *with our understanding of the hierarchy of such axioms as* **true axioms about the universe of sets** *that this prediction – the formal theory* ZF + AD *is consistent – is justified.*

As a consequence of my belief in this claim, I also made a prediction:

> *In the next 10,000 years there will be no discovery of an inconsistency in this theory.*

This is a specific and unambiguous prediction about the *physical universe* just as is the case for the analogous prediction in the previous section. Further, it is a prediction that does *not* arise by a reduction to a previously held truth (as, for example, is the case for the prediction that no counterexample to Fermat's Last Theorem will be discovered). This is a genuinely new prediction that I make in Woodin (2009) based on the development of set theory over the last 50 years and on my belief that the conception of the transfinite universe of sets is meaningful. I make this prediction independently of all speculation of what computational devices might be developed in the next 10,000 years (or whatever new sources of knowledge might be discovered) that would increase the effectiveness of research in mathematics.

Now the Skeptic might object that this prediction is not interesting or natural because the formal theories are not interesting or natural. But such objections are not allowed in physics; the ultimate physical theory should explain *all* (physical) aspects of the physical universe, not just those that we regard as natural. How can we apply a lesser standard for the ultimate mathematical theory? Of course, I also predict:

> *There will be no discovery* **ever** *of an inconsistency in this theory*;

and this prediction, if true, is arguably a physical law.

> **Skeptic's Retreat:** OK, I accept the challenge, noting that I only have to explain the predictions of formal consistency given by the large cardinal axioms. The formal theory

of set theory as given by the axioms ZFC is so "incomplete" that: *Any large cardinal axiom, in the natural formulation of such axioms, is either consistent with the axioms of set theory, or there is an elementary proof that the axiom cannot hold.*

To examine the Skeptic's Retreat and to assess how this, too, might be refuted, I need to briefly survey the basic template for large cardinal axioms in set theory.

4.3.2 Large Cardinal Axioms within Set Theory

A set N is *transitive* if every element of N is a subset of N. Transitive sets are fragments of V that are analogous to initial segments. For each ordinal α the set V_α is a transitive set.

The simplest (proper) class is the class of all ordinals. This class is a transitive class where a class $M \subseteq V$ is defined to be a transitive class if every element of M is a subset of M. The basic template for large cardinal axioms is as follows:

There is a transitive class M and an elementary embedding

$$j : V \to M$$

that is not the identity.

With the exception of the definition of a *Reinhardt cardinal*, which I shall discuss later, one can always assume that the classes M and j are classes that are logically definable from parameters by formulas of a fixed bounded level of complexity (Σ_2-formulas). Moreover, the assertion that j is an elementary embedding – that is, the assertion:

- For all formulas $\phi(x)$ and for all sets a,

$$V \vDash \phi[a]$$

if and only if $M \vDash \phi[j(a)]$

– is equivalent to the assertion:

- For all formulas $\phi(x)$, for all ordinals α, and for all sets $a \in V_\alpha$,

$$V_\alpha \vDash \phi[a]$$

if and only if $j(V_\alpha) \vDash \phi[j(a)]$.

Therefore, this template makes no essential use of the notion of a class. It is simply for convenience that I refer to classes (and this is the usual practice in set theory).

Suppose that M is a transitive class and that

$$j : V \to M$$

is an elementary embedding that is not the identity. Suppose that $j(\alpha) = \alpha$ for all ordinals α. Then one can show by transfinite induction that for all ordinals α, the embedding, j, is the identity on V_α. Therefore, because j is not the identity, there must exist an ordinal α such that $j(\alpha) \neq \alpha$. The least such ordinal is the *critical point* of j.

This must be a cardinal. The *critical point* of j is the large cardinal, and the existence of the transitive class M and the elementary embedding j are the witnesses for this.

A cardinal κ is a *measurable cardinal* if there exists a transitive class M and an elementary embedding

$$j : V \to M$$

such that κ is the critical point of j.

It is by requiring M to be *closer* to V that one can define large cardinal axioms far beyond the axiom, "There is a measurable cardinal." In general, the closer one requires M to be to V, the stronger the large cardinal axiom. The natural maximum axiom was proposed ($M = V$) by Reinhardt in his PhD thesis (see Reinhardt 1970). The associated large cardinal axiom is that of a *Reinhardt cardinal*.

Definition 3. A cardinal κ is a *Reinhardt cardinal* if there is an elementary embedding

$$j : V \to V$$

such that κ is the critical point of j.

The definition of a Reinhardt cardinal makes essential use of classes, but the following variation does not, and this variation (which is not a standard notion) is only formulated in order to facilitate this discussion. The definition requires a logical notion. Suppose that α and β are ordinals such that $\alpha < \beta$. Then

$$V_\alpha \prec V_\beta$$

if for all formulas, $\phi(x)$, for all $a \in V_\alpha$,

$$V_\alpha \vDash \phi[a]$$

if and only if $V_\beta \vDash \phi[a]$. Thus, $V_\alpha \prec V_\beta$ if and only if

$$I : V_\alpha \to V_\beta$$

is an elementary embedding where I is the identity map.

Definition 4. A cardinal κ is a *weak Reinhardt cardinal* if there exist $\gamma > \lambda > \kappa$ such that

(1) $V_\kappa \prec V_\lambda \prec V_\gamma$,

(2) there exists an elementary embedding

$$j : V_{\lambda+2} \to V_{\lambda+2}$$

such that κ is the critical point of j.

The definition of a weak Reinhardt cardinal only involves sets. The relationship between Reinhardt cardinals and weak Reinhardt cardinals is unclear, but one would naturally conjecture that, at least in terms of consistency strength, Reinhardt cardinals are stronger than weak Reinhardt cardinals, and hence my choice in terminology. The following theorem is an immediate corollary of the fundamental inconsistency results of Kunen (1971).

Theorem 5 (Kunen).   *There are no weak Reinhardt cardinals.*

The proof is elementary, so this does not refute the Skeptic's Retreat, but Kunen's proof makes essential use of the Axiom of Choice. The problem is open without this assumption. Further, there is really no known interesting example of a strengthening of the definition of a weak Reinhardt cardinal that yields a large cardinal axiom that can be refuted without using the Axiom of Choice. The difficulty is that without the Axiom of Choice it is extraordinarily difficult to prove anything about sets.

Kunen's proof leaves open the possibility that the following large cardinal axiom might be consistent with the Axiom of Choice. This, therefore, is essentially the strongest large cardinal axiom not known to be refuted by the Axiom of Choice (see Kanamori [1994] for more on this, as well as for the actual statement of Kunen's theorem).

Definition 6.   A cardinal κ is a *strongly $(\omega + 1)$-huge cardinal* if there exist $\gamma > \lambda > \kappa$ such that

(1) $V_\kappa \prec V_\lambda \prec V_\gamma$,

(2) there exists an elementary embedding

$$j : V_{\lambda+1} \to V_{\lambda+1}$$

such that κ is the critical point of j.

The issue of whether the existence of a weak Reinhardt cardinal is consistent with the axioms ZF is an important issue for the Set Theorist because by the results of Woodin (2009) the theory

ZF + "There is a weak Reinhardt cardinal"

proves the formal consistency of the theory

ZFC + "There is a proper class of strongly $(\omega + 1)$-huge cardinals."

This number theoretic statement is a theorem of number theory. As indicated previously, the notion of a strongly $(\omega + 1)$-huge cardinal is essentially the strongest large cardinal notion that is not known to be refuted by the Axiom of Choice.

Therefore, the number theoretic assertion that the theory

ZF + "There is a weak Reinhardt cardinal"

is consistent is a *stronger* assertion than the number theoretic assertion that the theory

ZFC + "There is a proper class of strongly $(\omega + 1)$-huge cardinals"

is consistent. More precisely, the former assertion implies, *but is not implied by*, the latter assertion, unless, of course, the theory

ZFC + "There is a proper class of strongly $(\omega + 1)$-huge cardinals"

is formally inconsistent. This raises an interesting question:

> *How could the Set Theorist ever be able to argue for the prediction that the existence of weak Reinhardt cardinals is consistent with axioms of set theory without the Axiom of Choice?*

Moreover, this *one* prediction implies *all* the predictions (of formal consistency) the Set Theorist can currently make based on the *entire* large cardinal hierarchy as presently conceived (in the context of a universe of sets that satisfies the Axiom of Choice). My point is that by appealing to the Skeptic's Retreat, one could reasonably claim that the theory

$$ZF + \text{"There is a weak Reinhardt cardinal"}$$

is formally consistent, and in making this *single* claim, one would subsume *all* the claims of consistency that the Set Theorist can make on the basis of our current understanding of the universe of sets (without abandoning the Axiom of Choice).

Before presenting a potential option to deal with this, I describe an analogous option of how the Set Theorist *can* claim that the theory

$$ZF + AD$$

is consistent, even though, as I have indicated, AD also refutes the Axiom of Choice. The explanation requires some definitions, which I shall require anyway. Gödel defined a very special transitive class $L \subseteq V$ and showed that all the axioms of ZFC hold when interpreted in L. The definition of L does not require the Axiom of Choice, so one obtains the seminal result that if the axioms ZF are consistent, then so are the axioms ZFC. Gödel also proved that the continuum hypothesis holds in L, thereby showing that one cannot formally refute the continuum hypothesis from the axioms ZFC (unless, of course, these axioms are inconsistent).

The definition of L is simply given by replacing the operation $\mathcal{P}(X)$ in the definition of $V_{\alpha+1}$ by the operation $\mathcal{P}_{\text{Def}}(X)$, which associates to the set X the set of all subsets $Y \subseteq X$ such that Y is logically definable in the structure $(X, \in)$ from parameters in X. For any infinite set X, $\mathcal{P}_{\text{Def}}(X) \subset \mathcal{P}(X)$ and $\mathcal{P}_{\text{Def}}(X) \neq \mathcal{P}(X)$.

Thus, one defines L_α by induction on the ordinal α, setting $L_0 = \emptyset$, setting

$$L_{\alpha+1} = \mathcal{P}_{\text{Def}}(L_\alpha),$$

and taking unions at limit stages. The class L is defined as the class of all sets a such that $a \in L_\alpha$ for some ordinal α. It is perhaps important to note that while there must exist a proper class of ordinals α such that

$$L_\alpha = L \cap V_\alpha,$$

this is not true for all ordinals α.

Relativizing the definition of L to $V_{\omega+1}$, we obtain the class $L(V_{\omega+1})$, which is more customarily denoted by $L(\mathbb{R})$; here one defines

$$L_0(\mathbb{R}) = V_{\omega+1}$$

and proceeds by induction exactly as above to define $L_\alpha(\mathbb{R})$ for all ordinals α. The class $L(\mathbb{R})$ is the class of all sets a such that $a \in L_\alpha(\mathbb{R})$ for some ordinal α.

Unlike the case for L, one cannot prove that the Axiom of Choice holds in $L(\mathbb{R})$, although one can show that all of the other axioms of ZFC hold in $L(\mathbb{R})$. The following theorem, which is related to Theorem 2, not only establishes the consistency of ZF + AD from simply the existence of large cardinals, but also establishes that $L(\mathbb{R}) \vDash AD$

(as a new truth about sets). See Kanamori (1994) for more on the history of this theorem and the attempts to establish that $L(\mathbb{R}) \vDash$ AD from large cardinal axioms.

Theorem 7 (Martin, Steel, Woodin). *Suppose there is a proper class of Woodin cardinals. Then*

$$L(\mathbb{R}) \vDash \text{AD}.$$

For the reasons I have indicated, one cannot hope to argue for the consistency of the theory

ZF + "There is a weak Reinhardt cardinal"

on the basis of *any* large cardinal axiom not known to refute the Axiom of Choice. The experience with the theory ZF + AD suggests that, as an alternative, one should seek both a generalization of $L(\mathbb{R})$ and some structural principles for this fragment such that the axiom that asserts both the existence of this fragment and that the structural principles hold in this fragment implies the formal consistency of the axiom that asserts the existence of a weak Reinhardt cardinal or, even better, that implies that the latter axiom actually holds in this fragment.

In fact, there *are* compelling candidates for generalizations of $L(\mathbb{R})$ and axioms for these fragments generalizing AD. But at present there is simply no plausible candidate for such a generalization of $L(\mathbb{R})$ in which the axiom that there is a weak Reinhardt cardinal can even hold; nor is there a plausible candidate for a fragment together with structural principles for that fragment that would imply the formal consistency of the existence of a weak Reinhardt cardinal. This is explored more fully in Woodin (2009).

There is another potential option that is suggested by a remarkable theorem of Vopenka. To explain this further, I must give another definition that I shall also require in the subsequent discussion. This is the definition of the class HOD that originates in remarks of Gödel at the Princeton Bicentennial Conference in December 1946. The first detailed reference appears to be Lévy (1965) (see the review of Lévy [1965] by G. Kreisel).

Definition 8 (ZF)

(1) For each ordinal α, HOD_α is the set of all sets a such that there exists a transitive set $M \subset V_\alpha$ such that $a \in M$ and such that for all $b \in M$, b is definable in V_α from ordinal parameters.

(2) HOD is the class of all sets a such that $a \in \text{HOD}_\alpha$ for some α.

I caution that just as is the case for L_α, in general

$$\text{HOD}_\alpha \neq \text{HOD} \cap V_\alpha,$$

although for a proper class of ordinals α it is true that $\text{HOD}_\alpha = \text{HOD} \cap V_\alpha$.

The class HOD is quite interesting for a number of reasons, one of which is illustrated by the following observation of Gödel, which, as indicated, is stated within just the theory ZF, in other words, without assuming the Axiom of Choice.

Theorem 9 (ZF). $\text{HOD} \vDash \text{ZFC}$.

This theorem gives a completely different approach to showing that if the theory ZF is formally consistent, then so is the theory ZFC.

One difficulty with HOD is that the definition of HOD is not absolute; for example, in general HOD is not even the same as defined within HOD. As a consequence, almost any set theoretic question one might naturally ask about HOD is formally unsolvable. Two immediate such questions are whether $V = \text{HOD}$ and, more simply, whether HOD contains all the real numbers. Both of these questions are formally unsolvable but are of evident importance because they specifically address the complexity of the Axiom of Choice. If $V = \text{HOD}$, then there is no mystery as to why the Axiom of Choice holds, but, of course, one is left with the problem of explaining why $V = \text{HOD}$.

I end this section with the remarkable theorem of Vopenka alluded to earlier. The statement involves Cohen's method of *forcing* adapted to produce extensions in which the Axiom of Choice can fail; these are called *symmetric generic extensions*. For each ordinal α, there is a minimum extension of the class HOD that contains both HOD and V_α and in which the ZF axioms hold. This minimum extension is denoted by $\text{HOD}(V_\alpha)$.

Theorem 10 (ZF; Vopenka).   *For all ordinals α, $\text{HOD}(V_\alpha)$ is a symmetric generic extension of* HOD.

The alternative conception of truth for set theory that is suggested by this theorem and that could provide a basis for the claim that weak Reinhardt cardinals are consistent is the subject of the next section.

4.4 The Generic Multiverse of Sets

The challenge presented in the previous section – the challenge to account for the prediction that the existence of a weak Reinhardt cardinal is formally consistent with ZF axioms – suggests that one should consider a multiverse conception of the universe of sets. The point, of course, is that while the existence of a weak Reinhardt cardinal is not possible, granting the Axiom of Choice, this does not rule out that there may be (symmetric) generic extensions of V in which there are weak Reinhardt cardinals. Such a multiverse approach to the conception of the universe of sets would also mitigate the difficulties associated with the formal unsolvability of fundamental problems such as that of the continuum hypothesis, and the latter feature is the primary motivation for such an approach. This section is based on Woodin (2004).

Let the *multiverse* (of sets) refer to the collection of possible universes of sets. The truths of set theory according to the multiverse conception of truth are the sentences that hold in each universe of the multiverse. Cohen's method of *forcing*, which is the fundamental technique for constructing nontrivial extensions of a given (countable) model of ZFC, suggests a natural candidate for a multiverse; the *generic multiverse* is generated from each universe of the collection by closing under generic extensions (enlargements) and under generic refinements (inner models of a universe of which the given universe is a generic extension). To illustrate the concept of the generic multiverse, suppose that M is a countable transitive set with the property that

$$M \vDash \text{ZFC}.$$

Let $\mathbb{V}_M$ be the smallest set of countable transitive sets such that $M \in \mathbb{V}_M$ and such that for all pairs, (M_1, M_2), of countable transitive sets such that

$$M_1 \vDash \text{ZFC},$$

and such that M_2 is a generic extension of M_1, if either $M_1 \in \mathbb{V}_M$ or $M_2 \in \mathbb{V}_M$, then both M_1 and M_2 are in $\mathbb{V}_M$. It is easily verified that for each $N \in \mathbb{V}_M$,

$$\mathbb{V}_N = \mathbb{V}_M,$$

where $\mathbb{V}_N$ is defined using N in place of M. $\mathbb{V}_M$ is the generic multiverse generated in V from M.

The *generic multiverse conception of truth* is the position that a sentence is true if and only if it holds in each universe of the generic multiverse generated by V. This can be formalized within V in the sense that for each sentence ϕ there is a sentence ϕ^*, recursively depending on ϕ, such that ϕ is true in each universe of the generic multiverse generated by V if and only if ϕ^* is true in V. The sentence ϕ^* is explicit given ϕ and does not depend on V. For example, given *any* countable transitive set, M, such that $M \vDash \text{ZFC}$,

$$M \vDash \phi^*$$

if and only if $N \vDash \phi$ for all $N \in \mathbb{V}_M$ (the proof is given in Woodin [in press]). This is an important point in favor of the generic-multiverse position because it shows that, as far as assessing truth is concerned, the generic-multiverse position is not that sensitive to the meta-universe in which the generic multiverse is being defined.

Is the generic-multiverse position a reasonable one? The refinements of Cohen's method of *forcing* in the decades since his initial discovery of the method and the resulting plethora of problems shown to be unsolvable have, in a practical sense, almost compelled one to adopt the generic-multiverse position. This has been reinforced by some rather unexpected consequences of large cardinal axioms, which I discuss later in this section.

The purpose of this section is *not* to argue against *any* possible multiverse position, but to examine more carefully the generic-multiverse position within the context of modern set theory. In brief, I argue that modulo the Ω Conjecture (which I define in the next section), the generic-multiverse position outlined earlier is not plausible. The essence of the argument against the generic-multiverse position is that assuming the Ω Conjecture is true (and that there is a proper class of Woodin cardinals), then this position is simply a brand of formalism that denies the transfinite by reducing truth about the universe of sets to truth about a simple fragment such as the integers or, in this case, the sets of real numbers. The Ω Conjecture is invariant between V and any generic extension of V, and so the generic-multiverse position must either declare the Ω Conjecture to be true or false.

It is a fairly common (informal) claim that the quest for truth about the universe of sets is analogous to the quest for truth about the physical universe. However, I am claiming an important distinction. While physicists would rejoice in the discovery that the conception of the physical universe reduces to the conception of some simple fragment or model, the Set Theorist rejects this possibility. I claim that by the very nature of its conception, the set of all truths of the transfinite universe (the universe of sets) cannot be reduced to the set of truths of some explicit fragment of the universe of sets. Taking into account the iterative conception of sets, the set of all truths of an explicit fragment of the universe of sets cannot be reduced to the truths of an explicit *simpler* fragment. The latter is the basic position on which I shall base my arguments.

An assertion is Π_2 if it is of the form

"For every infinite ordinal α , $V_\alpha \vDash \phi$,"

for some sentence, ϕ. A Π_2 assertion is a *multiverse truth* if the Π_2 assertion holds in each universe of the multiverse. A key point:

Remark 11.   Arguably, the generic-multiverse view of truth is only viable for Π_2-sentences and not, in general, even for Σ_2-sentences (these are sentences expressible as the negation of a Π_2-sentence). This is because of the restriction to *set forcing* in the definition of the generic multiverse. Therefore, one can quite reasonably question whether the generic-multiverse view can possibly account for the predictions of consistency given by large cardinal axioms. At present there is no reasonable candidate for the definition of an expanded version of the generic multiverse that allows *class forcing* extensions and yet preserves the existence of large cardinals across the multiverse.

In the context in which there is a Woodin cardinal, let us use "δ_0" to denote the least Woodin cardinal. Hence, I am fixing a notation, just as "ω_1" is fixed as the notation for the least uncountable ordinal. Both ω_1 and δ_0 can change in passing from one universe of sets to an extension of that universe.

The assertion

"δ is a Woodin cardinal"

is equivalent to the assertion

$$V_{\delta+1} \vDash \text{"δ is a Woodin cardinal"}$$

and so $\delta = \delta_0$ if and only if

$$V_{\delta+1} \vDash \text{"$\delta = \delta_0$."}$$

Therefore, assuming that there is a Woodin cardinal, for each sentence ϕ, it is a Π_2 assertion to say that

$$V_{\delta_0+1} \vDash \phi$$

and it is a Π_2 assertion to say that $V_{\delta_0+1} \nvDash \phi$. Thus, in any one universe of the multiverse, the set of all sentences ϕ such that $V_{\delta_0+1} \vDash \phi$ – that is, the *theory* of V_{δ_0+1} as computed in that universe – is recursive in the set of Π_2–sentences (assertions) that hold in that universe. Further, by Tarski's theorem on the undefinability of truth, the latter set cannot be recursive in the former set.

These comments suggest the following multiverse laws, which I state in reference to an arbitrary multiverse position (although assuming that the existence of a Woodin cardinal holds throughout the multiverse).

First Multiverse Law

The set of Π_2 assertions that are multiverse truths is not recursive in the set of multiverse truths of V_{δ_0+1}.

The motivation for this multiverse law is that if the set of Π_2 multiverse truths is recursive in the set of multiverse truths of V_{δ_0+1}, then as far as evaluating Π_2 assertions is concerned, the multiverse is equivalent to the reduced multiverse of just the fragments V_{δ_0+1} of the universes of the multiverse. This amounts to a rejection of the transfinite beyond V_{δ_0+1} and constitutes, in effect, the unacceptable brand of formalism alluded to earlier. This claim would be reinforced should the multiverse position also violate a second multiverse law, which I now formulate.

A set $Y \subset V_\omega$ is definable in V_{δ_0+1} across the multiverse if the set Y is definable in the structure V_{δ_0+1} of each universe of the multiverse (possibly by formulas that depend on the parent universe). The second multiverse law is a variation of the first multiverse law.

Second Multiverse Law

The set of Π_2 assertions that are multiverse truths is not definable in V_{δ_0+1} across the multiverse.

Again, by Tarski's theorem on the undefinability of truth, this multiverse law is obviously a reasonable one *if* one regards the only possibility for the multiverse to be the universe of sets such that the set of multiverse truths of V_{δ_0+1} is simply the set of all sentences that are true in V_{δ_0+1} and the set of Π_2 assertions that are multiverse truths is simply the set of Π_2 assertions that are true in V. Likewise, the second multiverse law would have to hold if one modified the law to simply require that the set of Π_2 assertions that are multiverse truths is not uniformly definable in V_{δ_0+1} across the multiverse (i.e., by a single formula).

Assuming that both Ω Conjecture and the existence of a proper class of Woodin cardinals hold in each (or one) universe of the generic multiverse generated by V, both the first multiverse law and the second multiverse law are violated by the generic multiverse position. This is the basis for the argument I am giving against the generic multiverse position in this chapter. In fact, the technical details of how the generic multiverse position violates these multiverse laws provide an even more compelling argument against the generic multiverse position because the analysis shows that, in addition, the generic multiverse position is truly a form of formalism because of the connections to Ω logic. The argument also shows that the violation of the first multiverse law is explicit; that is, assuming the Ω Conjecture, there is an explicit recursive reduction of the set of Π_2 assertions that are generic multiverse truths to the set of generic-multiverse truths of V_{δ_0+1}.

There is a special case that I can present without any additional definitions and that is not contingent on any conjectures.

Theorem 12. *Suppose that M is a countable transitive set*

$$M \vDash \text{ZFC} + \text{"There is a proper class of Woodin cardinals"}$$

and that $M \cap \text{Ord}$ is as small as possible. Then $\mathbb{V}_M$ violates both multiverse laws.

4.4.1 Ω-logic

The generic multiverse conception of truth declares the continuum hypothesis to be neither true nor false and declares, granting large cardinals, assertion,

$$L(\mathbb{R}) \vDash AD,$$

to be true (see Theorem 7). I note that for essentially all current large cardinal axioms, the existence of a proper class of large cardinals holds in V if and only if it holds in $V^{\mathbb{B}}$ for all complete Boolean algebras, $\mathbb{B}$. In other words, in the generic multiverse position the existence of a proper class of, say, Woodin cardinals is either true or false because it either holds in every universe of the generic multiverse or holds in no universe of the generic multiverse (Hankins and Woodin 2000).

I am going to analyze the generic multiverse position from the perspective of Ω-logic, which I first briefly review. I will use the standard modern notation for Cohen's method of forcing; potential extensions of the universe, V, are given by complete Boolean algebras $\mathbb{B}$, $V^{\mathbb{B}}$ denotes the corresponding Boolean valued extension, and for each ordinal α, $V_\alpha^{\mathbb{B}}$ denotes V_α as defined in that extension.

Definition 13.   Suppose that T is a countable theory in the language of set theory and ϕ is a sentence. Then

$$T \vDash_\Omega \phi$$

if for all complete Boolean algebras, $\mathbb{B}$, for all ordinals, α, if $V_\alpha^{\mathbb{B}} \vDash T$, then $V_\alpha^{\mathbb{B}} \vDash \phi$.

If there is a proper class of Woodin cardinals, then the relation $T \vDash_\Omega \phi$ is generically absolute. This fact, which arguably was a completely unanticipated consequence of large cardinals, makes Ω-logic interesting from a metamathematical point of view. For example, the set

$$\mathcal{V}_\Omega = \{\phi \mid \emptyset \vDash_\Omega \phi\}$$

is generically absolute in the sense that for a given sentence, ϕ, the question whether or not ϕ is logically Ω-valid i.e., whether or not $\phi \in \mathcal{V}_\Omega$ is absolute between V and all of its generic extensions. In particular, the method of forcing *cannot* be used to show the formal independence of assertions of the form $\emptyset \vDash_\Omega \phi$.

Theorem 14.   *Suppose that T is a countable theory in the language of set theory, ϕ is a sentence, and there exists a proper class of Woodin cardinals. Then for all complete Boolean algebras, $\mathbb{B}$, $V^{\mathbb{B}} \vDash$ "$T \vDash_\Omega \phi$" if and only if $T \vDash_\Omega \phi$.*

There are a variety of technical theorems that show that one cannot hope to prove the generic invariance of Ω-logic from any large cardinal hypothesis weaker than the existence of a proper class of Woodin cardinals – for example, if $V = L$, then definition of $\mathcal{V}_\Omega$ is not absolute between V and $V^{\mathbb{B}}$, for *any* nonatomic complete Boolean algebra, $\mathbb{B}$.

It follows easily from the definition of Ω-logic that for any Π_2-sentence, ϕ,

$$\emptyset \vDash_\Omega \phi$$

if and only if for all complete Boolean algebras, $\mathbb{B}$,

$$V^{\mathbb{B}} \vDash \phi.$$

Therefore, by the theorem above, assuming that there is a proper class of Woodin cardinals, for each sentence, ψ, the assertion

For all complete Boolean algebras, $\mathbb{B}$, $V^{\mathbb{B}} \vDash$ "$V_{\delta_0+1} \vDash \psi$"

is itself absolute between V and $V^{\mathbb{B}}$ for all complete Boolean algebras $\mathbb{B}$. This remarkable consequence of the existence of a proper class of Woodin cardinals actually seems to be evidence for the generic multiverse position. In particular, this shows that the generic multiverse position, at least for assessing Π_2 assertions, and so for assessing all assertions of the form

$$V_{\delta_0+1} \vDash \phi,$$

is equivalent to the position that a Π_2 assertion is true if and only if it holds in $V^{\mathbb{B}}$ for all complete Boolean algebras $\mathbb{B}$. Notice that if $\mathbb{R} \not\subset L$ and if V is a generic extension of L, then this equivalence is *false*. In this situation, the Π_2 sentence that expresses $\mathbb{R} \not\subset L$ holds in $V^{\mathbb{B}}$ for all complete Boolean algebras, $\mathbb{B}$, but this sentence fails to hold across the generic multiverse generated by V (because L belongs to this multiverse).

To summarize, suppose that there exists a proper class of Woodin cardinals in each universe of the generic multiverse (or equivalently that there is a proper class of Woodin cardinals in at least one universe of the generic multiverse). Then for each Π_2 sentence ϕ, the following are equivalent:

(1) ϕ holds across the generic multiverse;
(2) "$\emptyset \vDash_\Omega \phi$" holds across the generic multiverse;
(3) "$\emptyset \vDash_\Omega \phi$" holds in at least one universe of the generic multiverse.

For any Σ_2-sentence ϕ, the assertion, "$\emptyset \not\vDash_\Omega (\neg\phi)$", is, by the definitions, equivalent to the assertion that for some complete Boolean algebra, $\mathbb{B}$, $V^{\mathbb{B}} \vDash \phi$. Therefore, assuming that there exists a proper class of Woodin cardinals, for any Σ_2-sentence, if in one universe of the generic multiverse generated by V the sentence ϕ is true, then in *every* universe of the generic multiverse generated by V the sentence ϕ can be forced to be true by passing to a generic extension of that universe. This same remarkable fact applies to symmetric forcing extensions as well.

Therefore, assuming that there is a proper class of Woodin cardinals, it seems that the generic multiverse view of truth can account for the prediction that weak Reinhardt cardinals are consistent with ZF; the relevant Σ_2-sentence is the sentence that asserts that there exists a complete Boolean algebra $\mathbb{B}$ and there exists a term $\tau \in V^{\mathbb{B}}$ for a transitive set such that (with Boolean value 1)

$$V(\tau) \vDash \text{ZF} + \text{``\textit{There is a weak Reinhardt cardinal.''}}$$

If this sentence is true in one universe of the generic multiverse generated by V, then it must be true in *every* universe of the generic multiverse generated by V. Further, by Theorem 10, if there is a weak Reinhardt cardinal, then in the generic multiverse generated by HOD this Σ_2-sentence is actually declared as true.

But on close inspection, one realizes that this is not really a justification at all. The sentence above is *meaningful* in the generic multiverse view of truth, but there is no explanation of why it is true. This is exactly as is the case for the sentence that asserts that the formal theory

$$\text{ZF} + \text{``There is a weak Reinhardt cardinal.''},$$

is consistent.

To evaluate more fully the generic multiverse position, one must understand the logical relation $T \vDash_\Omega \phi$. In particular, a natural question arises: is there a corresponding proof relation?

4.4.2 The Ω Conjecture

I define the proof relation, $T \vdash_\Omega \phi$. This requires a preliminary notion that a set of reals be *universally Baire* (Feng, Magidor, and Woodin 1992). In fact, I shall define $T \vdash_\Omega \phi$, assuming the existence of a proper class of Woodin cardinals and exploiting the fact that there are a number of (equivalent) definitions. Without the assumption that there is a proper class of Woodin cardinals, the definition is a bit more technical (Woodin 2004). Recall that if S is a compact Hausdorff space, then a set $X \subseteq S$ has the *property of Baire* in the space S if there exists an open set $O \subseteq S$ such that symmetric difference,

$$X \triangle O,$$

is meager in S (contained in a countable union of closed sets with empty interior).

Definition 15. A set $A \subset \mathbb{R}$ is *universally Baire* if for all compact Hausdorff spaces, S, and for all continuous functions,

$$F : S \to \mathbb{R},$$

the preimage of A by F has the property of Baire in the space S.

Suppose that $A \subseteq \mathbb{R}$ is universally Baire. Suppose that M is a countable transitive model of ZFC. Then M is *strongly A-closed* if for all countable transitive sets N such that N is a generic extension of M,

$$A \cap N \in N.$$

Definition 16. Suppose there is a proper class of Woodin cardinals. Suppose that T is a countable theory in the language of set theory and ϕ is a sentence. Then $T \vdash_\Omega \phi$ if there exists a set $A \subset \mathbb{R}$ such that

(1) A is universally Baire,

(2) for all countable transitive models, M, if M is strongly A-closed and $T \in M$, then

$$M \vDash \text{``} T \vdash_\Omega \phi.\text{''}$$

Assuming there is a proper class of Woodin cardinals, the relation, $T \vdash_\Omega \phi$, is generically absolute. Moreover, *Soundness* holds as well.

Theorem 17. *Assume there is a proper class of Woodin cardinals. Then for all (T, ϕ) and for all complete Boolean algebras, $\mathbb{B}$, $T \vdash_\Omega \phi$ if and only if $V^{\mathbb{B}} \vDash "T \vdash_\Omega \phi."$*

Theorem 18 (Soundness). *Assume that there is a proper class of Woodin cardinals and that $T \vdash_\Omega \phi$. Then $T \vDash_\Omega \phi$.*

I now come to the Ω Conjecture, which in essence is simply the conjecture that the Gödel completeness theorem holds for Ω-logic; see Woodin (2009) for a more detailed discussion.

Definition 19 (Ω Conjecture). Suppose that there exists a proper class of Woodin cardinals. Then for all sentences ϕ, $\emptyset \vDash_\Omega \phi$ if and only if $\emptyset \vdash_\Omega \phi$.

Assuming the Ω Conjecture, one can analyze the generic multiverse view of truth by computing the logical complexity of Ω-logic. The key issue, of course, is whether the generic multiverse view of truth satisfies the two multiverse laws. This is the subject of the next section. We end this section with a curious connection between the Ω Conjecture, HOD, and the universally Baire sets. This requires a definition.

Definition 20. A set $A \subseteq \mathbb{R}$ is OD if there exists an ordinal α and a formula ϕ such that

$$A = \{x \in \mathbb{R} \mid V_\alpha \vDash \phi[x]\}.$$

Theorem 21. *Suppose that there is a proper class of Woodin cardinals and that for every set $A \subseteq \mathbb{R}$, if A is OD, then A is universally Baire. Then*

$$\text{HOD} \vDash "\Omega \text{ Conjecture."}$$

4.4.3 The Complexity of Ω-logic

Let $\mathcal{V}_\Omega$ (as defined prior to Theorem 14) be the set of sentences ϕ such that

$$\emptyset \vDash_\Omega \phi,$$

and (assuming there is a proper class of Woodin cardinals) let $\mathcal{V}_\Omega(V_{\delta_0+1})$ be the set of sentences, ϕ, such that

$$\text{ZFC} \vDash_\Omega "V_{\delta_0+1} \vDash \phi."$$

Assuming there is a proper class of Woodin cardinals, then the set of generic multiverse truths that are Π_2 assertions is of the same Turing complexity as $\mathcal{V}_\Omega$ (i.e., each set is recursive in the other). Further (assuming there is a proper class of Woodin cardinals), the set $\mathcal{V}_\Omega(V_{\delta_0+1})$ is precisely the set of generic multiverse truths of V_{δ_0+1}. Thus, the requirement that the generic multiverse position satisfies the first multiverse law, as discussed previously, reduces to the requirement that $\mathcal{V}_\Omega$ not be recursive in the set $\mathcal{V}_\Omega(V_{\delta_0+1})$.

The following theorem is a corollary of the basic analysis of Ω-logic in the context that there is a proper class of Woodin cardinals.

Theorem 22. *Assume there is a proper class of Woodin cardinals and that the* Ω *Conjecture holds. Then the set* $\mathcal{V}_\Omega$ *is recursive in the set* $\mathcal{V}_\Omega(V_{\delta_0+1})$.

Therefore, assuming that the existence of a proper class of Woodin cardinals and that the Ω Conjecture both hold across the generic multiverse generated by V, the generic multiverse position violates the first multiverse law. What about the second multiverse law? This requires understanding the complexity of the set $\mathcal{V}_\Omega$. From the definition of $\mathcal{V}_\Omega$ it is evident that this set is definable in V by a Π_2 formula: if $V = L$, then this set is recursively equivalent to the set of all Π_2 sentences that are true in V. However, in the context of large cardinal axioms the complexity of $\mathcal{V}_\Omega$ is more subtle.

Theorem 23. *Assume there is a proper class of Woodin cardinals and that the* Ω *Conjecture holds. Then the set* $\mathcal{V}_\Omega$ *is definable in* V_{δ_0+1}.

Therefore, if the Ω Conjecture holds and there is a proper class of Woodin cardinals, then the generic multiverse position that the only Π_2 assertions that are true are those that are true in each universe of the generic multiverse also violates the second multiverse law, for this set of assertions is itself definable in V_{δ_0+1} across the generic multiverse. I make one final comment here. The *weak multiverse laws* are the versions of the two multiverse laws I have defined where V_{δ_0+1} is replaced by $V_{\omega+2}$. Assuming the Ω Conjecture and that there is a proper class of Woodin cardinals, the generic multiverse position actually violates the weak first multiverse law, and augmented by a second conjecture, it also violates the weak second multiverse law; more details can be found in Woodin (in press). The example of Theorem 12 violates both weak multiverse laws.

4.5 The Infinite Realm

If the Ω Conjecture is true, then what is the only plausible alternative to the conception of the universe of sets is arguably ruled out. The point is that any multiverse conception based on a (reasonable) multiverse not smaller than the generic multiverse also violates the two multiverse laws. Here a multiverse, $\mathbb{V}_M^*$, associated to a countable transitive model M is *not smaller* than the generic multiverse if for each $N \in \mathbb{V}_M^*$, $\mathbb{V}_N \subseteq \mathbb{V}_M^*$. Similarly, $\mathbb{V}_M^*$ is *smaller* than the generic multiverse (generated by M) if $\mathbb{V}_M^* \subseteq \mathbb{V}_M$. The following theorem gives a more precise version of this claim.

Theorem 24. *Suppose that M is a countable transitive set*

$$M \vDash \text{ZFC} + \text{``There is a proper class of Woodin cardinals''}$$

and that

$$M \vDash \text{``The } \Omega \text{ Conjecture.''}$$

Suppose $\mathbb{V}_M^*$ *is a multiverse generated by M that is not smaller than the generic multiverse and that contains only transitive sets N such that*

$$N \vDash \text{ZFC} + \text{``There is a proper class of Woodin cardinals''}$$

and such that

$$N \vDash \text{``The } \Omega \text{ Conjecture.''}$$

Then the multiverse view of truth given by $\mathbb{V}_M^*$ violates both multiverse laws.

Finally, there is no compelling candidate for a multiverse view of truth based on a multiverse generated by V that is smaller than the generic multiverse, other than the multiverse that just contains V. But there certainly are candidates. For example, restricting the generic multiverse to the multiverse generated by only allowing forcing notions that are *homogeneous* could (as far as I know) give a multiverse view that does not violate the multiverse laws. Further, given one motivation I cited for a generic multiverse view (based on Theorem 10), such a restriction is a natural one.

At this point, I just do not see any argument for such a restricted multiverse view. The fundamental problem with a multiverse view based on a multiverse that is smaller than the generic multiverse is that there must exist Π_2 sentences that are declared to be true and that are not true in the generic multiverse view. At present there is simply no natural candidate for such a collection of Π_2 sentences that is not simply the set of all Π_2 sentences true in V.

The determined advocate for the generic multiverse conception of truth might simply accept the failure of the multiverse laws as a deep fact about truth in set theory and seek salvation in the multiverse truths that are beyond Π_2 sentences. But then this advocate must either explain the restriction to set forcing extensions in the definition of the generic multiverse or specify exactly how the multiverse is to be defined. Restricting to set generic extensions resurrects the specter of unsolvable problems. For example, consider the question of whether there exists a cardinal λ such that the generalized continuum hypothesis holds for all cardinals above λ. The answer to this question is invariant across the generic multiverse. Alternatively, specifying the class forcing extensions that are to be allowed seems utterly hopeless at the present time if the corresponding multiverse conception of truth is to declare the question above to be meaningless *and* yet preserve large axioms as meaningful. Further, there is the issue of whether truth in this expanded generic multiverse is reducible to truth in the universes of that multiverse, as is the case of the generic multiverse.

Of course, one could just conclude from all of this that the Ω Conjecture is *false* and predict that the solution to specifying the true generic multiverse will be revealed by the nature of the failure of the Ω Conjecture. But the Ω Conjecture is invariant across the generic multiverse. Thus, it is not unreasonable to expect that both the Ω Conjecture has an answer and, further, if that answer is false, then the Ω Conjecture be *refuted* from some large cardinal hypothesis. Many of the metamathematical consequences of the Ω Conjecture follow from the nontrivial Ω-satisfiability of the Ω Conjecture; this is the assertion that for some universe V^* of the generic multiverse generated by V, there exists an ordinal α such that

$$V_\alpha^* \vDash \text{ZFC} + \text{``There is a proper class of Woodin cardinals''}$$

and such that

$$V_\alpha^* \vDash \text{``The } \Omega \text{ Conjecture.''}$$

This assertion is itself a Σ_2-assertion, and so assuming there is a proper class of Woodin cardinals, this assertion must also be invariant across the generic multiverse generated by V. While the claim that if the Ω Conjecture is false then the Ω Conjecture must be refuted from some large cardinal hypothesis is debatable, the corresponding claim for the nontrivial Ω-satisfiability of the Ω Conjecture (in the sense just defined) is much harder to argue against. The point here is that while there are many examples of sentences that are provably absolute for set forcing and that cannot be decided by any large cardinal axiom, there are no known examples where the sentence is Σ_2. In fact, if the Ω Conjecture is true, then there really can be no such example. Finally, it seems unlikely that there is a large cardinal axiom that proves the nontrivial Ω-satisfiability of the Ω Conjecture and yet there is no large cardinal axiom that proves the Ω Conjecture.

4.5.1 Probing the Universe of Sets: The Inner Model Program

The *Inner Model Program* is the detailed study of large cardinal axioms. The first construction of an inner model is attributed to (Gödel 1938, 1947). This construction founded the Inner Model Program; the transitive class constructed is denoted by L, and I have given the definition. The question of whether $V = L$ is an important one for set theory. The answer has profound implications for the conception of the universe of sets.

Theorem 25. *(Scott 1961) Suppose there is a measurable cardinal. Then* $V \neq L$.

The *Axiom of Constructibility* is the axiom that asserts "$V = L$"; more precisely, this is the axiom that asserts that for each set a there exists an ordinal α such that $a \in L_\alpha$. Scott's theorem provided the first indication that the Axiom of Constructibility is independent of the ZFC axioms. At the time there was no compelling reason to believe that the existence of a measurable cardinal was consistent with the ZFC axioms, so one could not make the claim that Scott's theorem established the formal independence of the Axiom of Constructibility from the ZFC axioms. Of course, it is an immediate corollary of Cohen's results that the Axiom of Constructibility is formally independent of the ZFC axioms. The modern significance of Scott's theorem is more profound; I would argue that Scott's theorem establishes that the Axiom of Constructibility is *false*. This claim (that $V \neq L$) is not universally accepted, but in my view no one has come up with a credible argument against this claim.

The Inner Model Program seeks generalizations of L for the large cardinal axioms; in brief, it seeks generalizations of the Axiom of Constructibility that are compatible with large cardinal axioms (such as the axioms for measurable cardinals and beyond). It has been a very successful program, and its successes have led to the realization that the large cardinal hierarchy is a very "robust" notion. The results that have been obtained provide some of our deepest glimpses into the universe of sets. Despite the rather formidable merits indicated previously, there is a fundamental difficulty with the prospect of using the Inner Model Program to counter the Skeptic's Retreat. The problem is in the basic methodology of the Inner Model Program. To explain this, I must give a (brief) description of the (technical) template for inner models.

The inner models that are the goal and focus of the Inner Model Program are defined layer by layer working up through the hierarchy of large cardinal axioms, which in turn is naturally revealed by the construction of these inner models. Each layer provides the foundation for the next, and L is the first layer.

Roughly (and in practice), in constructing the inner model for a specific large cardinal axiom, one obtains an exhaustive analysis of all weaker large cardinal axioms. There can be surprises here in that seemingly different notions of large cardinals can coincide in the inner model. Finally, as one ascends through the hierarchy of large cardinal axioms, the construction generally becomes more and more difficult.

However, there is a fundamental problem with appealing to the Inner Model Program to counter the Skeptic's Retreat. Suppose, for example, that a hypothetical large cardinal axiom "Φ" provides a counterexample to the Skeptic's Retreat and this is accomplished by the Inner Model Program. *To use the Inner Model Program to refute the existence of an "Φ-cardinal," one must first be able to successfully construct the inner models for all smaller large cardinals, and this hierarchy would be fully revealed by the construction.*

Perhaps this could happen, but it can only happen *once*. This is the problem. Having refuted the existence of an "Φ-cardinal," how could one then refute the existence of any *smaller* large cardinals, for one would have solved the inner model problem for these smaller large cardinals? The fundamental problem is that *the Inner Model Program seems inherently unable, by virtue of its inductive nature, to provide a framework for an evolving understanding of the boundary between the possible and the impossible (large cardinal axioms).*

Thus, it would seem that the Skeptic's Retreat is, in fact, a powerful counterattack, but there is something wrong here, and the answer lies in understanding large cardinal axioms that are much stronger than those within reach of the current hierarchy of inner models.

4.5.2 Supercompact Cardinals and Beyond

Paraphrasing a standard definition: an *extender* is an elementary embedding

$$E : V_{\alpha+1} \cap M \rightarrow V_{\beta+1} \cap M$$

where M is a transitive class such that $M \vDash \text{ZFC}$. Necessarily $\alpha \leq \beta$ and $E(\alpha) = \beta$. It is not difficult to show that if for all ordinals $\xi \leq \alpha$, $E(\xi) = \xi$, then for all $a \in V_{\alpha+1}$, $E(a) = a$. Thus, if E is nontrivial, there must exist an ordinal $\gamma \leq \alpha$, such that $E(\gamma) \neq \gamma$ and the least such ordinal γ is the *critical point* of E and is denoted $\text{CRT}(E)$.

Extenders are the building blocks for the Inner Model Program, which seeks enlargements of L that are transitive classes N such that N contains *enough* extenders to witness that the targeted large cardinal axiom holds in N. The complication is in specifying just which extenders are to be included in N.

The definition of a supercompact cardinal is due to Reinhardt and Solovay (see Kanamori 1994) for more on the history of the axiom. Below is a reformulation of the definition attributed to Magidor in terms of extenders.

Definition 26.   A cardinal δ is a *supercompact cardinal* if for each ordinal $\beta > \delta$ there exists an extender

$$E : V_{\alpha+1} \to V_{\beta+1}$$

such that $E(\kappa) = \delta$, where $\kappa = \mathrm{CRT}(E)$.

Slightly stronger is the notion that δ is an *extendible cardinal*: for all $\alpha > \delta$ there exists an extender

$$E : V_{\alpha+1} \to V_{\beta+1}$$

such that $\mathrm{CRT}(E) = \delta$.

As I have already indicated, the strongest large cardinal axioms not known to be inconsistent with the Axiom of Choice are the family of axioms asserting the existence of strongly $(\omega + 1)$-huge cardinals. These axioms have seemed so far beyond any conceivable inner model theory that they simply are not understood.

The possibilities for an inner model theory at the level of supercompact cardinals and beyond have been essentially a complete mystery until recently. The reason lies in the nature of extenders. Again for expository purposes, let me define an extender

$$E : V_{\alpha+1} \to V_{\beta+1}$$

to be a *suitable* extender if $E(\mathrm{CRT}(E)) > \alpha$. Thus, an extender

$$E : V_{\alpha+1} \cap M \to V_{\beta+1} \cap M$$

is a suitable extender if it is not *too long* and if $V_{\beta+1} \subset M$. For example, suppose that κ is a strongly $(\omega + 1)$-huge cardinal as defined in Definition 6. Then there exists $\gamma > \lambda > \kappa$ such that

(1) $V_\kappa \prec V_\lambda \prec V_\gamma$,
(2) there exists an elementary embedding

$$j : V_{\lambda+1} \to V_{\lambda+1}$$

such that κ is the critical point of j.

Thus, j is an extender, but *not* a suitable extender. In particular, the existence of a strongly $(\omega + 1)$-huge cardinal *cannot* be witnessed by a suitable extender.

The Inner Model Program at the level of supercompact cardinals and beyond seeks enlargements N of L such that there are enough extenders E such that $E|N \in N$ to witness that the targeted large cardinal axiom holds in N. For the weaker large cardinal axioms, this has been an extremely successful program. For example, Mitchell and Steel (1994) have defined enlargements at the level of Woodin cardinals. In fact, they define the basic form of such enlargements of L up to the level of *superstrong cardinals*, which are just below the level of supercompact cardinals. The Mitchell-Steel models are constructed from sequences of extenders. The basic methodology is to construct N

from a sequence of extenders that includes enough extenders to directly witness that the targeted large cardinal axiom holds in N.

For the construction of the Mitchell-Steel models there is a fundamental requirement that the extenders on sequence from which the enlargement of L is constructed be derived from extenders

$$E : V_{\alpha+1} \cap M \to V_{\beta+1} \cap M$$

such that $V_{\alpha+1} \subset M$. Following this basic methodology, the enlargements of L at the level of supercompact cardinals and beyond must be constructed from extender sequences, which now include extenders that are *restrictions* of extenders of the form

$$E : V_{\alpha+1} \to V_{\beta+1},$$

and Steel has shown that the basic methodology of analyzing extender models encounters serious obstructions once there are such extenders on the sequence, particularly if the extenders are not suitable.

But by some fairly recent theorems something completely unexpected and remarkable happens. Suppose that N is a transitive class, for some cardinal δ,

$$N \vDash \text{"}\delta \text{ is a supercompact cardinal,"}$$

and that this is witnessed by the class of all $E|N$ such that $E|N \in N$ and such that E is a suitable extender. Then the transitive class N is close to V, and N inherits essentially all large cardinals from V. The amazing thing is that this must happen *no matter how N is constructed*. This would seem to undermine my earlier claim that inner models should be constructed from extender sequences that contain enough extenders to witness that the targeted large cardinal axiom holds in the inner model. It does not, and the reason is that by simply requiring that $E|N \in N$ for enough suitable extenders from V to witness that the large cardinal axiom, "There is a supercompact cardinal," holds in N, one (and this is the surprise) necessarily must have $E|N \in N$ for a *much* larger class of extenders, $E : V_{\alpha+1} \to V_{\beta+1}$. Therefore, the *principle* that there are enough extenders in N to witness that the targeted large cardinal axiom holds in N *is preserved* (as it must be). The *change*, in the case that N is constructed from a sequence of extenders that includes restrictions of suitable extenders, is that these extenders do not have to be on the sequence from which N is constructed. In particular in this case, large cardinal axioms can be witnessed to hold in N by "phantom" extenders; these are extenders of N that are not on the sequence and that *cannot* be witnessed to hold by *any* extender on the sequence. This includes large cardinal axioms at the level of strongly $(\omega + 1)$-huge cardinals. As a consequence of this, one can completely avoid the cited obstacles because *one does not need to have the kinds of extenders on the sequence that give rise to the obstacles*. Specifically, one can restrict consideration to extender sequences of just extenders derived from suitable extenders, and this is a paradigm shift in the whole conception of inner models.

The analysis yields still more. Suppose that there is a positive solution (in ZFC) to the inner model problem for just one supercompact cardinal. Then as a corollary one would obtain a proof of the following conjecture:

Conjecture (ZF). *There are no weak Reinhardt cardinals.*

Suppose this conjecture is actually true and is proved according to the scenario that I have just described. This would, in a convincing fashion, refute the Skeptic's Retreat, providing for the *first time* an example of a natural large cardinal axiom proved to be inconsistent as a result of a deep structural analysis.

In fact, it is possible to isolate a specific conjecture that must be true if there is a positive solution to the inner model problem for one supercompact cardinal and that itself suffices for this inconsistency result. This conjecture (which is the HOD Conjecture of Woodin [2009]) concerns HOD, and we refer the interested (and dedicated) reader to Woodin (2009) for details. Actually, a corollary of the HOD Conjecture suffices to prove the inconsistency conjectured earlier, and this corollary we can easily state. First, a cardinal κ is a *regular cardinal* if every subset $X \subset \kappa$ with $|X| < \kappa$ is bounded in κ. Thus, ω is a regular cardinal as is ω_1 (assuming the Axiom of Choice). The corollary of the HOD Conjecture is the following: there is a proper class of regular cardinals that are not measurable cardinals in HOD.

I mention this for two reasons. First, this is a specific and precise conjecture that does not involve the Inner Model Program at all and so offers an independent route to proving the conjectured inconsistency cited earlier. Second, it identifies specific combinatorial consequences of having a successful solution to the inner model problem for one supercompact cardinal and so provides a potential basis for establishing that there is no solution to the inner model problem for one supercompact cardinal.

The extension of the Inner Model Program to the level of one supercompact cardinal (the definition of "ultimate L") will come with a price. The successful extension of the Inner Model Program to a large cardinal axiom can no longer serve as the basis for the claim of the formal consistency of that axiom. The reason is that, as the previous discussion indicates, in extending the Inner Model Program to the level of one supercompact cardinal, one will have extended the Inner Model Program to essentially all known large cardinals. The ramifications are discussed at length in Woodin (2009). In brief, further progress in understanding (and even discovering) large cardinal axioms would have to depend on *structural* considerations of "ultimate L."

4.6 Conclusions

The development of the mathematical theory of infinity has led to a number of specific predictions. These predictions assert that certain technical axioms concerning the existence of large cardinals are not formally inconsistent with the axioms of set theory. As I have indicated, these predictions are actually predictions about the physical universe. To date there is no known (and credible) explanation for these predictions except that they are true because the corresponding axioms are true in the universe of sets. As the arguments of the first section indicate, these same issues arise even for the conception of large finite sets.

As discussed in the second section, there is a serious challenge to this claim, even ignoring the often-cited challenge: the ubiquity of unsolvable problems in set theory. The challenge arises from the fact that there are formal axioms of infinity that are arguably a serious foundational issue for set theory for two reasons. First, these axioms are known to refute the Axiom of Choice, and second, these axioms are known to be "stronger" than essentially all the notions of infinity believed to be formally consistent with the Axiom of Choice. Here the metric for strength is simply the inference relation for the corresponding predictions (of formal consistency). The issues raised by this are twofold. First (regarding the debate between the Set Theorist and the Skeptic), there is no need to explain the success of a single prediction; it is a succession of ever-stronger successful predictions that demands explanation. But this one prediction of consistency subsumes all the predictions made to date, so there is no series of predictions that requires explanation. Second, for the Set Theorist to account for this one prediction, it would seem that a different conception of the universe of sets is required.

The conception of a universe of sets in which the Axiom of Choice fails creates more difficulties than it solves, so this does not seem to be a viable option. However, any large cardinal axiom (which is expressible by a Σ_2-sentence) that can hold in a universe of sets satisfying all of the axioms except for the Axiom of Choice can hold in a generic extension of a universe of sets that does satisfy the Axiom of Choice. Therefore, this challenge, as well as the challenge posed by formerly unsolvable problems such as that of the continuum hypothesis, might be addressed (but perhaps not completely in a satisfactory manner) by adopting the conception of a multiverse of sets. Here the Ω Conjecture emerges as a key conjecture. If this conjecture is true, then what is arguably the only candidate for a multiverse view for the infinite realm that can address these challenges also fails to be a viable alternative (accepting the requirement that the multiverse laws of Section 4.4 be satisfied). Therefore, if the multiverse view is correct, the Ω Conjecture must be false.

The attempt to understand how the Ω Conjecture might be refuted leads directly to the Inner Model Program. The Inner Model Program is the attempt to generalize the definition of L to yield transitive classes M in which large cardinal axioms hold. If the Inner Model Program as described in the fifth section can be extended to the level of a single supercompact cardinal, then no known large cardinal axiom can refute the Ω Conjecture. Further, one would also obtain as corollary the verification of a series of conjectures. These conjectures imply that the large cardinal axioms, such as the axiom that asserts the existence of a weak Reinhardt cardinal, which pose such a challenge to the conception of the universe of sets, are formally inconsistent. These inconsistency results would be the first examples of inconsistency results for large cardinal axioms obtained only through a very detailed analysis.

Finally, the extension of the Inner Model Program to the level of one supercompact cardinal will yield examples (where *none* are currently known) of a *single* formal axiom that is compatible with all the known large cardinal axioms and that provides an axiomatic foundation for set theory that is immune to independence by Cohen's method. This axiom will not be unique, but there is the very real possibility that among these axioms there is an optimal one (from structural and philosophical considerations), in which case we will have returned, against all odds or reasonable expectation, to the view of truth for set theory that was present at the time when the investigation of set theory began.

References

Feng, Qi, Magidor, Menachem, and Woodin, W. Hugh. 1992. Universally Baire sets of reals. Judan et al., eds., *Mathematics Sciences Research Institute Publication*, vol. 26, pp. 203–42. Springer Verlag.

Gödel, Kurt. 1938. Consistency-proof for the generalized continuum-hypothesis. *Proceedings of the National Academy of Sciences USA* 25: 220–24.

Gödel, Kurt. 1940. *The Consistency of the Continuum Hypothesis. Annals of Mathematics Studies*, no. 3. Princeton: Princeton University Press.

Hamkins, Joel David, and Woodin, W. Hugh. 2000. Small forcing creates neither strong nor Woodin cardinals. *Proceedings of the American Mathematical Society* 128 (10): 3025–29.

Kanamori, Akihiro. 1994. *The Higher Infinite. Perspectives in Mathematical Logic*. Berlin: Springer-Verlag.

Kunen, Kenneth. 1971. Elementary embeddings and infinitary combinatorics. *Journal of Symbolic Logic* 36: 407–13.

Lévy, Azriel. 1965. Definability in axiomatic set theory. I. In *Logic, Methodology and Philosophical Sciences, Proceedings of the 1964 International Congress*, pp. 127–51. Amsterdam: North-Holland.

Mitchell, William J., and Steel, John R. 1994. *Fine Structure and Iteration Trees*. Berlin: Springer-Verlag.

Mycielski, Jan, and Steinhaus, H. 1962. A mathematical axiom contradicting the axiom of choice. *Bulletin of the Polish Academy of Sciences, Mathematics, Astronomy, and Physics* 10: 1–3.

Reinhardt, William N. 1970. Ackermann's set theory equals ZF. *Annals of Mathematics Logic* 2 (2): 189–249.

Scott, Dana. 1961. Measurable cardinals and constructible sets. *Bulletin of the Polish Academy of Science, Mathematics, Astronomy, and Physics* 9: 521–24.

Woodin, W. Hugh. 1998. The tower of Hanoi. In *The Second International Meeting on Truth in Mathematics*, pp. 329–51. Oxford: Oxford University Press.

Woodin, W. Hugh. 2004. *Set Theory after Russell: The Journey Back to Eden*, vol. 6. In *de Gruyter Series in Logic and Its Applications*. Berlin: Walter de Gruyter.

Woodin, W. Hugh. In press. *The Continuum Hypothesis: The Generic-Multiverse of Sets and the Ω Conjecture*.

Woodin, W. Hugh. In press. The transfinite universe. To appear in Gödel. Centenary, Cambridge University Press.

Woodin W. Hugh. Suitable extender sequences. Preprint, pp. 1–677, July.

Zermelo, Ernst. 1930. Uber Grenzzahlen und Mengenbereiche: Neue Untersuchungenuberdie Grundlagen der Mengenlehre. *Fundamenta Mathematicae*: 16: 29–47.

CHAPTER 5

A Potential Subtlety Concerning the Distinction between Determinism and Nondeterminism

W. Hugh Woodin

5.1 The Coding of Information into Time

It is well known that the property of randomness for finite binary sequences based on information content is not decidable (Li and Vitányi 1997). We produce a dramatic version of this, but it is not our goal to simply reproduce this undecidability result; rather, our intention in this chapter is to illustrate a potentially subtle aspect of the distinction between determinism and nondeterminism. This subtlety is the possibility of coding arbitrary information into time in such a way that a *specific* deterministic process computes precisely that information (as additional output).

Our basic argument begins as follows. Given any specific set of physical laws, we produce a Turing program e_0 with the following feature. First, the output of the program e_0 (by virtue of its format) must be a finite binary sequence s, but there may be no output generated. Let t be any nonempty finite binary sequence (e.g., flip a quantum coin 10^{100} times, and let t be the outcome). Then by "extending time" and preserving all the specified physical laws, one can arrange that in the "new" universe the program e_0 generates as additional output exactly the chosen string t (so that in the case in which no output was initially generated by the program, in the new universe the total output generated is exactly t; otherwise, the total output generated is exactly s appended by the sequence t). From the perspective of the inhabitants of the universe (i.e., our perspective), passing from the initial universe to the extension is not an observable change because all laws have been preserved; more precisely, the intial universe is not an observable entity in the extension. The program e_0 is explicit; it only depends (as it must) on the specification of the physical laws – if that specification is simple, then so is the program. We are implicitly assuming that time is infinite in the idealized universe and so the "extension" is a nonstandard extension (which again is the only possibility). In fact, for the construction of e_0 we give, the program e_0 generates no output within our (idealized) universe. We emphasize that the property for the potential output that we have indicated that the program e_0 has applies to *any* abstract version of our universe that obeys the specified laws and is not an artifact of our universe: it holds *across the entire multiverse* (of mathematical versions of the universe) defined by the specified laws.

Our construction of e_0 actually gives a Turing program e_0^* that witnesses a more dramatic version of the property just discussed, and it is this program that is the basis for our claim of a potentially subtle aspect to the distinction of determinism versus nondeterminism. We describe e_0^* first more formally and then informally.

If in a given universe no output is generated by e_0^* with input 0, then by extending time one can arrange that any specified finite binary sequence is now the output of e_0^* acting on input 1. Otherwise, in the given universe, e_0^* computes a finite (positive) integer N_0 from input 0, and for any specified finite binary sequence, by extending time one can arrange that for some input $i < N_0$, the output of e_0^* is the given sequence. In this situation, the indicated bound N_0 on potential inputs *cannot change* once it has been calculated, and for any input $n < N_0$, if e_0^* generates output from this input, then e_0^* must generate output acting on all inputs $k < n$ and *none* of these outputs can be changed by extending time.

Informally, in any given universe the program e_0^* produces either no output or an ordered library of binary sequences together with an upper bound for the size of the library. In the former case, by extending time one can arrange that *any* binary sequence is added to the library as the first (and only) entry. In the latter case, by extending time, the upper bound cannot change, sequences cannot be deleted from the library, and *any* binary sequence can be added to the library both as the last entry and as the unique additional entry.

How does e_0^* relate to the distinction of determinism versus nondeterminism? One can view the inputs to e_0^* as the "initial conditions" needed to derive the outcome of a physical process based on the physical evolution of that process. For *any* physical process there is an input from which e_0^* can compute the outcome of that physical process given sufficient time. Moreover, and this is the key point, the size of the input space is *bounded* and that bound is *independent* of the physical process to be analyzed. In particular, modulo a fixed offset, the input space is *much simpler* than the physical process to be analyzed.

If one attempts to counter any possible implication of the existence of e_0^* for the distinction of determinism versus nondeterminism by rejecting all meaning for infinite time even in an idealized sense, then one clearly has far more serious issues to contend with than the distinction between determinism and nondeterminism in the physical universe, more specifically, what is the meaning of nondeterminism in a specific finite setting?

5.2 The Program e_0

For the formal construction of the program e_0 in a specific instance, we regard the idealized physical universe as simply the idealized mathematical universe of *number theory*; this is the formal mathematical structure,

$$(\mathbb{N}, +, \cdot, <)$$

of the nonnegative integers with the operations of addition and multiplication and with the associated total order. We adopt as our base theory (i.e., the specified set of "physical laws") the standard Peano axioms as implemented in formal logic; these are the formal axioms, PA. In fact, all of our considerations apply to any (recursive) extension of these axioms.

We restrict consideration to the collection of Turing programs that, by the format of the program, generate a finite binary sequence in a specific fashion: the program attempts to compute a finite sequence,

$$\langle (s_i, N_i) : i \leq m \rangle$$

by successively computing each pair (s_i, N_i), such that for all $i + 1 \leq m$, $N_{i+1} > N_i$, s_i, s_{i+1} are each (nonempty) binary sequences, and s_i is an initial segment of s_{i+1}. If the program fails to compute such a sequence – in particular, if the program fails to compute (s_0, N_0) – then there is no output generated; otherwise, the sequence s_m is the (total) *output* of the program. Increasing the running time of the program can only alter the output by lengthening the output sequence (i.e., by creating additional output), but this can only happen if $N_m > 0$. Thus, once the program succeeds in calculating (s_0, N_0), the output of the program can only increase at most N_0 times as the running time increases. Our convention is that for such Turing programs e, if no output is generated, then the output is empty. This is unambiguous because, by the format we have described, the program can never generate the empty sequence as output ($s_m \neq \emptyset$). We could just as easily require, by altering the format, that output is always produced, but the technical details would be less natural. The point is that whatever the format, the output generated in the standard universe must be an initial segment of the output generated in any possible nonstandard universe, and it is only the potential for additional output that we are interested in.

By combining the construction of a Gödel sentence with the Kleene recursion theorem and appealing to the Friedman isomorphism theorem, we construct an index e_0 for a Turing program in the format just described and with the following property. For any countable model $\mathcal{M} \vDash$ PA, if t is *any* (internally) finite binary sequence of $\mathcal{M}$ that properly extends the sequence that is the output of the program e_0 as implemented within $\mathcal{M}$ (if any such output exists), then there is a countable model, $\mathcal{N} \vDash$ PA, such that

(1) $\mathcal{M}$ is an initial segment of $\mathcal{N}$,
(2) t is exactly the (total) output of the program e_0 generated within $\mathcal{N}$.

Any index e_0 satisfying this condition (and in the format indicated above) clearly witnesses the informal claims we have made, and e_0 easily gives the existence of a program e_0^* as claimed. We are now faced with a rather precise mathematical problem (to find e_0 with the indicated property).

For the construction of e_0 we are assuming that PA is formally consistent, and for the specific e_0 we construct, if

$$\mathcal{M} \vDash \text{PA}$$

is any model in which PA is consistent, there is no output of the program e_0 generated within the model $\mathcal{M}$. Therefore, if t is any finite binary sequence, there must exist a model

$$\mathcal{M} \vDash \text{PA}$$

in which the output of the program e_0 is exactly the given binary sequence t. This much weaker property of e_0 corresponds to the theorem that the problem of determining for a given sequence whether the sequence is random on the basis of information content is not decidable. Arguably, this weaker property alone does not illustrate any subtlety in the distinction between determinism and nondeterminism in the physical universe. The reason is that this weaker property of e_0 only holds for those idealized universes (models of PA) in which an additional law holds ("PA is consistent"), and this additional law is not on the specified list (PA). More precisely, the actual program e_0 that we produce (in the proof of Theorem 5) has the property that for *any* model,

$$\mathcal{M} \vDash \text{PA},$$

the following are equivalent:

(1) $\mathcal{M} \vDash$ "PA *is inconsistent.*"
(2) The program e_0 generates output within $\mathcal{M}$.

Finally, for any consistent, recursive theory T extending PA, there is an analogous index e_T that has the analogous property of e_0 but relative to models $\mathcal{M} \vDash T$.

5.3 The Existence of e_0

The remainder of this chapter is simply concerned with the proof that e_0 exists, and we will assume familiarity with the basic notions of formal mathematical logic. We begin by fixing some notation. We let $\mathcal{L}_0$ denote the formal language for number theory; this is the formal language with two binary function symbols (one for "+" and one for "·") and a binary relation symbol (for "<"). Suppose $\mathcal{M} = (M, +, \cdot, <)$ is an $\mathcal{L}_0$-structure such that

$$\mathcal{M} \vDash \text{PA}.$$

Then the relation $<$ of the structure $\mathcal{M}$ is the natural order defined in $\mathcal{M}$ from the operations of "+" and "·"; this is the order: $a < b$ if and only if $a \neq b$ and $b = a + c$ for some $c \in M$.

If $\mathcal{M} \cong (\mathbb{N}, +, \cdot, <)$, then $\mathcal{M}$ is *standard* and we identify $\mathcal{M}$ with $(\mathbb{N}, +, \cdot, <)$. Otherwise, $\mathcal{M}$ is *nonstandard* and we shall always view the standard structure as an initial segment of $\mathcal{M}$. More generally, if

$$(N, +, \cdot, <) \vDash \text{PA}$$

we have the natural notion that $(M, +, \cdot, <)$ is a *proper initial segment* of $(N, +, \cdot, <)$: $M \subset N$, M is an initial segment of N in the total order $(N, <)$, *and* the operations of

"+" and "·" of the two structures, $(M, +, \cdot, <)$ and $(N, +, \cdot, <)$, agree on the elements of M. We can now state the Friedman isomorphism theorem (Kaye 1991).

Theorem 1 *(Friedman isomorphism theorem). Every countable model of* PA *that is nonstandard is isomorphic to a proper initial segment of itself.*

We shall need a variation of this theorem, and the statement requires some more notation. Suppose $(M, +, \cdot, <) \vDash$ PA. A set $A \subset M$ is an *internal set* if A is bounded in M and if A is logically definable in $(M, +\cdot, <)$ from parameters and similarly for subsets of $M \times M$. Thus, a sequence

$$\langle a_i : i \in I \rangle$$

where $I \subset M$ is an *internal sequence* if the set of pairs,

$$\{(a, i) \mid i \in I, a = a_i\}$$

is an internal set. Notice that if $\mathcal{M}$ is a proper initial segment of $\mathcal{N}$ (with $\mathcal{M}, \mathcal{N} \vDash$ PA), then every internal set of $\mathcal{M}$ is an internal set of $\mathcal{N}$.

Suppose $\mathcal{M} = (M, +, \cdot, <)$ is an $\mathcal{L}_0$-structure, $\mathcal{M} \vDash$ PA, and that $\mathcal{M}$ is nonstandard. Then SS($\mathcal{M}$) denotes the *Scott set* of $\mathcal{M}$, which is the set of all $A \cap \mathbb{N}$ such that A is an internal set of $\mathcal{M}$ (Scott 1962).

Finally, let $\mathcal{L}_0^+$ be the formal language obtained from $\mathcal{L}_0$ by adding a constant symbol c. If $\mathcal{M} = (M, +, \cdot, <)$ is an $\mathcal{L}_0$-structure if $a \in M$, then we let $(\mathcal{M}; a)$ denote the $\mathcal{L}_0^+$-structure obtained from $\mathcal{M}$ by interpreting c by a.

If $\phi(x)$ is a formula of $\mathcal{L}_0$, then $\phi(c)$ denotes the sentence of $\mathcal{L}_0^+$ obtained by substituting the constant c for all the free occurrences of the variable x in ϕ. Thus, if $(M, +, \cdot, <)$ is an $\mathcal{L}_0$-structure and if $a \in M$, then

$$(M, +, \cdot, <) \vDash \phi[a]$$

if and only if $((M, +, \cdot, <); a) \vDash \phi(c)$.

For each $\mathcal{L}_0$-structure, $\mathcal{M} = (M, +, \cdot, <)$, such that $\mathcal{M} \vDash$ PA, and for each $a \in M$, we let $\Sigma_1^0(\mathcal{M}; a)$ denote the set of all Σ_1^0-sentences, ψ, of $\mathcal{L}_0^+$ such that

$$(\mathcal{M}; a) \vDash \psi$$

Suppose that $(M, +, \cdot, <)$, $(N, +, \cdot, <)$ are each models of PA, $(M, +, \cdot, <)$ is a proper initial segment of $(N, +, \cdot, <)$, $(M, +, \cdot, <)$ is nonstandard, and $a \in M$. By the definitions, the following must hold where $\mathcal{M} = (M, +, \cdot, <)$ and $\mathcal{N} = (N, +, \cdot, <)$:

(1) SS($\mathcal{M}$) = SS($\mathcal{N}$).
(2) $\Sigma_1^0(\mathcal{M}; a) \subseteq \Sigma_1^0(\mathcal{N}; a)$.

Another version of the Friedman isomorphism theorem is that if $\mathcal{M}$ and $\mathcal{N}$ are countable nonstandard models of PA, then $\mathcal{M}$ is isomorphic to a proper initial segment of $\mathcal{N}$ if and only if (1) and (2) hold with $a = 0$. The following is a corollary of the proof of the Friedman isomorphism theorem. For the statement of the theorem we are regarding the formulas and symbols of the language $\mathcal{L}_0^+$ as elements of $\mathbb{N}$ (through Gödel numbering [Gödel 1931]), and we shall give a very brief sketch of the proof by reducing the proof to the key lemma that underlies Theorem 1.

Theorem 2.   *Suppose that $\mathcal{M} = (M, +, \cdot, <)$ is a countable nonstandard model of PA, $a \in M$, and $\phi(x_0)$ is a formula of $\mathcal{L}_0$. Suppose T is a consistent theory of $\mathcal{L}_0^+$ such that*

(i) PA $\subseteq T$ *and* $\phi(c) \in T$,

(ii) $T \in SS(\mathcal{M})$,

(iii) $\Sigma_1^0(\mathcal{M}; a) \subset T$.

Then there exists a countable model $\mathcal{N}$ such that $\mathcal{M}$ is an initial segment of $\mathcal{N}$ and such that $(\mathcal{N}; a) \vDash T$.

PROOF.   By the theory of nonstandard models of PA (see Kaye 1991), for each $b \in M$ such that b is nonstandard, there exists an initial segment M^* of $(M, <)$ such that

(1.1) $(M^*, +, \cdot, <) \vDash$ PA,

(1.2) $b \notin M^*$ and $(M^*, +, \cdot, <)$ is nonstandard,

where $(M^*, +, \cdot, <)$ is the $\mathcal{L}_0$-structure obtained by restricting the operations of $\mathcal{M}$ to M^* (and so M^* is closed under these operations).

Therefore, by "over-spill" and because T is consistent, there exists a nonstandard model $\mathcal{M}^*$ such that $\mathcal{M}^* \vDash$ PA, $\mathcal{M}^*$ is an initial segment of $\mathcal{M}$ and such that there is an internal set S of $\mathcal{M}^*$ such that $T \subset S$ and such that

$$\mathcal{M}^* \vDash \text{ "}S \text{ is a finite consistent } \mathcal{L}_0^+\text{-theory."}$$

By applying the Gödel completeness theorem within $\mathcal{M}^*$ to S, we obtain a nonstandard model, $\mathcal{N}_0 = (N_0, +, \cdot, <)$ and $a_0 \in N_0$ such that

(2.1) $(\mathcal{N}_0; a_0) \vDash T$,

(2.2) $SS(\mathcal{M}) = SS(\mathcal{M}^*) = SS(\mathcal{N}_0)$.

Now by the proof of the Friedman isomorphism theorem (again, see [Kaye 1991]), and because

$$(\mathcal{N}_0; a_0) \vDash \Sigma_1^0(\mathcal{M}; a),$$

there exists an initial segment of $\mathcal{N}_0$ containing a_0 that is isomorphic to $\mathcal{M}$ by an isomorphism that sends a_0 to a, and so $\mathcal{N}$ exists as required.   □

Again by a fixed Gödel numbering we regard the set of Turing programs in the specified format for producing as output a finite binary sequence as exactly the set $\mathbb{N}$. For each $e \in \mathbb{N}$ and for each model

$$\mathcal{M} \vDash \text{PA}$$

we let Output$(e:\mathcal{M})$ denote the output of the program e generated within the universe, $\mathcal{M}$. Note that Output$(e:\mathcal{M})$ is an internal sequence of $\mathcal{M}$ and if $\mathcal{M}$ is a proper initial segment of $\mathcal{N}$, then either

$$\text{Output}(e:\mathcal{M}) = \text{Output}(e:\mathcal{N})$$

or Output$(e:\mathcal{M})$ is a proper initial segment of Output$(e:\mathcal{N})$. We let Output$(e : \mathbb{N})$ denote the output of e generated within the standard model of PA (i.e., within our idealized universe). Notice that with our convention, the assertion

$$\text{“Output}(e:\mathbb{N}) = \emptyset\text{”}$$

is simply the assertion that the program e generates no output. We also note that by our conventions the assertion

$$\text{“Output}(e:\mathbb{N}) = s\text{”}$$

is by complexity a Π_1^0-assertion if $s = \emptyset$ and it is a Σ_2^0-assertion if $s \neq \emptyset$ it is not in general a Σ_1^0-assertion. However, the assertion

$$\text{“}s \text{ is a proper initial segment of Output}(e:\mathbb{N})\text{”}$$

is easily verified to be a Σ_1^0-assertion. The next two theorems are versions of the Kleene recursion theorem adapted to our setting (Kleene 1938).

Theorem 3   *(Kleene recursion theorem). Suppose that*

$$\pi : \mathbb{N} \to \mathbb{N}$$

is a recursive function. Then there exists $e \in \mathbb{N}$ such that

$$\text{Output}(e:\mathbb{N}) = \text{Output}(\pi(e):\mathbb{N}).$$

We shall need the uniform version of the Kleene recursion theorem.

Theorem 4   *(uniform Kleene recursion theorem). Suppose that*

$$\pi : \mathbb{N} \to \mathbb{N}$$

is a recursive function. Then there exists $e \in \mathbb{N}$ such that

$$\text{Output}(e:\mathcal{M}) = \text{Output}(\pi(e):\mathcal{M})$$

for all $\mathcal{L}_0$-structures $\mathcal{M}$ such that $\mathcal{M} \vDash$ PA.

We now come to our main theorem, which we can now precisely formulate using the notation we have fixed.

Theorem 5.   *There exists $e_0 \in \mathbb{N}$ such that for all countable models,*

$$\mathcal{M} \vDash \text{PA}$$

if $s = \text{Output}(e_0:\mathcal{M})$, and if t is an internal binary sequence of $\mathcal{M}$ such that s is a proper initial segment of t, then there exists a countable model $\mathcal{N} \vDash$ PA such that $\mathcal{M}$ is a proper initial segment of $\mathcal{N}$ and such that

$$\text{Output}(e_0:\mathcal{N}) = t.$$

PROOF.   We shall abuse notation, and for any $\mathcal{L}_0$-structure $\mathcal{M}$ such that

$$\mathcal{M} \vDash \text{PA},$$

we identify the internal binary sequences of $\mathcal{M}$ with their corresponding canonical codes. Thus, for each internal binary sequence s of $\mathcal{M}$, we interpret $\Sigma_1^0(\mathcal{M}; s)$ as the set of sentences $\phi(c)$ of $\mathcal{L}_0^+$ such that $\phi(x)$ is a Σ_1^0-formula of $\mathcal{L}_0$ and such that

$$\mathcal{M} \vDash \phi[a]$$

where a is the canonical code of s. More generally, for any formula $\phi(x)$ of $\mathcal{L}_0$, we write

$$(\mathcal{M}, s) \vDash \phi(c)$$

to indicate that $\mathcal{M} \vDash \phi[a]$, where again a is the canonical code of s.

With this convention, for all $\mathcal{L}_0$-structures, $\mathcal{M}$, such that $\mathcal{M} \vDash$ PA, for all $e \in \mathbb{N}$, and for all internal finite binary sequences s of $\mathcal{M}$ (including the empty sequence), the following are equivalent:

(1.1) Output$(e{:}\mathcal{M}) = s$.

(1.2) $(\mathcal{M}; s) \vDash$ "Output$(e{:}\mathbb{N}) = c$."

Given our fixed Gödel numbering of the symbols and formulas of $\mathcal{L}_0^+$, a formal proof p from $\mathcal{L}_0^+$ is a finite sequence

$$p = \langle n_i : i \leq m \rangle$$

where each $n_i \in \mathbb{N}$. We define the *size* of p to be $n \cdot m$, where $n = \max\{n_i \mid i \leq m\}$. There is a natural order on triples, (p, s, n), where p is a formal $\mathcal{L}_0$-proof of size k for some $k \leq n$, s is a binary sequence of length k for some $k \leq n$, and $n \in \mathbb{N}$:

$$(p^*, s^*, n^*) \leq (p, s, n)$$

if $n^* < n$ or if $n^* = n$ and $(p^*, s^*) \leq (p, s)$ in any fixed (recursive) order of all pairs (p, s) of finite sequences (for any fixed n there are only finitely many possible pairs (p, s) for which (p, s, n) is a *legal* triple).

Fix $e \in \mathbb{N}$. Let e^* be the Turing program whose output is as follows (we are describing a specific computation). Let (p_0, s_0, n_0) be least (in the order defined above) such that the following hold:

(2.1) s_0 is a (nonempty) binary sequence of length k for some $k \leq n_0$,

(2.2) p_0 is a proof from PA $\cup \Sigma_1^0(\mathbb{N}; s_0)$ that Output$(e{:}\mathbb{N}) \neq c$,

(2.3) p_0 has size k for some $k \leq n$,

(2.4) for each Σ_1^0 formula $\phi(x)$ of $\mathcal{L}_0$ such that $\phi(c)$ occurs in p_0, the least witness that $\phi[s_0]$ holds is below n.

Suppose that (p_i, s_i, n_i) has been defined. Let $(p_{i+1}, s_{i+1}, n_{i+1})$ be least such that the following hold:

(3.1) s_{i+1} is a binary sequence of length k for some $k \leq n_{i+1}$ and s_i is a proper initial segment of s_{i+1},

(3.2) p_{i+1} is a proof from PA $\cup \Sigma_1^0(\mathbb{N}; s_{i+1})$ that Output$(e{:}\mathbb{N}) \neq c$,

(3.3) p_{i+1} has size k for some $k \leq n_{i+1}$ and the size of p_{i+1} is strictly less than the size of p_i,

(3.4) for each Σ_1^0 formula $\phi(x)$ of $\mathcal{L}_0$ such that $\phi(c)$ occurs in p_{i+1}, the least witness that $\phi[s_{i+1}]$ holds is below n_{i+1}.

If (p_0, s_0, n_0) is not defined, then there is no output of e^*, and so with our convention

$$\text{Output}(e^*{:}\mathbb{N}) = \emptyset.$$

Otherwise, $\text{Output}(e^*{:}\mathbb{N}) = s_m$, where m is largest such that (p_m, s_m, n_m) is defined. Viewing e^* as generating the sequence

$$\langle (s_i, N_i) : i \leq m \rangle,$$

where for all $i \leq m$, $N_i = \text{size}(p_i)$, clearly e^* can be required to have the proper format. Thus, there is a recursive function

$$\pi : \mathbb{N} \to \mathbb{N}$$

such that for each $e \in \mathbb{N}$, $\pi(e)$ is a Turing program that computes the binary sequence using e as described above.

By the uniform Kleene recursion theorem, there exists $e_0 \in \mathbb{N}$ such that for all $\mathcal{L}_0$-structures, $\mathcal{M}$, if

$$\mathcal{M} \vDash \text{PA}$$

then $\text{Output}(e_0{:}\mathcal{M}) = \text{Output}(\pi(e_0){:}\mathcal{M})$.

We finish by proving that e_0 witnesses the theorem. Suppose that $\mathcal{M}$ is a countable $\mathcal{L}_0$-structure and that

$$\mathcal{M} \vDash \text{PA}.$$

Let $s = \text{Output}(e_0; \mathcal{M})$, and let t be an internal binary sequence of $\mathcal{M}$ such that s is a proper initial segment of t.

Let $T = \text{PA} \cup \Sigma_1^0(\mathcal{M}; t) \cup \phi_0(c)$, where $\phi_0(x)$ is the formula of $\mathcal{L}_0$ that expresses (in all models of PA)

$$\text{“Output}(e_0{:}\mathbb{N}) = x\text{.”}$$

Thus, $T \in \text{SS}(\mathcal{M})$. We claim that T is consistent. We consider only the case that $\text{Output}(e_0, \mathcal{M}) \neq \emptyset$ (i.e., the case that e_0 produces output in $\mathcal{M}$); the case that

$$\text{Output}(e_0, \mathcal{M}) = \emptyset$$

is similar (and simpler). Let

$$\langle (p_i, s_i, n_i) : i \leq b \rangle$$

be the internal sequence of $\mathcal{M}$ generated by the program $\pi(e_0)$ within $\mathcal{M}$. Because

$$\text{Output}(e_0{:}\mathcal{M}) = \text{Output}(\pi(e_0){:}\mathcal{M})$$

necessarily Output$(\pi(e_0):\mathcal{M}) \neq \emptyset$ and so the sequence $\langle (p_i, s_i, n_i) : i \leq b \rangle$ must both exist and have the property that $s_b = s$. For each $i \leq b$, let N_i be such that N_i is the size of p_i as calculated in $\mathcal{M}$. We first show that $N_b \notin \mathbb{N}$, (i.e., N_b is nonstandard). Assume toward a contradiction that N_b is standard. Therefore, p_b is an $\mathcal{L}_0^+$-proof witnessing

$$\text{PA} \cup \Sigma_1^0(\mathcal{M}; s_b) \vdash \text{``Output}(e_0:\mathbb{N}) \neq c\text{''}$$

But we have that the following hold given that $s = s_b$:

(4.1) $(\mathcal{M}; s) \vDash \text{PA} \cup \Sigma_1^0(\mathcal{M}; s_b)$,

(4.2) Output$(e_0:\mathcal{M}) = $ Output$(\pi(e_0):\mathcal{M}) = s_b$;

hence, this is a contradiction. Therefore, N_b is nonstandard. If T is not consistent, then there exists a proof, p, witnessing

$$(\mathcal{M}; t) \cup \Sigma_1^0(\mathcal{M}; t) \vdash \text{``Output}(e_0:\mathbb{N}) \neq c\text{''}$$

and this proof p necessarily has size strictly less than N_b. But then within $\mathcal{M}$ it is possible to define $(p_{b+1}, s_{b+1}, n_{b+1})$ because one can choose p_{b+1} to have size at most the size of p and thereby satisfy the requirement that p_{b+1} have size strictly less than the size of p_b, and this again is a contradiction.

Therefore, T is a consistent theory, and so by Theorem 2, because

$$T \in \text{SS}(\mathcal{M})$$

and that

$$T = \text{PA} \cup \Sigma_1^0(\mathcal{M}; t) \cup \phi_0(c)$$

where $\phi_0(x)$ is the formula of $\mathcal{L}_0$, "Output$(e_0:\mathbb{N}) = x$," there exists $\mathcal{N}$ such that

(5.1) $\mathcal{N} \vDash \text{PA}$,

(5.2) $\mathcal{M}$ is an initial segment of $\mathcal{N}$,

(5.3) $\mathcal{N} \vDash \phi_0[t]$.

But then Output$(e_0:\mathcal{N}) = t$, and so $\mathcal{N}$ is as required.   □

The proof of Theorem 5 easily adapts to prove the following more general version.

Theorem 6.   *Suppose that T is a recursive $\mathcal{L}_0$-theory such that* $\text{PA} \subseteq T$. *There exists $e_T \in \mathbb{N}$ such that for all countable models,*

$$\mathcal{M} \vDash T$$

if $s = $ Output$(e_T:\mathcal{M})$ and if t is an internal binary sequence of $\mathcal{M}$ such that s is a proper initial segment of t, then there exists a countable model $\mathcal{N} \vDash T$ such that $\mathcal{M}$ is a proper initial segment of $\mathcal{N}$ and such that

$$\text{Output}(e_T:\mathcal{N}) = t.$$

References

Gödel, Kurt. 1931. Über formal unentscheidbare sätze der principia mathematica und verwandter systeme I. *Monatshefte für Mathematik u Physik* 38: 349–60.

Kaye, Richard. 1991. *Models of Peano Arithmetic*, vol. 15. In *Oxford Logic Guides*. New York: Oxford University Press.

Kleene, S. C. 1938. On notation for ordinal numbers. *Journal Symbolic Logic* 3: 150–55.

Li, Ming, and Vitányi, Paul. 1997. *An Introduction to Kolmogorov Complexity and Its Applications. Graduate Texts in Computer Science*. 2nd ed. New York: Springer-Verlag.

Scott, Dana. 1962. Algebras of sets binumerable in complete extensions of arithmetic. In *Recursive Function Theory*, vol. 5. In *Proceedings of the Symposium on Pure Mathematics*, pp. 117–21. Providence, RI: American Mathematical Society.

CHAPTER 6

Concept Calculus: Much Better Than

Harvey M. Friedman

My contribution to this volume is the initial publication in concept calculus. This is the term I used to describe a new development that seeks to connect two structures that are normally thought to have little or no connection.

On the one hand, there is ordinary commonsense thinking. Generally speaking, ordinary commonsense thinking is rather unstructured and has little in the way of the careful exactness that drives science.

Nevertheless, philosophers have long been committed to taking many aspects of ordinary commonsense thinking at face value and to searching and analyzing fundamental principles. For instance, consider the development of ordinary language philosophy (see, e.g., http://en.wikipedia.org/wiki/Ordinary_language_philosophy).

On the other hand, there is mathematical thinking. Here there has been a tremendously successful and productive development in laying out and analyzing fundamental principles.

Several different approaches apply to the analysis of ordinary mathematical thinking. The most well known and well studied of these are in terms of set theory. Others include class theory and category theory.

It has been established that all of these different approaches to the analysis of ordinary mathematical thinking result in foundational systems that are, in an appropriate and precise sense, equivalent. In other words, each one can be interpreted in all of the others.

In concept calculus, we seek foundational systems for commonsense concepts, as opposed to mathematical concepts. The fundamental principles that are uncovered are plausible or compelling to varying extents. We are feeling our way in identifying fundamental principles, at this very early stage.

What we have found is this: the foundational systems that we have uncovered for the commonsense notions of "better than" and "much better than" are, in fact, equivalent, in the same basic sense of mutual interpretability, with the foundational systems for mathematical thinking.

This establishes the beginning of a deep and unexpected connection between commonsense thinking and mathematical thinking. In particular, this provides an interpretation of mathematical thinking in commonsense terms.

We believe that concept calculus will continue to expand fruitfully into a very large number of realms of commonsense thinking. We have already obtained preliminary results concerning a realm of commonsense thinking we call "abstract cosmology." We will report on these developments elsewhere.

6.1 Introduction

We have discovered an unexpectedly close connection between the logic of mathematical concepts and the logic of informal concepts from commonsense thinking. Our results indicate that they are, in a certain precise sense, equivalent. This connection is new, and there is the promise of establishing similar connections involving a very wide range of informal concepts.

We call this development "concept calculus." In this chapter, we focus on just one context for concept calculus. We use two particular informal concepts from commonsense thinking. These are the informal binary relations "better than" and "much better than"

As discussed briefly in Section 6.10, these relations can be looked at mereologically, using the part/whole and the infinitesimal part/whole relation.

Sections 6.2 and 6.3 contain background information about interpretability between theories, which should be informative for readers not familiar with this fundamentally important concept credited to Alfred Tarski. We are now preparing a book on this topic (Friedman and Visser in preparation).

In Section 6.4 we present our basic axioms involving "better than," "much better than," and identity. These axioms are of a simple character and range from obvious to intriguingly plausible.

We anticipate that concept calculus will be extremely flexible, so that axioms can be chosen to accommodate many diverse points of view, while still maintaining the mutual interpretability with systems such as Z and ZF that we establish here.

For instance, the axioms investigated here preclude there being a best object. There are important viewpoints where a best object is an essential component. We anticipate a formulation accommodating a best object that stands in relation to the system here as does class theory to set theory.

In Section 6.4, you will find three groups of axioms:

Basic

Diverse Exactness
Strong Diverse Exactness
Very Strong Diverse Exactness
Super Strong Diverse Exactness

Unlimited Improvement
Strong Unlimited Improvement

We put primary emphasis on the system MBT (much better than) = Basic + Diverse Exactness + Strong Unlimited Improvement. We prove that MBT is mutually interpretable with ZF (and, hence, ZFC, as ZF and ZFC are mutually interpretable).

However, there are other meritorious combinations that we show are mutually interpretable with ZF. In fact, we show that if we choose one axiom from each of the three groups (thus, Basic must be included), then we get a system mutually interpretable with ZF, with exactly one exception: Basic + Diverse Exactness + Unlimited Improvement is interpretable in ZF/P (ZF without the power set) and may be much weaker still.

Zermelo set theory (Z) is a particularly important relatively strong fragment of ZF of substantial foundational significance. In particular, ZC (Z with the axiom of choice) forms a very smooth and workable foundation for mathematics that is nearly as comprehensive, in practice, as ZFC. Of course, known exceptions to this are particularly interesting and noteworthy. See Friedman (in preparation) for a discussion.

We show that Basic plus any of the last three forms of Diverse Exactness form a system that is mutually interpretable with Z.

We close with Section 6.10, in which we give a very brief discussion of some further developments.

A corollary of the results here is a proof of the equivalence of the consistency of MBT and the consistency of ZF(C), within a weak fragment of arithmetic such as EFA = exponential function arithmetic. In particular, this provides a proof of the consistency of mathematics (as formalized by ZFC), assuming the consistency of MBT. The same holds for the variants discussed earlier that are mutually interpretable with ZF.

We have also obtained a number of results in concept calculus involving a variety of other informal concepts and a variety of formal systems, including ZF and beyond. We are planning a comprehensive book on concept calculus.

6.2 Interpretation Power

The notion of interpretation plays a crucial role in concept calculus. Interpretability between formal systems was first precisely defined by Alfred Tarski. We work in the usual framework of first-order predicate calculus with equality.

Definition An interpretation of S in T consists of:

i. A one-place relation defined in T that is meant to carve out the domain of objects that S is referring to, from the point of view of T.

ii. A definition of the constants, relations, and functions in the language of S by formulas in the language of T, whose free variables are restricted to the domain of objects that S is referring to (in the sense of i).

iii. It is required that every axiom of S, when translated into the language of T by means of i and ii, becomes a theorem of T.

It is now standard to allow quite a lot of flexibility in i through iii. Specifically:

a. Parameters are allowed in all definitions.

b. The domain objects can be tuples.

c. The equality relation in S need not be interpreted as equality but, instead, as an equivalence relation. The interpretations of the domain, constants, and relations must respect this equivalence relation. Functions are interpreted as "functional" relations that respect this equivalence relation.

A detailed discussion of interpretations between theories will appear in a forthcoming book (Friedman and Visser in preparation).

We caution the reader that interpretations may not preserve truth. They only preserve provability. Two illustrative examples follow:

S consists of the axioms for linear order, together with "there is a least element."

i. $\neg(x < x)$.
ii. $(x < y \wedge y < z) \rightarrow x < z$.
iii. $x < y \vee y < x \vee x = y$.
iv. $(\exists x)(\forall y)(x < y \vee x = y)$.

T consists of the axioms for linear order, together with "there is a greatest element."

i. $\neg(x < x)$.
ii. $(x < y \wedge y < z) \rightarrow x < z$.
iii. $x < y \vee y < x \vee x = y$.
iv. $(\exists x)(\forall y)(y < x \vee x = y)$.

S, T are theories in first-order predicate calculus with equality, in the same language: just $<$.

CLAIM. S is interpretable in T and vice versa. They are mutually interpretable.

Obvious interpretation of S in T: In T, take the objects of S to be everything (according to T). Define $x < y$ of S to be $y < x$ in T. Define $x = y$ of S to be $x = y$ in T.

Interpretation of the axioms of S formally yields the following:

i′. $\neg(x < x)$.
ii′. $(y < x \wedge z < y) \rightarrow z < x$.
iii′. $y < x \vee x < y \vee x = y$.
iv′. $(\exists x)(\forall y)(y < x \vee x = y)$.

These are obviously theorems of T.

The obvious interpretation of T in S is the same! The interpretation of the axioms of T formally yields the following:

i″. $\neg(x < x)$.
ii″. $(y < x \wedge z < y) \rightarrow z < x$.
iii″. $y < x \vee x < y \vee x = y$.
iv″. $(\exists x)(\forall y)(x < y \vee x = y)$.

These are obviously theorems of S.

We now discuss a much more sophisticated example. Let P = Peano arithmetic be the theory in 0, S, +, $\bullet$ with successor axioms, defining equations for +, $\bullet$, and the scheme of induction for all formulas in this language.

Now consider "finite set theory." By this, we mean ZF with the axiom of infinity replaced by its negation, that is, ZF\I + $\neg$I.

Theorem (well known)   P, ZF\I + $\neg$I are mutually interpretable.

The theorem is usually attribution to Tarski.

To interpret P in ZF\I + $\neg$I, nonnegative integers are interpreted as the finite von Neumann ordinals in ZF\I + $\neg$I. 0, S, +, $\bullet$, = are interpreted in the normal way on the finite von Neumann ordinals in ZF\I + $\neg$I.

To interpret ZF\I + $\neg$I in P, sets are coded by the natural numbers in P. A common method writes $n = 2^{m_1} + \cdots + 2^{m_k}$ and has n coding the set of sets coded by the m's. This uses all of the natural numbers in P, with = interpreted as =.

In many examples of mutual interpretability, the considerably stronger relation of synonymy holds. The strongest notion of synonymy normally considered is that of having a common definitional extension. There are some important weaker notions.

Notions of synonymy and other topics concerning interpretability are treated systematically and extensively in a forthcoming book (Friedman and Visser in preparation).

Synonymy and its natural variants exhibit many delicate phenomena. It is obvious that S, T cited previously are synonymous. It is proved in Kaye and Wong (2007) that P and ZF\I + $\neg$I are synonymous, if we formulate the axiom of foundation in ZF as a scheme.

However, it is proved in Enayat, Schmerl, and Visser (2008) that P and ZF\I + $\neg$I are not synonymous, if foundation is formulated in the more usual way as a single sentence.

6.3 Basic Facts about Interpretation Power

Every theory is interpretable in any inconsistent theory. Thus, the most powerful level of interpretation power is inconsistency. The following fundamental fact there is no maximal interpretation power, short of inconsistency. From Feferman (1960) demonstrates that

Theorem 3.1 (in ordinary predicate calculus with equality)   *Let S be a consistent recursively axiomatized theory. There exists a consistent finitely axiomatized system T such that S is interpretable in T but T is not interpretable in S.*

This is proved using Gödel's second incompleteness theorem. Consider $T = $ EFA + Con(S), where EFA is exponential function arithmetic. If T is interpretable in S, then EFA proves Con(S) → Con(EFA + Con(S)). By Gödel's second incompleteness theorem, EFA + Con(S) is inconsistent, which is a contradiction.

Comparability(?).   Let S, T be recursively axiomatized theories. Then S is interpretable in T or T is interpretable in S?

There are plenty of natural and interesting examples of incomparability for finitely axiomatized theories that are rather weak.

Let T_1 be the theory of strict linear orderings, where every element has an immediate predecessor and immediate successor.

Let T_2 be the theory of strict linear orderings, where between any two elements there is a third, and there are no least or greatest elements.

We use $(Q, <)$, where Q is the set of all rational numbers, which forms a model of T_2.

Lemma 3.2 *There is no model $(D, R, \equiv)$ of T_1 definable in $(Q, <)$.*

PROOF Let $(D, R, \equiv)$ be a model of T_1 that is definable in $(Q, <)$. Let P be the set of parameters used to define $(D, R, \equiv)$. Here D is the domain, R is the interpreted linear order relation, and $\equiv$ is the interpreted equality relation. Let k be such that every element of D is a tuple of length at most k.

Let $x \in D$. Consider $[x], [x] + 1, [x] + 2, \ldots$, the equivalence classes under $\equiv$ that are the successive immediate successors of $[x]$ in $(D, R, \equiv)$. Let $\alpha_0, \alpha_1, \ldots$ be such that each $\alpha_i \in [x] + i$. Given that the lengths of the α's are bounded by k, we see that by Ramsey's theorem we can find $i < j < k$ such that $(\alpha_i, \alpha_j), (\alpha_i, \alpha_k), (\alpha_j, \alpha_k)$ are all of the same order type over P. Hence, there is an automorphism of $(Q, <)$ that is the identity on P, which sends (α_i, α_j) to (α_i, α_k). Hence, there is an automorphism of $(D, R, \equiv)$ that sends (α_i, α_j) to (α_i, α_k). But $(D, R, \equiv)$ satisfies

$$\text{The distance from } \alpha_i \text{ to } \alpha_j \text{ is } j - i.$$
$$\text{The distance from } \alpha_i \text{ to } \alpha_k \text{ is } k - i.$$

This is a contradiction. □

We now use $(Z, <)$, where Z is the set of all integers, which forms a model of T_1. It is well known that T_2 has elimination of quantifiers, if we add the $+1$ and $\neg 1$ functions.

Let $p, r \geq 1$. The p, r types of the $x \in Z^r$ are the set of all true formulas $\varphi(x)$ in $(Z, <)$, where φ has at most p quantifiers, and no parameters are allowed.

Lemma 3.3 *Let $(E, S, \sim)$ be a model of T_2, definable in $(Z, <)$. There is a nondegenerate interval I in $(E, S, \sim)$, $r \geq 1$, a partial r tuple u from Z, and a p, r type σ, such that the following holds. For all nondegenerate subintervals J of I and integers t, J contains an extension of u, of p, r type σ, where the new coordinates are all $> t$.*

PROOF Let $E, S, \sim$ be as given. If we partition E into finitely many pieces, one of the pieces must be somewhere dense in $(E, S, \sim)$. Therefore, we can assume without loss of generality that σ is a p, r type, I_1 is a nondegenerate interval of $(E, S, \sim)$, and the $x \in D^r$ of p, r type σ are dense in I_1.

If possible, fix an integer t_1 and a coordinate position $1 \leq r_i \leq r$ such that the $x \in D^r$ of p, r type σ whose r_1-th coordinate is t_1 are dense in some nondegenerate subinterval I_2 of I_1.

If possible, choose an integer t_2 and another coordinate position $1 \leq r_2 \leq r$ so that the $x \in D^r$ of p, r type σ whose r_1-th coordinate is t_1 and whose r_2-th coordinate is t_2 are dense in some nondegenerate subinterval $I_3 \subseteq I_2$ of $(E, S, \sim)$.

Continue in this way as long as possible. This results in a partial r tuple u and a nondegenerate subinterval I, such that the extensions of u of p, r type σ are dense in I.

Let J be a nondegenerate subinterval of I and t be an integer. Suppose J does not contain an extension of u, of p, r type σ, where the new coordinates are all $> t$. Then the union over the remaining coordinate positions $i \leq r$ of the extensions of u of p, r type σ, whose i-th coordinate is $\leq t$, is dense in J. Hence, one of these sets is dense in some nondegenerate subinterval of J. Therefore, we have $t' \leq t$ and a remaining coordinate position $i \leq r$ such that the extensions of u of p, r type σ, whose i-th coordinate is t', are dense in some nondegenerate subinterval of J. This contradicts that we could not continue the process. Hence, J does contain an extension of u, of p, r type σ, where the new coordinates are all $> t$.   □

Lemma 3.4 *There is no model $(E, S, \sim)$ of T_2, definable in $(Z, <)$.*

PROOF   Let $(E, S, \sim)$ be a model of T_2 definable in $(Z, <)$, using q quantifiers, with no parameters. Let $p \gg q$.

Let I, r, u be as given by Lemma 3.3. Let $J < K$ be nondegenerate subintervals of I in $(E, S, \sim)$. By Lemma 3.3, let $x \in J \cap D^r$ extend u, with p, r type σ, where all new coordinates of x are $\gg pmax(u)$. By Lemma 3.3, let $y \in K \cap D^r$ extend u, with p, r type σ, where all new coordinates of y are $\gg pmax(x)$. Then $x \, S \, y$. By Lemma 3.3, let z be an r tuple of p, r type σ, extending u, from the interval (x, y) in $(E, S, \sim)$, such that all new coordinates of z are $\gg pmax(x, y)$. By the quantifier elimination in $(Z, <)$, we see that (x, z) and (y, z) satisfy the same q quantifier formulas without parameters, in $(Z, <)$. Hence, $x \, S \, z \leftrightarrow y \, S \, z$. But $x \, S \, z \, S \, y$. This is a contradiction.   □

Theorem 3.5 *T_1 is not interpretable in T_2. T_2 is not interpretable in T_1.*

PROOF   By Lemmas 3.2 and 3.4.   □

Theorem 3.6 *Let S be a consistent recursively axiomatized theory. There exist consistent finitely axiomatized theories T_1, T_2, both in a single binary relation symbol, such that*

  i. *S is provable in T_1, T_2;*

  ii. *T_1 not interpretable in T_2;*

  iii. *T_2 is not interpretable in T_1.*

For the proof of a sharper result, see Theorem 2.7 in Friedman (2007).

However, are there examples of incomparability between natural theories that are metamathematically strong, for example, where P is interpretable?

STARTLING OBSERVATION. *Any two natural theories S, T, known to interpret P, are known (with small numbers of exceptions) to have: S is interpretable in T or T is interpretable in S. The exceptions are believed to also have comparability.*

As a consequence, there has emerged a rather large linearly ordered table of "interpretation powers" represented by natural formal systems. Several natural systems may occupy the same position.

We call this growing table the *Interpretation Hierarchy*. See Friedman (2007, sect. 7).

6.4 Better Than, Much Better Than

We use the following informal notions: better than ($>$) and much better than ($\gg$). These are binary relations. Passing from $>$ to $\gg$ is an example of what we call *concept amplification*. Equality is taken for granted.

We need to consider properties of things. The properties that we consider are to be given by first-order formulas. Their extensions are called "ranges of things."

When informally presenting axioms, we prefer to use "range of things" rather than "set of things," as we do not want to commit to set theory here.

We say that x is (much) better than a given range of things if and only if x is (much) better than every element of that range of things.

Note that by transitivity of better than (see Basic, which follows), if x is better than a given range of things, then it is also better than everything that something in that range of things is better than.

We say that x is exactly better than a given range of things if and only if x is better than every element of that range of things, and everything that something in that range of things is better than, and nothing else.

Thus, among the x that are better than a given range of things, the x that are exactly better are better than the fewest possible things.

We now introduce some important axiom groups.

BASIC (B).   Nothing is better than itself. If x is better than y and y is better than z, then x is better than z. If x is much better than y, then x is better than y. If x is much better than y and y is better than z, then x is much better than z. If x is better than y and y is much better than z, then x is much better than z. There is something that is much better than any given x, y. If x is much better than y, then x is much better than something better than y.

The following basic principle asserts that "we can find any given bounded level of goodness in a variety of ways."

DIVERSE EXACTNESS (DE).   Let x be better than a given range of things. There is something that is exactly better than the given range of things, that x is not better than.

In DE, ranges of things are given by formulas in $L(>, \gg, =)$, with side parameters allowed.

Our next basic principle asserts that "if x is much better than y, and bears a relation to y, then there is no limit to how much x can be improved while still maintaining that relation to y."

UNLIMITED IMPROVEMENT (UI).   Assume that x is much better than and related to y by a given binary relation. Then arbitrarily good x are related to y by the given binary relation.

In UI, binary relations are given by formulas in $L(>, =)$ with no side parameters.

In order to obtain mutual interpretability with strong set theories, we strengthen Unlimited Improvement in the following natural way.

STRONG UNLIMITED IMPROVEMENT (SUI). Let x and a ternary relation be given. There are arbitrarily good y such that x, y are each related, by the given ternary relation, to the same pairs of things that x is much better than.

In SUI, binary relations are given by formulas in $L(>, =)$ with no side parameters.

We place the greatest emphasis on the system MBT (much better than) = B + DE + SUI. We prove that MBT is mutually interpretable with ZF.

We isolate the following three strengthenings of Diverse Exactness:

STRONG DIVERSE EXACTNESS (SDE). Let x be much better than something better than a given range of things. Then x is better than some, but not all, things exactly better than the given range of things.

VERY STRONG DIVERSE EXACTNESS (VSDE). Let x be much better than something better than a given range of things. Then x is much better than some, but not all, things exactly better than the given range of things.

SUPER STRONG DIVERSE EXACTNESS (SSDE). Let x be much better than something, and a given range of things. Then x is better than some, but not all, things exactly better than the given range of things.

In these three, ranges of things are given by formulas in $L(>, \gg, =)$, with side parameters allowed.

We derive SDE from B + VSDE and from B + SSDE.

We also derive SDE and UI from MBT.

We show that B + SDE is mutually interpretable with Z. Here Z is Zermelo set theory, a fragment of ZF of considerable strength. This result holds for B + VSDE + SSDE.

We show that B + SDE + UI is mutually interpretable with ZF.

We also show that all seven of our axioms (axiom schemes), together, is mutually interpretable with ZF(C

To avoid any ambiguities, we now present these axioms formally.

Let φ be a formula in $L(>, \gg, =)$, where y is not free in φ. We define

$$y > \varphi \; iff \; (\forall x)(\varphi \to y > x).$$

$$y \gg \varphi \; iff \; (\forall x)(\varphi \to y \gg x).$$

$$y >_{ex} \varphi \; iff \; (\forall z)(y > z \leftrightarrow (\exists x)(\varphi \wedge (x = z \vee x > z))).$$

Note that in the preceding formulas, we think of $\varphi = \varphi(x)$. Thus, the variable x has a special status. The other variables, y, z, are allowed to be any distinct variables, other than x, that do not appear in φ.

BASIC (B). $\neg x > x$. $x > y \wedge y > z \to x > z$. $x \gg y \to x > y$. $x \gg y \wedge y > z \to x \gg z$. $x > y \wedge y \gg z \to x \gg z$. $(\exists z)(z \gg x \wedge z \gg y)$. $x \gg y \to (\exists z)(x \gg z \wedge z > y)$.

DIVERSE EXACTNESS (DE). $y > \varphi \to (\exists z)(\neg y > z \wedge z >_{ex} \varphi)$, where φ is a formula in $L(>, \gg, =)$ in which y, z are not free.

UNLIMITED IMPROVEMENT (UI). $x \gg y \wedge \varphi \to (\exists x)(x > w \wedge w))$, where φ is a formula of $L(>, =)$ with at most the free variables x, y.

STRONG UNLIMITED IMPROVEMENT (SUI). $(\exists y)(y > z \wedge (\forall v, w)(x \gg v, w \to (\varphi \leftrightarrow \varphi[x/y])))$, where φ is a formula of $L(>, =)$ with at most the free variables x, v, w, in which y does not appear.

STRONG DIVERSE EXACTNESS (SDE). $y \gg z > \varphi \to (\exists z)(y > z >_{ex} \varphi) \wedge (\exists z)(\neg y > z \wedge z >_{ex} \varphi)$, where φ is a formula in $L(>, \gg, =)$ in which y, z are not free.

VERY STRONG DIVERSE EXACTNESS (VSDE). $y \gg z > \varphi \to (\exists z)(y \gg z >_{ex} \varphi) \wedge (\exists z)(\neg y \gg z \wedge z >_{ex} \varphi)$, where φ is a formula in $L(>, \gg, =)$ in which y, z are not free.

SUPER STRONG DIVERSE EXACTNESS (SSDE). $y \gg \varphi \wedge (\exists x)(y \gg x) \to (\exists z)(y > z >_{ex} \varphi) \wedge (\exists z)(\neg y > z \wedge z >_{ex} \varphi)$, where φ is a formula in $L(>, \gg, =)$ in which y, z are not free.

We define

$$MBT = B + DE + SUI.$$

We take Z to be pairing, extensionality, union, power set, separation, and infinity. It is customary to take the axiom of infinity to be in either of two forms.

The first is the same weak form that is most commonly used in ZF. It asserts the existence of a set containing the element $\oslash$ and closed under the operation that sends x to $x \cup \{x\}$. It is well known that this form of Z does not suffice to prove the existence of $V(\omega)$.

The second form is the stronger form, based on the operation that sends x, y to $x \cup \{y\}$. This form of Z does suffice to prove the existence of $V(\omega)$.

What happens to Russell's paradox in this context? In sets, we start with

> There is a set whose elements are exactly the sets
> with a given property.

and obtain a contradiction that Frege missed and Russell saw. The corresponding principle here is

> There is something which is better than, exactly,
> the things with a given property and
> those things they are better than.

This immediately leads to a contradiction. Even the much weaker

> There is something which is better than the things
> with a given property.

gives an immediate contradiction, because there cannot be anything that is better than all things, by irreflexivity. In other words, nothing can be better than itself.

Thus, Russell's Paradox now becomes entirely transparent and never would have trapped anyone: it disappears as a paradox. Clearly, there is no residual feeling of mystery here, as there is in the context of sets and properties.

6.5 Some Implications

We establish a number of implications, some of which are needed for Section 6.9.

Theorem 5.1 *The following are provable in B. SSDE $\to$ SDE $\to$ DE. VSDE $\to$ SDE $\to$ DE.*

PROOF   Assume SSDE. Let $x \gg y > S$. Then $x \gg S$, and apply SSDE.

Assume SDE. Let $x > S$. Let $y \gg x > S$. By SDE, let $z >_{ex} S$, $\neg z < y$. Then $\neg z < x$. The last claim is immediate.   □

Recall that $MBT = B + DE + SUI$.
We say that a is $>$-equivalent to b if and only if $(\forall x)(a > x \leftrightarrow b > x)$.
We say that a is $>$-included in b if and only if $(\forall x)(a > x \to b > x)$.
We say that a, b are incomparable if and only if $a \neq b \wedge \neg a > b \wedge \neg b > a$.

Theorem 5.2 *MBT proves $B + SDE + UI$.*

PROOF   It is immediate that MBT proves $B + UI$. We have only to show that MBT proves SDE. The diversity in SDE follows from DE. Let

(1) $x \gg y > E$.

Without loss of generality, assume that

(2) E is $>$ transitive.

Assume that

(3) nothing $< x$ is $>_{ex} E$.

We derive a contradiction.
By DE, let

(4) $z >_{ex} E, \neg x > z, u >_{ex} \{x\}, v >_{ex} \{y, z\}, w >_{ex} \{u, v\}$.

Define

(5) $P(w', x', y', z', u', v') \leftrightarrow x' > y' \wedge y', z'$ are incomparable $\wedge x', z'$ are incomparable $\wedge u' >_{ex} \{x'\} \wedge v' >_{ex} \{y', z'\} \wedge w' >_{ex} \{u', v'\} \wedge z'$ is $>$-included in $y' \wedge (\forall a < b < w')(a$ is not $>$-equivalent to $z')$.

We claim that

(6) $P(w, x, y, z, u, v)$.

Suppose y, z are comparable. By (1), (4), $x > y \wedge \neg x > z$. Hence $z > y$. By (4), $y \in E$. By (1), $y > E$. Hence $y > y$, which is a contradiction. Hence y, z are incomparable.

Suppose x, z are comparable. Since y, z are incomparable and $x > y$, we have $x > z$. This contradicts (4).

By (1), (4), z is $>$-included in y. Hence, we have only to verify that $(\forall a < b < w)$ (a is not $>$-equivalent to z). By (4), we need only verify that $(\forall a < b < w)$ (a is not $>_{ex} E$).

Suppose $a < b < w \wedge a >_{ex} E$. By (4), $a < b \leq u \vee a < b \leq v$. Hence, $a < b \leq x \vee a < b \leq y \vee a < b \leq z$. Hence, $a < x \vee a < y \vee a < z$. By (3), the first two disjuncts imply that a is not $>_{\mathrm{ex}} E$. If $a < z$, then by (4) $a \in E$, and so not $a >_{\mathrm{ex}} E$.

We claim that

(7) $P(w', x', y', z', u', v') \to u', v'$ are incomparable.

Suppose $P(w', x', y', z', u', v')$. Suppose $u' \le v'$. By (5), $x' \le y' \lor x' \le z'$. This contradicts (5).

Suppose $v' \le u'$. Then $y', z' \le x'$. This contradicts (5).

We claim that

(8) $P(w', x', y', z', u', v') \land P(w', x'', y'', z'', u'', v'') \to (x', y', z', u', v') = (x'', y'', z'', u'', v'')$.

Suppose $P(w', x', y', z', u', v') \land P(w', x'', y'', z'', u'', v'')$. Then $w' >_{\mathrm{ex}} \{u', v'\}$, $w' >_{\mathrm{ex}} \{u'', v''\}$. By (7), $u' \le u'' \lor u' \le v'', v' \le u'' \lor v' \le v'', u'' \le u' \lor u'' \le v', v'' \le u' \lor v'' \le v'$.

Case 1. $u' \le u''$. By (7), $u'' \le u', u' = u''$.

Case 2. $u' \le v''$. By (7), $v'' \le u', u' = v''$.

By symmetry, $v' = v'' \lor v' = u''$. Hence, $\{u', v'\} = \{u'', v''\}$.

Suppose $u' = v''$. Since y'', z'' are incomparable, we have $x' \le y'' \lor x' \le z'', y'' \le x', z'' \le x'$. Hence, $x' = y'' \le x' \lor x' \le z'' \le x'$. Therefore, $x' = y'' \lor x' = z''$. This contradicts the incomparability of y'', z''. Hence, $u' \ne v''$. Therefore, $u' = u'' \land v' = v''$. From the first conjunct, $x' = x''$.

By the incomparability of y', z', and of y'', z'', we have $\{y', z'\} = \{y'', z''\}$. Suppose $y' = z''$. Then $x' > z''$, and so $x'' > z''$, which is impossible. Hence, $y' = y'' \land z' = z''$.

We now define

(9) $R(w', b, c) \leftrightarrow (\exists x', y', z', u', v')(P(w', x', y', z', u', v') \land c = y' \land (b < z' \lor b = y'))$.

We claim that

(10) $(\forall b, c)(R(w, b, c) \leftrightarrow c = y \land (b \in E \lor b = y))$.

Let $R(w, b, c)$. Let $P(w, x', y', z', u', v') \land c = y' \land (b < z' \lor b = y')$. By (6), (8), (9), $c = y \land (b < z \lor b = y)$. Hence, $c = y \land (b < z \lor b = y)$.

Let $c = y \land (b \in E \lor b = y)$. We have $R(w, y, y)$, since $P(w, x, y, z, u, v)$ and (9). Let $b \in E$. Then $R(w, b, y)$, since $P(w, x, y, z, u, v), b < z$, and (9).

By (10),

(11) $(\forall b, c)(R(w, b, c) \to b, c \ll w)$.

By SUI, let

(12) $w' > w \land (\forall b, c \ll w)(R(w, b, c) \leftrightarrow R(w', b, c))$.

By (11), (12),

(13) $R(w', y, y)$.

Let

(14) $P(w', x', y', z', u', v')$.

By (8), (9),

(15) $(\forall b)(R(w', b, b) \leftrightarrow b = y')$.

By (13),

(16) $y = y'$.

By (8), (9),

(17) $y = y'$.

By (5), (15),

(18) z' is $>$-included in y.

By (9),

(19) $(\forall b, c)(R(w', b, c) \rightarrow b, c \ll w)$.

By (12), (19),

(20) $(\forall b, c)(R(w, b, c) \leftrightarrow R(w', b, c))$.

By (9), (16),

(21) $(\forall b, c)((c = y \wedge (b < z' \vee b = y)) \leftrightarrow (c = y \wedge (b < z \vee b = y)))$. $(\forall b)$ $(b < z' \leftrightarrow z < z)$. z', z are $>$-equivalent.

By (5), (21),

(22) $(\forall a < b < w')(a$ is not $>$-equivalent to $z)$.

By (4), (9), (22),

(23) $z < w < w' \wedge z$ is not $>$-equivalent to z.

This is a contradiction.   □

6.6 Interpretation of MBT in ZF

We will use the following axioms for ZF here; the primitives are $\in, =$:

EXTENSIONALITY. $(\forall x)(x \in y \leftrightarrow x \in z) \rightarrow y = z$.
PAIRING. $(\exists x)(y \in x \wedge z \in x)$.
UNION. $(\exists x)(\forall y)(\forall z)(y \in z \wedge z \in w \rightarrow y \in x)$.
SEPARATION. $(\exists x)(\forall y)(y \in x \leftrightarrow y \in z \wedge \varphi)$, where x is not free in φ.
POWER SET. $(\exists x)(\forall y)((\forall z)(z \in y \rightarrow z \in w) \rightarrow y \in x)$.
INFINITY. $(\exists x)(\emptyset \in x \wedge (\forall y, z)(y \in x \wedge z \in x \rightarrow y \cup \{z\} \in x))$.
FOUNDATION. $x \neq \emptyset \rightarrow (\exists y)(y \in x \wedge \neg(\exists z)(z \in y \wedge z \in x))$.
COLLECTION. $(\forall y)(y \in x \rightarrow (\exists z)(\varphi)) \rightarrow (\exists w)(\forall y)(y \in x \rightarrow (\exists z)(z \in w \wedge \varphi))$, where w is not free in φ.

Infinity is more usually formulated in the weaker form, in which the previously cited $y \cup \{z\}$ is replaced by $y \cup \{y\}$. Replacement is more commonly used than Collection, in which $(\exists z)(\varphi)$ in the antecedent is replaced by $(\exists!z)(\varphi)$. It is well known that when both these changes are made, the resulting system is logically equivalent to the axioms for ZF.

We will give an interpretation of MBT in ZF. In fact, the construction interprets MBT + VSDE + SSDE in ZF. Hence, by Theorems 5.1 through 5.3, we have interpreted every axiom considered here, simultaneously in ZF.

In light of the axiom in Basic that asserts that for any two things, there is something much better, it is natural to strengthen the last axiom in Basic to the following:

$$\#)\ x \gg y \wedge x \gg z \to (\exists w)(x \gg w \wedge w > y \wedge w > z).$$

Definition   Basic′ is Basic, where the last axiom is replaced by #).

Recall that MBT = B + DE + SUI. We will give an interpretation of B′ + VSDE + SSDE + SUI in ZF. Hence, by Theorem 5.1, this interprets every axiom considered here, simultaneously in ZF.

We work within ZF. We first form a transfinite hierarchy of structures $(D_\alpha, >_\alpha)$.

For any binary relation $>$, we define $x < y \leftrightarrow y > x$, $x \leq y \leftrightarrow x < y \vee x = y$, $x \geq y \leftrightarrow x > y \vee x = y$.

Definition   Let $(E, >)$ be transitive and irreflexive, and $K \subseteq E$. We say that x is an exact upper bound of K over $(E, >)$ if and only if for all $y \in E, x > y \leftrightarrow (\exists z \in K)(z \geq y)$. We say that K is $(E, >)$ transitive if and only if $K \subseteq E$ and for all $x, y, z \in K, x > y \wedge y > z \to x > z$.

Definition   We define pairs $(D_\alpha, >_\alpha)$, for all ordinals α. Define $(D_0, >_0) = (\emptyset, \emptyset)$. Suppose $(D_\alpha, >_\alpha)$ has been defined and is transitive and irreflexive. Define $(D_{\alpha+1}, >_{\alpha+1})$ to extend $(D_\alpha, >_\alpha)$ by adding an exact upper bound of every $(D_\alpha, >_\alpha)$ transitive set, even if this set already has an exact upper bound over $(D_\alpha, >_\alpha)$ lying in D_α. For limit ordinals λ, define $D_\lambda = \cup_{\alpha<\lambda}D_\alpha, >_\lambda = \cup_{\alpha<\lambda}>_\alpha$.

It will be convenient to have a form of this definition that uses a purely set theoretic construction. This is accomplished by giving an explicit definition of the exact upper bounds.

Definition′   We define pairs $(D_\alpha, >_\alpha)$, for ordinals α, where $>_\alpha \subseteq D_\alpha \times D_\alpha$. Define $(D_\alpha, >_\alpha) = (\emptyset, \emptyset)$. Suppose $(D_\alpha, >_\alpha)$ has been defined, where $>_\alpha$ is transitive and irreflexive. We set $D_{\alpha+1} = D_\alpha \cup \{(\alpha, A): A \text{ is } (D_\alpha, >_\alpha) \text{ transitive}\}$. We define $x >_{\alpha+1} y \leftrightarrow x >_\alpha y \vee (\exists A)(x = (\alpha, A) \wedge y \in A)$. For limit ordinals λ, set $D_\lambda = \cup_{\beta<\lambda}D_\beta, >_\lambda = \cup_{\beta<\lambda}>_\beta$.

Lemma 6.1   *Let $\alpha, \beta < \lambda$.*

i. $\alpha \leq \beta \to D_\alpha \subseteq D_\beta \wedge >_\alpha \subseteq >_\beta$.

ii. $x \in D_{\alpha+1} \backslash D_\alpha \to x = (\alpha, \{y : x >_{\alpha+1} y\})$.

iii. $x \in D_\alpha \to (\exists \beta < \alpha)(x = (\beta, \{y : x >_{\beta+1} y\}))$.

iv. $x >_\alpha y \to (\exists \beta < \alpha)(y \in D_\beta)$.

v. $x, y \in D_\alpha \to (x >_\beta y \to x >_\alpha y)$.

PROOF   Let $\alpha, \beta < \lambda$. Claim i is proved by fixing α and applying transfinite induction to $\beta \geq \alpha$. For claim ii, let $x \in D_{\alpha+1} \backslash D_\alpha$. Let $x = (\alpha, A)$, where A is $(D_\alpha, >_\alpha)$ transitive. Hence, $A \subseteq D_\alpha$. We claim that $A = \{y: x >_{\alpha+1} y\}$. To see this, suppose $x >_{\alpha+1} y$. Now $x >_\alpha y$ is impossible, given that $x \notin D_\alpha$. Hence, $y \in A$. Conversely, suppose $y \in A$. Then $x >_{\alpha+1} y$.

For claim iii, let $x \in D_\alpha$. Let α' be least such that $x \in D_{\alpha'}$. Then $\alpha' \leq \alpha$ and $x \in D_{\alpha'} \backslash D_{\alpha'-1}$. By ii, $x = (\alpha'-1, \{y : x >_{\alpha'} y\})$. Set $\beta = \alpha' - 1$.

We prove claim iv by transfinite induction on α. For $\alpha = 0$, the statement is vacuously true given that $D_0 = \emptyset$. Suppose

$$(\forall x, y)(x >_\alpha y \rightarrow (\exists \beta < \alpha)(y \in D_\beta)).$$

Let $x >_{\alpha+1} y$. If $x >_\alpha y$, then $(\exists \beta < \alpha)(y \in D_\beta)$, and so $(\exists \beta < \alpha + 1)(y \in D_\beta)$.
Otherwise, let $x = (\alpha, A)$, $y \in A$, where A is $(D_\alpha, >_\alpha)$ transitive. Then $y \in D_\alpha$, and so $(\exists \beta < \alpha + 1)(y \in D_\beta)$.

Finally, suppose λ is a limit ordinal, and for all $\alpha < \lambda$,

$$(\forall x, y)(x >_\alpha y \rightarrow (\exists \beta < \alpha)(y \in D_\beta)).$$

Let $x >_\lambda y$. Then $y \in D_\lambda$, and so $(\exists \beta < \lambda)(y \in D_\beta)$.

We prove claim v by fixing α and applying transfinite induction to $\beta \geq \alpha$. For $\beta = \alpha$, the statement is trivial. Let $\beta \geq \alpha$, and suppose

$$(\forall x, y)(x, y \in D_\alpha \rightarrow (x >_\beta y \rightarrow x >_\alpha y)).$$

Let $x, y \in D_\alpha, x >_{\beta+1} y$. If $x >_\beta y$, then $x >_\alpha y$. Otherwise, let $x = (\beta, A)$, $y \in A$, where A is $(D_\beta, >_\beta)$ transitive. By iii, $\beta < \alpha$, which is impossible.

Finally, suppose λ is a limit ordinal, and for all $\alpha < \lambda$,

$$(\forall x, y)(x, y \in D_\alpha(x >_\lambda y \rightarrow x >_\alpha y)).$$

Let $x >_\beta y, \beta < \lambda$. Then $x >_\alpha y$. $\square$

Lemma 6.2   *For all α, $(D_\alpha, >_\alpha)$ is irreflexive and transitive.*

PROOF   We prove this by transfinite induction on α. Clearly, $(D_0, >_0)$ is irreflexive and transitive. Suppose $(D_\alpha, >_\alpha)$ is irreflexive and transitive. Suppose $x >_{\alpha+1} x$. By Lemma 5.1, $x \in D_\alpha$, and so $x >_\alpha x$. This violates the irreflexivity of $>_\alpha$.

To see that $(D_{\alpha+1}, >_{\alpha+1})$ is transitive, let $x >_{\alpha+1} y, y >_{\alpha+1} z$, where $x, y, z \in D_{\alpha+1}$. Then $y \in D_\alpha, z \in D_\alpha$. If $x \in D_\alpha$, then $x >_\alpha z, x >_{\alpha+1} z$. Suppose $x \in D_{\alpha+1} \backslash D_\alpha$. Let $x = (\alpha, A)$, where A is $(D_\alpha, >_\alpha)$ transitive. Then $y \in A, y >_\alpha z$. Hence, $z \in A$, and so $x >_{\alpha+1} z$. Hence, $(D_{\alpha+1}, >_{\alpha+1})$ is irreflexive and transitive.

Suppose for all $\beta < \lambda$, $(D_\beta, >_\beta)$ is irreflexive and transitive. Then obviously $(D_\lambda, >_\lambda)$ is irreflexive. Let $x >_\lambda y \wedge y >_\lambda z$, where $x, y, z \in D_\lambda$. Let $x, y, z \in D_\beta, \beta < \lambda$. Then $x >_\beta y \wedge y >_\beta z$. Therefore, $x >_\beta z$. Hence, $(D_\lambda, >_\lambda)$ is transitive. $\square$

Definition   Fix S to be a nonempty set of limit ordinals, with no greatest element, whose union is the limit ordinal λ. We define $M[S]$ to be the following structure $(D_\lambda, >_\lambda, \gg_S)$. $D_\lambda, >_\lambda$ have already been defined. We define $x \gg_S y$, if and only if

$$x, y \in D_\lambda \wedge (\exists \alpha, \beta \in S)(\alpha < \beta \wedge y \in D_\alpha \wedge (\forall w \in D_\beta)(x >_\lambda w)).$$

Remark We could have defined $x \gg_S y$, if and only if

$$x, y \in D_\lambda \wedge (\exists \alpha \in S)(\alpha < \beta \wedge y \in D_\alpha \wedge (\forall w \in D_\alpha)(x >_\lambda w)),$$

but then it can be shown that under this definition, Strong Diverse Exactness would fail in $M[S]$.

Lemma 6.3 *$M[S]$ satisfies Basic$'$ + VSDE.*

PROOF Because $(D_\lambda, >_\lambda)$ is irreflexive and transitive, we have the first two statements in Basic. The third is obvious.

Suppose $x \gg_S y$, $y >_\lambda z$. Let $\alpha < \beta$ be from S, $y \in D_\alpha$, $(\forall w \in D_\beta)(x >_\lambda w)$. Then $z \in D_\alpha$, and so $x \gg_S z$.

Suppose $x >_\lambda y$, $y \gg_S z$. Let $\alpha < \beta$ be from S, $z \in D_\alpha$, $(\forall w \in D_\beta)(y >_\lambda w)$. Then $(\forall w \in D_\beta)(x >_\lambda w)$, and so $x \gg_S z$.

Let $y, z \in D_\lambda$. Let $y, z \in D_\alpha$. Let $\beta < \gamma$ be from S, where $\alpha < \beta$. Then $(\gamma, D_\gamma) \gg_S y, z$.

Let $x \gg_S y, z$. Let $\alpha_1 < \beta_1$ be from S, where $y \in D_{\alpha_1} \wedge (\forall w \in D_{\beta_1})(x \gg w)$. Let $\alpha_2 < \beta_2$ be from S, where $z \in D_{\alpha_2} \wedge (\forall w \in D_{\beta_2})(x \gg w)$. Let $\alpha = max(\alpha_1, \alpha_2)$, $\beta = max(\beta_1, \beta_2)$. Then $y, z \in D_\alpha \wedge (\forall w \in D_\beta)(x \gg w)$. Since α is a limit ordinal, let $y, z \in D_\gamma$, $\gamma < \alpha$. Then $(\gamma, D_\gamma) >_\lambda y, z$. Also $x \gg_S (\gamma, D_\gamma)$.

For VSDE, let $y \gg_S z >_\lambda \varphi$. Let $\alpha < \beta$ be from S, where $z \in D_\alpha$ and $(\forall w \in D_\beta)(y >_\lambda w)$. Let $B = \{x : \varphi(x)$ holds in $M[S]\}$. Let $z \in D_\gamma$, $\gamma < \alpha$. Then (γ, B) is an exact upper bound for B. Hence, $y \gg_S (\gamma, B) >_{ex} B$. Also $(\delta, B) >_{ex} B$ and $\neg(\delta, B) >_\lambda (\gamma, B)$, provided $\delta > \gamma$. □

For SSDE, we need a further condition on S. This is because SSDE will fail in $M[S]$ if some element of S is a limit of elements of S.

For SSDE, we need a further condition on S. This is because SSDE will fail in $M[S]$ if some element of S is a limit of elements of S.

Lemma 6.4 *Suppose S is of order type ω. Then $M[S]$ satisfies SSDE.*

PROOF Let S be as given. Let $y \gg_S \varphi$, and $y \gg_S u$ for some u. Let $B = \{x : \varphi(x)$ holds in $M[S]\}$. First assume $B = \varnothing$. Let $\alpha < \beta$ be from S, where $u \in D_\alpha$, $(\forall w \in D_\beta)(y >_\lambda w)$. Then $y \gg_S (1, \varnothing)$.

Now assume $B \neq \varnothing$. For each $x \in B$, let $\alpha_x < \beta_x$ be from S such that $x \in D_{\alpha_x} \wedge (\forall w \in D_{\beta_x})(y >_\lambda w)$. Then the β_x are bounded below λ, and so the β_x have a max, β. Also, the α_x are bounded below λ, and also have a max, α. Obviously $\alpha < \beta$ and $(\forall w \in D_\beta)(y >_\lambda w)$. Clearly (α, B) is an exact upper bound for B. Since $(\alpha, B) \in D_\beta$, clearly $y >_\lambda (\alpha, B)$.

Let $y \in D_\gamma$, $\gamma < \lambda$. Note that (γ, y) is also an exact upper bound for B. Clearly $\neg y >_\lambda (\beta, z)$. □

We have given an interpretation of B$'$ + VSDE + SSDE in ZF. Actually, we only need $V(\omega^2)$ for this construction, taking $S = \{\omega, \omega \times 2, \omega \times 3, \ldots\}$, $\lambda = \omega^2$. Thus, we have not provided an interpretation of even B + SDE in Z, or Zermelo set theory. In

Section 6.6 we will modify $M[S]$ so that it does provide an interpretation of $B' +$ VSDE + SSDE in Z.

We now come to Strong Unlimited Improvement. Here we need further conditions on S, even for Unlimited Improvement.

Lemma 6.5 *For all $k \geq 1$ there exists $r \geq 1$ such that the following holds. Suppose S has order type ω, where for all α from S, $V(\alpha)$ is an elementary submodel of $V(\lambda)$ for formulas with at most r quantifiers. Then $M[S]$ is a model of k-SUI, which is SUI in which the formula φ has at most k quantifiers.*

PROOF By Lemmas 6.3 and 6.4, it suffices to verify that k-SUI holds in $M[S]$. Let k, r, S be as above. Unless stated otherwise, φ is always evaluated in $(D_\lambda, >_\lambda)$. Let $\varphi(u, y, z)$ define the ternary relation for k-SUI, for "u is related to y, z". Choose r to be $100(k + 1)$. We will evaluate φ in $M[S]$.

Let x be given. We can assume that $(\exists v)(x \gg_S v)$. Let $\beta \in S$ be greatest such that $(\forall w \in D_\beta)(x >_\lambda w)$. Then β is not the least element of S. Let α be the preceding element of S. Let $E = \{< y, z >: x \gg_S y, z \wedge \varphi(x, y, z)\}$. Then $E = \{<y, z> \in V(\alpha) : \varphi(x, y, z)\}$. Then $E \in V(\beta)$. We have

(1) $(\exists x)((\forall w \in D_\beta)(x >_\lambda w) \wedge (\forall y, z \in E)(\varphi(x, y, z)))$.

It suffices to show that β can be replaced by any ordinal $<\lambda$ in (1). Suppose this is false, and let γ be least such that β cannot be replaced by γ in (1). Then γ is defined from $E \in V(\beta)$ in $V(\lambda)$, with at most r quantifiers, and so by elementarity, $\gamma < \beta$. But $\gamma \geq \beta$. $\square$

Theorem 6.6 *The system T, consisting of Basic, the four versions of Diverse Exactness, and the two versions of Unlimited Improvement, is interpretable in ZF. We can even use Basic'.*

PROOF We claim that each finite fragment of T is provably consistent in ZF. This is clear by Lemmas 6.3 through 6.5, and Theorem 5.1.

We now define the interpretation of T in ZF as follows. If $Con(T)$, then use the construction of a model of T via the completeness theorem for predicate calculus. Otherwise, let T' be the largest initial segment of the axioms of T that is consistent. Use the construction of a model of T' via the completeness theorem for predicate calculus. In any case, the model, provably in ZF, satisfies any particular axiom of T, by the preceding claim.

6.7 Interpretation of B + VSDE + SSDE in Z

Z is the system in $\in, =$ with the following axioms:

EXTENSIONALITY. $(\forall x)(x \in y \leftrightarrow x \in z) \to y = z$.
PAIRING. $(\exists x)(y \in x \wedge z \in x)$.
UNION. $(\exists x)(\forall y)(\forall z)(y \in z \wedge z \in w \to y \in x)$.
SEPARATION. $(\exists x)(\forall y)(y \in x \leftrightarrow y \in z \wedge \varphi)$, where x is not free in φ.
POWER SET. $(\exists x)(\forall y)((\forall z)(z \in y \to z \in w) \to y \in x)$.
INFINITY. $(\exists x)(\varnothing \in x \wedge (\forall y, z)(y \in x \wedge z \in x \to y \cup \{z\} \in x))$.

The preceding axioms are the first six axioms (schemes) for ZF.

We now give an interpretation of B + VSDE + SSDE in Z. As remarked in Section 6.6, the construction there requires all of the $V(\omega \times n), n < \omega$. Even $V(\omega \times 2)$ is not available as a set in Z, and $V(\omega \times 3)$ is not available as a class in Z.

Lemma 7.1 *There are definable operators* *, **, *with no parameters, such that the following is provable in Z. Let* $(E, >)$ *be an irreflexive transitive relation, where* $E \subseteq V(\omega + n), n < \omega$. *Then* $(E^*, >^{**})$ *is an irreflexive transitive relation such that:*

i. $E^* \subseteq V(\omega + n + 6)$.

ii. $E \subseteq E^*, > \subseteq >^{**}$.

iii. $x \in E \rightarrow (x >^{**} y \leftrightarrow x > y)$.

iv. Any finite subset of E^* *has a strict upper bound in* $(E^*, >^{**})$.

PROOF   Let $(E, >) \in V(\omega + n), n < \omega$ be irreflexive and transitive. We let E^* be E together with all ordered pairs $(E, (x_1, \ldots, x_k))$, where $k \geq 1$ and each $x_i \in E$. For $\alpha, \beta \in E*$, we define $\alpha >^{**} \beta$ as follows:

Case 1. $\alpha, \beta \in E$. Then $\alpha >^{**} \beta \leftrightarrow R(\beta, \alpha)$.

Case 2. $\alpha \notin E, \beta \in E$. Let $\alpha = (E, (x_1, \ldots, x_k))$. Then $\alpha >^{**} \beta \leftrightarrow (\exists i)(\beta = x_i)$.

Case 3. $\alpha, \beta \notin E$. Let $\alpha = (E, (x_1, \ldots, x_k)), \beta = (E, (y_1, \ldots, y_n))$. Then $\alpha >^{**} \beta \leftrightarrow$ $(x_1, \ldots, x_k)$ is a (not necessarily consecutive) subsequence of $(y_1, \ldots, y_n)$.

Case 4. $\alpha \in E, \beta \notin E$. Then $\neg \alpha >^{**} \beta$.

We represent a finite sequence $(x_1, \ldots, x_k)$ from E by a function with domain k. Thus, the elements of the finite sequence are of the form $\{\{i\}, \{i, x\}\}$, where $x \in A$, and so lie in $V(\omega + n + 2)$. Hence, the finite sequences from E all lie in $V(\omega + n + 3)$. Therefore, each $(E, (x_1, \ldots, x_n)) = \{\{E\}, \{E, (x_1, \ldots, x_n)\}\}$ lies in $V(\omega + n + 5)$. Hence, $E^* \in V(\omega + n + 6)$.   □

We are now ready to define pairs $(E_n, >_n)$, for all integers $n < \omega$. Define $E_0 = \varnothing, >_0 = \varnothing$. Suppose $(E_n, >_n)$ has been defined. We now define $(E_{n+1}, >_{n+1})$.

We define $E_{n+1} = E_n^* \cup \{(n, 0, A) : A$ is $(E_n^*, >_n^{**})$ transitive$\}$. We define $x >_{n+1}$ $y \leftrightarrow x >_n^{**} y \vee (\exists A)(x = (n, 0, A) \wedge y \in A)$.

We use $(n, 0, A)$ instead of (n, A) because $(n, 0, A)$ is an ordered triple, and so we cannot duplicate ordered pairs in the construction.

We can now create $E_\omega = \cup_{n<\omega} E_n, >_\omega = \cup_{n<\omega} >_n$, as proper classes in Z.

We define $\gg$ by $x \gg y \leftrightarrow (\exists n)(y \in E_n^* \wedge (\forall w \in E_{n+1})(x >_\omega w))$.

We claim that Z proves that $(E_\omega, >_\omega, \gg)$ forms a model of B + VSDE + SSDE. More precisely, for each axiom of B + VSDE + SSDE, Z proves that the axiom holds in $(E_\omega, >_\omega, \gg)$.

The verification of Basic$'$ is straightforward, except for the last assertion.

Lemma 7.2 $(E_\omega, >_\omega, \gg)$ *satisfies Basic$'$*.

PROOF   This follows the proof of Lemma 6.3, except for the last axiom of Basic$'$. Let $x \gg y, z$. Let $y \in E_n^*, (\forall w \in E_{n+1})(x >_\omega w)$. Let $z \in E_m*$,

$(\forall w \in E_{m+1})(x >_\omega w)$. Let $r = max(n, m)$. Then $y, z \in E_r^*$, $(\forall w \in E_{r+1})(x >_\omega w)$. Let $u \in E_r^*$, $u >_r^{**} y, z$. Then $u >_\omega y, z$ and $x \gg u$. $\qquad \square$

Lemma 7.3 $(E_\omega, >_\omega, \gg)$ *satisfies SSDE.*

PROOF Let $y \gg \varphi$. Let $B = \{x : (\exists y)(x \geq_\omega y \wedge \varphi(x)$ holds in $(E_\omega, >_\omega, \gg))\}$. For each $x \in B$, let $f(x)$ be least such that $x \in E_{f(x)}^*$, $(\forall w \in E_{f(x)+1})(y >_\omega w)$. Let n be the maximum of the $f(x)$. Then $(\forall w \in E_{n+1})(y >_\omega w)$. Clearly, $(n, 0, B)$ is an exact upper bound for all x such that $\varphi(x)$ holds in $(E_\omega, >_\omega, \gg)$, over $(E_\omega, >_\omega)$. Because $(n, 0, B) \in E_{n+1}$, we have $y >_\omega (n, 0, B)$.

Let $y \in E_t$. Clearly, $(t, 0, B)$ is another exact upper bound for B, and $\neg y >_\omega (t, 0, B)$. $\qquad \square$

Lemma 7.4 $(E_\omega, >_\omega, \gg)$ *satisfies VSDE.*

PROOF Let $y \gg z > \varphi$. Let $z \in E_{n+1}^*$, $(\forall w \in E_{n+2})(y >_\omega w)$. Let $B = \{x : (\exists y)(x \geq_\omega y \wedge \varphi(x)$ holds in $(E_\omega, >_\omega, \gg))\}$. Then $B \subseteq E_n^*$, and so $(n, 0, B)$ is an exact upper bound for all x such that $\varphi(x)$ holds in $(E_\omega, >_\omega, \gg)$, over $(E_\omega, >_\omega)$. Because $(n, 0, B) \in E_{n+1}$, we have $y \gg (n, 0, B)$. $\qquad \square$

Theorem 7.5 $B + VSDE + SSDE$ *is interpretable in Z. We can even use* B'.

PROOF By Lemmas 7.2 and 7.3. $\qquad \square$

6.8 Interpretation of Z in B + SDE

In this section, we give an interpretation of Z in B + SDE. We use the same interpretation in Section 6.9. We will work entirely in B + SDE.

As in Section 6.6, we formulate Z as follows:

EXTENSIONALITY. $(\forall x)(x \in y \leftrightarrow x \in z) \to y = z$.
PAIRING. $(\exists x)(y \in x \wedge z \in x)$.
UNION. $(\exists x)(\forall y)(\forall z)(y \in z \wedge z \in w \to y \in x)$.
SEPARATION. $(\exists x)(\forall y)(y \in x \leftrightarrow y \in z \wedge \varphi)$, where x is not free in φ.
POWER SET. $(\exists x)(\forall y)((\forall z)(z \in y \to z \in w) \to w \in x)$.
INFINITY. $(\exists x)(\varnothing \in x \wedge (\forall y, z)(y \in x \wedge z \in x \to y \cup \{z\} \in x))$.

Definition We define

$x < y \leftrightarrow y > x$.

$x \leq y \leftrightarrow x < y \vee x = y$.

$x \ll y \leftrightarrow y \gg x$.

$x \geq y \leftrightarrow x > y \vee x = y$.

$x \neq y \leftrightarrow \neg x = y$.

x inc $y \leftrightarrow \neg x \leq y \wedge \neg y \leq x$.

$x \equiv y \leftrightarrow (\forall z)(z < x \leftrightarrow z < y)$.

$x \sim \varnothing \leftrightarrow (\forall y)(\neg x > y)$.

$y >_{ex} \{x_1, \ldots, x_n\}$ if $(\forall x)(x < y \leftrightarrow x \leq x_1 \lor \ldots \lor x \leq x_n)$.

$x <^* y \leftrightarrow x < y \land \neg(\exists z)(x < z < y)$.

$A(x, y) \leftrightarrow y \sim \oslash \land \neg y \leq x$.

$A(x, y_1, \ldots, y_n) \leftrightarrow A(x, y_1) \land \ldots \land A(x, y_n) \land y_1, \ldots, y_n$ are distinct.

In the above, $n \geq 1$. The "ex" in $>_{ex}$ means "exact."

Lemma 8.1 $(\exists y)(x_1, \ldots, x_n \ll y)$. $(\exists y)(y >_{ex} \{x_1, \ldots, x_n\})$. $(\forall x)(\exists y_1, \ldots, y_n)$ $(A(x, y_1, \ldots, y_n))$.

PROOF   The first claim is proved by induction on n using $(\forall x, y)(\exists z)(x, y \ll z)$.

For the second claim, let $x_1, \ldots, x_n$. By the first claim, let $y > x_1, \ldots, x_n$. By DE, $(\exists z)(z >_{ex} \varphi)$, where φ is $x = x_1 \lor \ldots \lor x = x_n$.

The third claim is proved by induction on n. Let x be given, and let $A(x, y_1, \ldots, y_{n-1})$, where this is considered vacuously true if $n = 1$. Let $x' > x, y_1, \ldots, y_{n-1}$. Obviously, $x' > \varphi$, where φ is $x \neq x$. By DE,

$$(\exists y_n)(\neg y_n < x' \land y_n >_{ex} \varphi),$$
$$(\exists y_n)(\neg y_n \leq x, y_1, \ldots, y_{n-1} \land y_n \sim \oslash),$$
$$(\exists y_n)(\neg y_n \leq x \land y_n \neq y_1, \ldots, y_{n-1} \land y_n \sim \oslash),$$
$$A(x, y_1, \ldots, y_n). \qquad \qquad \square$$

Lemma 8.2 *Let φ be a formula of $L(>, \gg, =)$ in which u, v do not appear. Suppose $y >_{ex} \varphi$, where $(\forall u, v)(\varphi[x/u] \land \varphi[x/v] \to \neg u < v)$. Then $(\forall x)$ $(x <^* y \leftrightarrow \varphi)$.*

PROOF   Let y, φ be as given. Then

$$(\forall z)(z < y \leftrightarrow (\exists x)(\varphi \land z \leq x)).$$

Suppose φ. Then $x < y$. Also $\varphi[x/w] \to \neg x < w < y$. Hence, $x <^* y$. On the other hand, suppose $x <^* y$. Then $x < y$, and so let $\varphi[x/u] \land x \leq u$. Clearly, $u < y$, and so $x \leq u < y$. By $x <^* y$, we have $x = u$. Hence, φ. $\qquad \square$

Definition $z >_{ex^*} \varphi \leftrightarrow (\forall x)(x <^* z \leftrightarrow \varphi)$.

Lemma 8.3 *Suppose $w \gg y > \varphi \land (\forall u, v)(\varphi[x/u] \land \varphi[x/v] \to \neg u < v)$. There exists $z < w$ such that $z >_{ex^*} \varphi$.*

PROOF   Let w, y, φ be as given. By SDE, let $w > z >_{ex} \varphi$. By Lemma 8.2, $z >_{ex^*} \varphi$. $\qquad \square$

Lemma 8.4 *Suppose $A(z, w) \land u, v \leq z \land x >_{ex} \{u, w\} \land y >_{ex} \{v, w\}$. Then $\neg x < y$. Furthermore, if $x = y$, then $u = v$. Also, $(\forall x')(x' <^* x \leftrightarrow x' = u \lor x' = w)$.*

PROOF   Let z, w, u, v, x, y be as given. Assume $x < y$. Then $x \leq v \lor x \leq w, x \leq v, x < y$, which is a contradiction.

Suppose $x = y$. Then $u < x, u < y, u \leq v \lor u \leq w, u \leq v \lor u = w, u \leq v \lor w \leq z, u \leq v$. Similarly, $v < y, v < x, v \leq u \lor v \leq w, v \leq u \lor v = w, v \leq u \lor w \leq z, v \leq u$. Hence, $u = v$.

Suppose $x' <^* x$. Then $x' \leq u \vee x' \leq w, x' \leq u \vee x' = w$. If $x' < u$, then $x' < u < x$, violating $x' <^* x$. Hence, $x' = u \vee x' = w$.

To see that $u <^* x$, note that $u < x$, and let $u < b < x$. Then $b \leq u \vee b \leq w, b \leq w$, which is a contradiction.

To see that $w <^* x$, note that $w < x$, and let $w < b < x$. Then $b \leq u \vee b \leq w, b \leq u, w < u, w < z$, which is a contradiction.   □

Lemma 8.5 *Let φ be a formula of $L(>, \gg, =)$ in which y, z, u, v, c do not appear. Assume $A(y, z), v \gg u \gg u' > y, z$. Then $y > \varphi \rightarrow (\exists c < v)(\forall x)(\varphi \leftrightarrow x < y \wedge (\exists x')(x, z <^* x' <^* c))$.*

PROOF Let φ, y, z, u, v be as given. Let $v \gg u \gg u' > y, z$. Assume $y > \varphi$.

By SDE, if φ then there exists $x' >_{\text{ex}} \{x, z\}$ with $x' < u$. Furthermore, by Lemma 8.4, these various $x' < u$ are incomparable under $<$ (i.e., no x' is $<$ any other x'). Hence, by Lemma 8.3, there exists c such that these various $x' < u$ are the $x' <^* c$. By Lemma 8.3, we can choose $c < v$. Then $x, z <^* x' <^* c$.

Now suppose $x < y \wedge x, z <^* x' <^* c$. Then $x' < u$. We want to show that φ. Let x^* be such that $x' >_{\text{ex}} \{x^*, z\}, \varphi[x/x^*]$. By Lemma 8.4, because $x <^* x'$, we have $x = x^* \vee x = z, x = x^* \vee z < y, x = x^*, \varphi$.   □

Definition (scheme) The "sets" are constructs of the form $\{x : \varphi\}$, where φ is a formula in $L(>, \gg, =)$, with the property that there exists y such that $(\forall x)(\varphi \rightarrow x < y)$. In other words, we require that "sets" be bounded above. (Here y is not in φ, but φ may have any other variables). A "set" is $< y$ if and only if every element is $< y$.

Fortunately, the preceding schematic definition is merely an important expositional convenience because of the following definition and Lemma.

Definition We say that a "set" B is coded by c using y, z, if and only if $A(y, z) \wedge B = \{x < y : (\exists x')(x, z <^* x' <^* c)\}$. We say that a "set" B is fully coded by c, y, z if and only if it is coded by c using y, z.

Lemma 8.6 *Let $A(y, z), v \gg u \gg u' \gg y, z$. Every "set" $< y$ is coded by some $c < v$ using y, z. Every "set" has a full code.*

PROOF The first claim is by Lemma 8.5. The second claim follows from the first claim using Lemma 8.1 and Basic.   □

From Lemma 8.6, we see that we can quantify over "sets, " by quantifying over full codes.

Definition Let $n \geq 2$. We say that $x_1, \ldots, x_n$ is coded by c using $y, a_1, \ldots, a_n$ if and only if $x_1, \ldots, x_n < y \wedge A(y, a_1, \ldots, a_n) \wedge (\exists u_1, \ldots, u_n)(u_1 >_{\text{ex}} \{x_1, a_1\} \wedge \ldots \wedge u_n >_{\text{ex}} \{x_n, a_n\} \wedge c >_{\text{ex}} \{u_1, \ldots, u_n\})$.

Lemma 8.7 *Let $n \geq 2, A(y, a_1, \ldots, a_n), v \gg u \gg u' \gg y, a_1, \ldots, a_n$. Every $x_1, \ldots, x_n < y$ is coded by some $c < v$ using $y, a_1, \ldots, a_n$.*

PROOF Let n, y, $a_1, \ldots, a_n$, v, u be as given. Let $x_1, \ldots, x_n < y$. By SDE, let $u_1, \ldots, u_n < u$ be such that each $u_i >_{ex} \{z_i, a_i\}$. By SDE, $(\exists c < v)(c >_{ex} \{u_1, \ldots, u_n\})$. $\square$

Lemma 8.8 *Let $n \geq 2$. Suppose $x_1, \ldots, x_n$ is coded by c using y, $a_1, \ldots, a_n$, and $x'_1, \ldots, x'_n$ is coded by c using y, $a_1, \ldots, a_n$. Then each $x_i = x'_i$.*

PROOF Let n, y, $a_1, \ldots, a_n$, $x_1, \ldots, x_n$, $x'_1, \ldots, x'_n$, c be as given. Then $A(y, a_1, \ldots, a_n) \wedge x_1, \ldots, x_n \leq y \wedge x'_1, \ldots, x'_n < y$. Let $u_i >_{ex} \{x_i, a_i\}$, $u'_i >_{ex} \{x'_i, a_i\}$, $c >_{ex} \{u_1, \ldots, u_n\}$, $c >_{ex} \{u'_1, \ldots, u'_n\}$.

If $a_i < u_j$, then $a_i \leq x_j \vee a_i \leq a_j$, $a_i \leq w \vee a_i = a_j$, $a_i = a_j$, $i = j$. Similarly, $a_i < u'_j \rightarrow i = j$.

We have $u_i < x$, $u_i \leq u'_1 \vee \ldots \vee u_i \leq u'_n$. Now $a_i < u_i$. Hence, $u_i \leq u'_j$ implies $a_i < u'_j$, $a_i \leq x'_j \vee a_i \leq a_j$, $a_i \leq w \vee a_i \leq a_j$, $a_i \leq a_j$, $a_i = a_j$, $i = j$. Hence, $u_i \leq u'_i$.

We have $u'_i < x$, $u'_i \leq u_1 \vee \ldots \vee u'_i \leq u_n$. Now $a_i < u'_i$. Hence, $u'_i \leq u_j$ implies $a_i < u_j$, $a_i \leq x_j \vee a_i \leq a_j$, $a_i \leq w \vee a_i \leq a_j$, $a_i \leq a_j$, $a_i = a_j$, $i = j$. Hence, $u_i \leq u'_i$.

We have shown that $u_i = u'_i$. We have $x_i < u_i$, $x_i < u'_i$, $x_i \leq x'_i \vee x_i \leq a_i$, $x_i \leq x'_i \vee x_i = a_i$, $x_i \leq x'_i \vee a_i \leq w$, $x_i \leq x'_i$.

We have $x'_i < u'_i$, $x'_i < u_i$, $x'_i \leq x_i \vee x'_i \leq a_i$, $x'_i \leq x_i \vee x'_i = a_i$, $x'_i \leq x_i \vee a_i \leq w$, $x'_i \leq x_i$. Hence, $x_i = x'_i$. $\square$

Definition (scheme) Let $n \geq 2$. The n-ary "relations" are constructs of the form $\{<x_1, \ldots, x_n>: \varphi\}$, where φ is a formula in $L(>, \gg, =)$, with the property that there exists y such that $(\forall x_1, \ldots, x_n)(\varphi \rightarrow x < y)$. (Here y is not in φ, but φ may have any other variables). An n-ary "relation" is $< y$ if and only if it holds only of arguments $< y$.

Fortunately, the preceding schematic definition is merely an important expositional convenience because of the following definition and Lemma.

Definition Let $n \geq 2$. We say that an n-ary "relation" R is coded by c using $y, w, a_1, \ldots, a_{n+1}$ if and only if $R = \{<x_1, \ldots, x_n>: x_1, \ldots, x_n < y \wedge (\exists c')(c'$ codes $x_1, \ldots, x_n$ using $y, a_1, \ldots, a_n \wedge c'$ lies in the "set" coded by c using $w, a_{n+1})\}$. We say that an n-ary "relation" R is fully coded by $c, y, w, a_1, \ldots, a_{n+1}$ if and only if it is coded by c using $y, w, a_1, \ldots, a_{n+1}$.

Note that full codes for "sets" have three components. Note that for $n \geq 2$, full codes for n-ary "relations" have $n + 4$ components.

Lemma 8.9 *Let $n \geq 2$, $w \gg z \gg y, a_1, \ldots, a_n \wedge v \gg u \gg u' \gg w, a_{n+1} \wedge A(y, a_1, \ldots, a_n) \wedge A(w, a_{n+1})$. Every n-ary "relation" $R < y$ is coded by some $c < v$ using $y, w, a_1, \ldots, a_{n+1}$. Every n-ary "relation" has a full code.*

PROOF Let n, v, u, w, z, y, $a_1, \ldots, a_{n+1}$ be as given. Let R be an n-ary "relation" with $R < y$.

Let B be the "set" $\{c' < w: (\exists x_1, \ldots, x_n)(R(x_1, \ldots, x_n) \wedge c'$ codes $x_1, \ldots, x_n$ using $y, a_1, \ldots, a_n)\}$. By Lemma 8.6, let $c < v$ code the "set" B using w, a_{n+1}.

We claim that R is coded by c using $y, w, a_1, \ldots, a_{n+1}$. To see this, let $R(x_1, \ldots, x_n)$. Then $x_1, \ldots, x_n < y$. By Lemma 8.7, let $c' < w$ code $x_1, \ldots, x_n$ using y, $a_1, \ldots, a_n$. Then c' lies in the "set" coded by c using w, a_{n+1}, given that $c' \in B$. Conversely, suppose $x_1, \ldots, x_n < y \wedge c'$ codes $x_1, \ldots, x_n$ using $y, a_1, \ldots, a_n \wedge$ c' lies in the "set" coded by c using w, a_{n+1}. Then $(\exists x_1, \ldots, x_n)(R(x_1, \ldots, x_n)$ $\wedge c'$ codes $x_1, \ldots, x_n$ using $y, a_1, \ldots, a_n)$. By Lemma 8.8, $R(x_1, \ldots, x_n)$.

The second claim follows from the first claim using Lemma 8.1 and Basic. $\square$

From Lemma 8.9, we see that for each $n \geq 2$, we can quantify over n-ary "relations" by quantifying over full codes.

Definition We say that R is a prelinear ordering if and only if R is a 2-ary "relation" such that for all $x, y, z \in \mathit{fld}(R)$,

  i. $x\,R\,x$.

  ii. $x\,R\,y \wedge y\,R\,Z \rightarrow x\,R\,z$.

  iii. $x\,R\,y \vee y\,R\,x$.

Definition Let R be a prelinear ordering. We write $x <_R y \leftrightarrow x\,R\,y \wedge \neg y\,R\,x$, $x =_R y \leftrightarrow x\,R\,y \wedge y\,R\,x$, $x \neq_R y \leftrightarrow \neg x =_R y$.

Definition A prewellordering is a prelinear ordering where every nonempty "subset" of its field has a least element.

Definition Let R be a prelinear ordering and $x \in \mathit{fld}(R)$. We write $R|{<}x$ for the restriction of R to the $y \in \mathit{fld}(R)$ with $y <_R x$. Note that $R|< x$ is a prelinear ordering.

Definition Let R, S be prelinear orderings. A comparison relation between R and S is an isomorphism "relation" from R onto S, or from R onto some $S|< x$, or from S onto some $R|< x$.

Lemma 8.10 *Let R, S be prewellorderings. There is a unique comparison "relation" between R and S. In case it is onto $R|< x$, then x is unique up to $=_R$.*

PROOF Let R, S be prewellorderings. Let x be an upper bound on the field of R and the field of S. Let $y \gg y' \gg x$. Now argue in the usual way to develop the unique comparison relation. $\square$

Definition We write $R \leq_{\mathrm{PWO}} S \leftrightarrow R, S$ are prewellorderings $\wedge$ the comparison relation between R, S is from R. $R <_{\mathrm{PWO}} S \leftrightarrow R \leq_{\mathrm{PWO}} S \wedge \neg S \leq_{\mathrm{PWO}} R$. $R =_{\mathrm{PWO}} S \leftrightarrow R \leq_{\mathrm{PWO}} S \wedge S \leq_{\mathrm{PWO}} R$.

Lemma 8.11 *$\leq_{PWO}$ is reflexive, transitive, and connected. $<_{PWO}$ is irreflexive and transitive. $R \leq_{PWO} S \leftrightarrow R <_{PWO} S \vee R =_{PWO} S$. $R =_{PWO} S \leftrightarrow R, S$ are prewellorderings $\wedge$ the comparison relation between R, S is from R onto S. $R <_{PWO} S \leftrightarrow R, S$ are prewellorderings $\wedge$ the comparison relation between R, S is from R onto some $S \mid < x$. Let R, S be prewellorderings. Then $R <_{PWO} S \vee S <_{PWO} R \vee R =_{PWO} S$, where the disjunct is unique.*

PROOF $\leq_{PWO}$ is obviously reflexive. It is transitive by using composition of binary "relations." It is connected by Lemma 8.11.

If $R <_{PWO} R$, then we have two different comparison relations between R, R, which is impossible. Hence, $<_{PWO}$ is irreflexive. S uppose $R <_{PWO} S \wedge S <_{PWO} T$. Then $R <_{PWO} T$ by composition of "relations."

Suppose $R \leq_{PWO} S$. If $\neg R <_{PWO} S$, then $S \leq_{PWO} R$, and so $R \leq_{PWO} S$. Conversely, obviously $R <_{PWO} S$ and $R =_{PWO} S$ separately imply $R \leq_{PWO} S$.

Suppose $R =_{PWO} S$. Then the comparison relation between R, S is from R, and the comparison relation between S, R is from S. If either comparison relation is not onto, then by composition, we get a comparison relation from R onto S that is not onto, or a comparison relation from S onto R that is not onto. This contradicts $R =_{PWO} S$ and the uniqueness of comparison relations. Hence, both comparison relations are onto. Conversely, if the comparison relation between R, S is from R onto S, then $R \leq_{PWO} S \wedge S \leq_{PWO} R$.

Suppose $R <_{PWO} S$. If the comparison relation between R, S is onto S, then $S \leq_{PWO} R$, which is a contradiction. If the comparison relation between R, S is from S, then $S \leq_{PWO} R$, which is a contradiction.

Suppose the comparison relation between R, S is from R onto some $S| < x$. Then $R \leq_{PWO} S$. If $S \leq_{PWO} R$, then the comparison relation between R, S is either from R onto S or from S onto some $R| < x$. This violates the uniqueness of the comparison relation.

Let R, S be prewellorderings. Then $R \leq_{PWO} S \vee S \leq_{PWO} R$. Hence, $R <_{PWO} S \vee S <_{PWO} R \vee R =_{PWO} S$. If more than one disjunct holds, then we violate the irreflexivity of $<_{PWO}$.   □

Lemma 8.12   *Let R be a nonempty 6-ary "relation" of full codes for prewellorderings. There is a prewellordering S with a full code in R such that every prewellordering T with a full code in R has $S \leq_{PWO} T$.*

PROOF   Let S be a prewellordering with a code in R. Look at the "set" B of all $x \in fld(S)$ such that the comparison relation between some S' with full code in R, and S, maps S' onto $S| < x$. If B is empty, then S is as required. Suppose B is nonempty. Let y be an S least element of B. Let T have full code in R, where the comparison relation between T, S maps T onto $S| < y$. Then T is as required.   □

Definition   A finite wellordering is a prewellordering whose equality relation is identity, where every nonempty "subset" of the field has a greatest element.

Definition   A finite "set" is the field of some finite wellordering R.

Definition   We write $x \leq_{fin} y$ if and only if x, y are finite "sets" such that there is a one-one function from x into y. $x <_{fin} y \leftrightarrow x \leq_{fin} y \wedge \neg y \leq_{fin} x$. $x =_{fin} y \leftrightarrow x \leq_{fin} y \wedge y \leq_{fin} x$.

Lemma 8.13   *Every one-one function from a finite "set" into itself is onto. Let x, y be finite "sets." Then $x <_{fin} y \leftrightarrow$ there is a one-one function from x into y that is not onto.*

PROOF   Let R be a finite wellordering. We first prove that for all $x \in A$, every $f : \{y : y \, R \, x\} \rightarrow \{y : y \, R \, x\}$ is onto. Let x be an R least counterexample. Let $f : \{y : y \, R \, x\} \rightarrow \{y : y \, R \, x\}$ be one-one and not onto. Then f is not empty. Let u be the R greatest element of $dom(f)$. We can obviously adjust f so that it is one-one and not onto, and $f(u) = u$. Now delete $f(u) = u$, to obtain $g : \{y : y \, R \, u\} \rightarrow \{y : y R u\}$, which is one-one and not onto. This is a contradiction.

Suppose $x <_{\text{fin}} y$. Let $f : x \rightarrow y$ be one-one. If f is onto, then by taking the inverse function, $y \leq_{\text{fin}} x$, which is impossible.

Suppose $f : x \rightarrow y$ is one-one but not onto. Then $x \leq_{\text{fin}} y$. If $y \leq_{\text{fin}} x$, then by composition, we obtain a one-one function from x into x that is not onto. This contradicts the first claim.   □

Lemma 8.14   *Let R, S be finite wellorderings. Then $fld(R) \leq_{fin} fls(S) \leftrightarrow R \leq_{PWO} S$, $fld(R) <_{fin} fls(S) \leftrightarrow R <_{PWO} S$, $fld(R) =_{fin} fls(S) \leftrightarrow R =_{PWO} S$. If A, B are finite "sets" and A is a proper "subset" of B, then $A <_{fin} B$.*

PROOF   Let R, S be finite wellorderings. Suppose $R <_{\text{PWO}} S$. Let f be the comparison map from R onto $S| < z$. Obviously, $fld(R) \leq_{\text{fin}} fls(S)$. If $fld(S) \leq_{\text{fin}} fld(R)$, then by composition, we obtain a one-one function from $fld(R)$ into $fld(R)$ that is not onto. This contradicts Lemma 8.13. Hence, $fld(R) <_{\text{fin}} fls(S)$.

Suppose $fld(R) <_{\text{fin}} fld(S)$. There is a one-one function from $fld(R)$ into $fld(S)$ that is not onto. There cannot be a one-one function from $fld(S)$ into $fld(R)$, because by composition we would violate the first claim of Lemma 8.14. Hence, $\neg S \leq_{\text{PWO}} R$, and so $R <_{\text{PWO}} S$.

Suppose $fld(R) \leq_{\text{fin}} fld(S)$. Then $fld(R) <_{\text{fin}} fld(S) \vee fld(S) \leq_{\text{fin}} fld(R)$. In the first case, $R <_{\text{PWO}} S$. In the second case, we have one-one $f : fld(R) \rightarrow fld(S)$, $g : fld(S) \rightarrow fld(R)$. The compositions are one-one, and therefore onto. Hence, f, g are onto, and so $R =_{\text{PWO}} S$.

Suppose $(x, R) \leq (y, S)$. Then evidently $x \leq_{\text{fin}} y$.

The fourth claim follows immediately from the first claim.

Now let A, B be finite "sets, " where A is a proper subset of B. Obviously, $A \leq_{\text{fin}} B$. If $B \leq_{\text{fin}} A$, then we obtain a one-one function from B into B that is not onto, violating Lemma 8.13.   □

Lemma 8.15   *Let R be a nonempty 3-ary "relation" of full codes for finite sets. There is a finite "set" x with a full code in R such that every finite "set" y with a full code in R has $x \leq_{fin} y$.*

PROOF   Let R be as given. Let u be a finite "set" with a full code in R. Let S be a finite wellordering. Let B be the "set" of all v in u such that some "set" x with a full code in R has $x \leq_{\text{fin}} \{w : w \, S \, v\}$. If B is empty, then u is as required. Otherwise, let v be the R least element of B. Let x have a full code in R, where $x \leq_{\text{fin}} \{w : w S v\}$. Then x is as required.   □

Lemma 8.16   *Let $x \gg y$. Let A be a finite "subset" of $\{z : z \ll x\}$. There exists $z \ll x$ such that z is not in A. There exists a finite "subset" B of $\{z : z \ll x\}$ such that $A <_{fin} B$.*

PROOF   Let x, y, A be as given. Let R be a finite wellordering with field A. We claim that

$$(\forall y \in A)(\exists z \ll x)\neg(\exists w)(w \; R \; y \wedge z \leq w).$$

This is proved by R induction on y. For the basis case, let y be R least. Choose $y < z \ll x$. Now let $y \in A$ and y' be the immediate R successor of y. Let $z \ll x$, $\neg(\exists w)(w \; R \; y \wedge z \leq w)$. If $\neg z \leq y'$, then keep the same z. If $z \leq y'z$, then use z^* such that $y' < z^* \ll x$.

Now set y to be R greatest. Using $w = y$, we see that the corresponding $z \ll x$ cannot lie in A.

For the second claim, choose B to be A together with any $z \ll x$ that lies outside A, and apply Lemma 8.14.   □

Lemma 8.17   *There exists a prewellordering R such that:*

i. R has no greatest element.

ii. For all $x \in$ fld(R) and nonempty "set" $B \subseteq \{y: y \; R \; x\}$ has an R greatest element.

PROOF   Let $x \gg y$. We work with the finite "subsets" of $\{z: z \ll x\}$. Let $A(x, z)$, $v \gg u \gg x, z$. By Lemma 8.6, every "set" $< x$ is coded by some $c < v$ using x, z. In particular, each finite "subset" of $\{z: z \ll x\}$ has a code $c < v$ using x, z. Let $R = \{<c, c'>: c, c' < v$, and there exists finite "subsets" B, C of $\{z: z \ll x\}$ that are coded by c, c' using x, z, respectively, where $B \leq_{\text{fin}} C\}$.   □

We are now ready to interpret Z. The interpreted sets will be certain systems (A, E, R, x), defined as follows:

Definition   We say that (A, E, R, x) is a system if and only if A is a "set, " E, R are binary "relations" on A, and $x \in A$, such that the following holds:

 i. E is an equivalence relation with field A.

 ii. $E(a, b) \wedge E(E(c, d)) \wedge R \; (E(a, c)) \to R(E(b, d))$.

 iii. Suppose $(\forall x \in A)(R(x, a) \leftrightarrow R(x, b))$. Then $E(a, b)$.

 iv. Let $B \subseteq A$ be a nonempty coded "set." $(\exists y \in B)(\forall z \in B)(\neg R(z, y))$.

 v. Let $B \subseteq A$ be a coded "set." Let $x \in B$, and B be closed under R, E predecessors. Then $A \subseteq B$.

Here B is closed under R, E predecessors means $(\forall y, z)(z \in B \wedge R(y, z) \to y \in B) \wedge (\forall y, z)(z \in B \wedge E(y, z) \to y \in B)$.

Also, (A, E, R, x) is not a quadruple, but rather a notational convenience that groups together the four items into a system.

Definition   For systems (A, E, R, x), and $y \in A$, define $y\#$ to be the least "subset" of A containing y, and closed under E, R predecessors. Define $(A, E, R, x)|y$ to be $(y\#, E|y\#, R|y\#, y)$. If we write $(A, E, R, x)|y$, then (A, E, R, x) is required to be a system and y is required to lie in A.

Definition Let (A, E, R, x) be a system. We say that c is a code for (A, E, R, x) using $y, w, a_1, \ldots, a_7$ if and only if this holds where (A, E, R, x) is treated as the 6-ary "relation" $\{<a, b, c, d, e, f>: a \in A \wedge E(b, c) \wedge R(d, e) \wedge f = x\}$. We say that $c, y, w, a_1, \ldots, a_7$ is a full code for (A, E, R, x) if and only if c is a code for (A, E, R, x) using $y, w, a_1, \ldots, a_7$.

Lemma 8.18 *Let (A, E, R, x) be a system. Then the R maximal elements of A are exactly the points E equivalent to x. Let $y \in A$. Then $(A, E, R, x)|y$ is a system.*

PROOF Let A, E, R, x, y be as given. It is obvious that $z\#$ is closed under E, R predecessors. Let $B = \{z: x \notin z\#\backslash[x]\}$, where $[x]$ is the E equivalence class of x. Then obviously $[x] \subseteq B$. Let $x \notin z\#\backslash[x]$, $E(w, z)$. Then $x \notin w\#\backslash[x]$ given that $z\# = w\#$. Let $R(w, z)$. Then clearly $w\# \subseteq z\#$, by considering $w\# \cap z\#$. Hence, $w\#\backslash[x] \subseteq z\#\backslash[x]$, and so $x \notin w\#\backslash[x]$. Hence, B contains $[x]$ and is closed under E, R predecessors. So $B = A$.

Thus, we have established that for all z in A, $x \notin z\#\backslash[x]$. Therefore, x is R maximal. Now let $C = \{z \in A: \neg E(z, x) \wedge z \text{ is not } R \text{ maximal}\}$. Then C contains $[x]$ and is closed under E, R predecessors. Hence, $C = A$. Therefore, every R maximal point is E equivalent to x.

For the second claim, let $y \in A$. To see that $(y\#, E|y\#, R|y\#, y)$ is a system, it suffices to show that:

   i. Every $C \subseteq y\#$ that contains $[y]$ and is closed under $E|y\#$, $R|y\#$ predecessors is C. This is immediate from the definition of $y\#$.

  ii. Every nonempty $C \subseteq y\#$ has an $R|y\#$ minimal element. Any R minimal element is an $R|y\#$ minimal element. $\qquad\qquad\square$

Definition We say that S is an isomorphism relation from (A, E, R, x) onto (A', E', R', x') if and only if:

    i. (A, E, R, x) and (A', E', R', x') are systems.

   ii. S is a binary "relation."

  iii. $S \subseteq A \times A'$.

  iv. $E(a, b) \wedge E'(c, d) \to (S(a, c) \leftrightarrow S(b, d))$.

   v. $S(a, b) \wedge S(a, c) \to E'(b, c)$.

  vi. $S(a, b) \wedge S(c, b) \to E(a, c)$.

 vii. $(\forall a \in A)(\exists b \in A')(S(a, b))$.

viii. $(\forall b \in A')(\exists a \in A)(S(a, b))$.

  ix. $S(a, b) \wedge S(c, d) \to (R(a, c) \leftrightarrow R'(b, d))$.

We write $(A, E, R, x) \approx (A', E', R', x')$ if and only if there is an isomorphism relation from (A, E, R, x) onto (A', E', R', x').

Lemma 8.19 $\approx$ *is an equivalence relation on systems. If S is an isomorphism from (A, E, R, x) onto (A', E', R', x'), then S^{-1} is an isomorphism from (A', E', R', x') onto (A, E, R, x).*

PROOF Left to the reader. $\square$

Lemma 8.20 *Let S be an isomorphism relation from (A, E, R, x) onto (A', E', R', x'). Let $a \in A, b \in A'$. Then $S(a, b) \leftrightarrow (\forall c)(R(c, a) \leftrightarrow (\exists d)(S(c, d) \wedge R'(d, b)))$. $S(a, b) \leftrightarrow (\forall d)(R'(d, b) \leftrightarrow (\exists c)(S(c, d) \wedge R(c, a)))$.*

PROOF Let $S, A, E, R, x, A', E', R', x'$ be as given. Let $a \in A, b \in A'$. The forward direction of the equivalence follows immediately from the definition. Now we assume

(1) $(\forall c)(R(c, a) \leftrightarrow (\exists d)(S(c, d) \wedge R'(d, b)))$.

We claim

(2) $(\forall b \in A')(\forall d)(R'(d, b) \leftrightarrow (\exists c)(S(c, d) \wedge R(c, a))$.

To see this, suppose $R'(d, b)$. Let $S(c, d)$. By (1), $R(c, a)$. Suppose $S(c, d) \wedge R(c, a)$. By (1), let $S(c, d') \wedge R'(d', b)$. Then $E(d, d'), R'(d, b)$. This establishes (2).

Let $S(a, b')$. By (2),

(3) $(\forall d)(R' \leftrightarrow d, b') ((\exists c)(S(c, d) \wedge R(c, a))$.

(4) $(\forall d)(R'(d, b) \leftrightarrow (\exists c)(S(c, d) \wedge R(c, a))$.

Hence, $E'(b, b'), S(a, b)$.

For the second claim, by Lemma 8.19, S^{-1} is an isomorphism from (A', E', R', x') onto (A, E, R, x). Apply the first claim to S^{-1}. $\square$

Lemma 8.21 *Let S be an isomorphism relation from (A, E, R, x) onto (A', E', R', x'). Then $S(x, x')$.*

PROOF Let $S, A, E, R, x, A', E', R', x'$ be as given. By Lemma 8.18, x is R maximal. Let $S(x, y)$. We claim that y is R' maximal. Suppose $R'(y, z)$. Let $S(w, z)$. Then $R(x, w)$, violating the R maximality x. By Lemma 8.18, $E'(y, x')$. Hence, $S(x, x')$. $\square$

Lemma 8.22 *Let S be an isomorphism relation from $(A, E, R, x)|y$ onto $(A', E', R', x')|y'$. Let S' be an isomorphism relation from $(A, E, R, x)|z$ onto $(A', E', R', x')|z'$. Then for all $a \in dom(S) \cap dom(S')$, $(\forall b)(S(a, b) \leftrightarrow (a, b))$.*

PROOF Let $S, A, E, R, x, y, A', E', R', x', y', S', z, z'$ be as given. Let a be R minimal such that $a \in dom(S) \cap dom(S')$, $(\exists b)(S(a, b) \leftrightarrow \neg S'(a, b))$. Fix b with $S(a, b) \leftrightarrow \neg S'(a, b)$. By Lemma 8.20,

(1) $S(a, b) \leftrightarrow (\forall c)(R(c, a) \leftrightarrow (\exists d)(S(c, d) \wedge R'(d, b)))$.
(2) $S'(a, b) \leftrightarrow (\forall c)(R(c, a) \leftrightarrow (\exists d)(S(c, d) \wedge R'(d, b)))$.

It suffices to prove that the right sides of (1) and (2) are equivalent. This is clear because if $R(c, a)$, then $c \in dom(S) \cap dom(S')$, and $(\forall d)(S(c, d) \leftrightarrow S'(c, d))$. $\square$

Lemma 8.23 *Let S be an isomorphism relation from (A, E, R, x) onto $(A', E', R', x')|y'$ and S' be an isomorphism relation from (A, E, R, x) onto $(A', E', R', x')|z'$. Then $E(y', z')$ and $S = S'$.*

PROOF Let $S, S', A, E, R, x, A', E', R', x', y', z'$ be as given. Apply Lemma 8.22 to $(A, E, R, x)|x$ and $(A, E, R, x)|x$, obtaining $S = S'$. By Lemma 8.21, $E(y', z'), S = S'$. □

Lemma 8.24 *Let (A, E, R, x) be a system. Then $y \in A \wedge \neg E(y, x) \leftrightarrow (\exists z)(R(z, x) \wedge y \in z\#)$.*

PROOF Let A, E, R, x be as given. For the forward direction, it suffices to show that $B = \{y \in A: E(y, x) \vee (\exists z)(R(z, x) \wedge y \in z\#)\}$ contains $[x]$ and is closed under E, R predecessors. Closure under E predecessors is obvious. Now let $R(u, y), y \in B$. If $E(y, x)$, then $R(u, x) \wedge u \in x\#$, and so $u \in B$. Suppose $R(z, x) \wedge y \in z\#$. Then $u \in z\#$, and so $u \in B$.

Conversely, let $R(z, x) \wedge y \in z\#$. Suppose $E(y, x)$. Then $R(z, y), y \in z\#, z \in z\#$, contradicting Lemma 8.18. □

Lemma 8.25 *Suppose $(\forall a)(R(a, x) \to (\exists a')(R'(a', x') \wedge (A, E, R, x)|a \approx (A', E', R', x')|a'))$, $(\forall a')(R'(a', x') \to (\exists a)(R(a, x) \wedge (A', E', R', x')|a' \approx (A, E, R, x)|a))$. Then $(A, E, R, x) \approx (A', E', R', x')$.*

PROOF Let $A, E, R, x, A', E', R', x'$ be as given. By Lemma 8.23, the a' in the first hypothesis is unique up to E'. For each a such that $R(a, x)$, let T_a be the unique isomorphism from $(A, E, R, x)|a$ onto $(A', E', R', x')|a'$, where a' is uniquely determined up to E'. The various T_a cohere, according to Lemma 8.22. Let T be the union of the various T_a. Let T^* be T extended by $\{\{<y, z>: E(x, y) \wedge E'(x', z)\}\}$.

By the coherence, T respects E, E' and is univalent up to E'. By the second hypothesis, T is onto the a' with $R'(a', x')$.

Note that $dom(T)$ is the union of the $a\#$, $R(a, x)$, which, by Lemma 8.24, is $A\backslash[x]$. Similarly, $rng(T) = A'\backslash[x']$. Hence, T^* also respects E, E', is univalent up to E', and is onto A'.

Suppose $u, v \in A\backslash[x]$. Let $T(u, u'), T(v, v')$. We claim that

$$R(u, v) \leftrightarrow R'(u', v').$$

To see this, let $u \in a\#, R(a, x), v \in b\#, R(b, x)$. Suppose $R(u, v)$. Then $u \in b\#$, and so $u, v \in dom(T_b)$. By coherence, $T_b(u, u'), T_b(v, v')$. Hence, $R'(u', v')$. The converse is analogous.

We also claim that

$$R(u, x) \leftrightarrow R'(u', x')$$
$$R(x, v) \leftrightarrow R'(x', v').$$

Suppose $R(u, x)$. Let u^* be such that $R'(u^*, x')$ and $(A, E, R, x)|u \approx (A', E', R', x')|u^*$. Then $T_u(u, u^*)$, and so $T(u, u^*)$. By coherence, $E'(u', u^*)$. Hence, $R(u', x')$.

Note that $x \notin a\#, x' \notin b\#$, making the second equivalence vacuously true. This is because if $x \in a\# \vee x' \in b\#$, then x is a maximal element of $a\# \vee x'$ is a maximal element of $b\#$, violating Lemma 8.18. From this we see that T is an isomorphism relation from (A, E, R, x) onto (A', E', R', x'). □

Definition   We write $(A, E, R, x) \in' (A', E', R', x')$ if and only if $(\exists y)(R'(y, x')$ $\wedge (A, E, R, x) \approx (A', E', R', x')|y)$.

Definition   The interpretation of $L(\in, =)$ is as follows. The sets are the systems (A, E, R, x). The interpretation of $\in$ is $\in'$. The interpretation of $=$ is $\approx$.

We now show that the interpretations of the axioms of Z hold. Recall that we are working in B + SDE. Hence, we are going to show that the interpretations of the axioms of Z are provable in B + SDE.

Lemma 8.26   *The interpretation of Extensionality holds.*

PROOF   Straightforward, using Lemma 8.25.   □

Lemma 8.27   *Let P be a property of full codes of systems whose fields are $< y$. (We do not require that P is bounded). There exists a system (A, E, R, x) such that every (A', E', R', x') with a full code with property P has $(A', E', R', x') \in (A, E, R, x)$, and every $(A, E, R, x)|u$, $R(u, x)$ is isomorphic to a system with a full code with property P.*

PROOF   Let P, y be as given. Let $w \gg z \gg y, a_1, \ldots, a_6 \wedge v \gg u \gg w, a_7$ $\wedge A(y, a_1, \ldots, a_6) \wedge A(w, a_7)$. By Lemma 8.9, every 6-ary "relation" $< y$ is coded by some $c < v$ using $y, w, a_1, \ldots, a_7$. In particular, each 6-ary "relation" representing a system whose field is $< y$ is coded by some $c < v$ using $y, w, a_1, \ldots, a_7$.

Let A be the "set" of all codes $c < v$ using $y, w, a_1, \ldots, a_7$, of systems $(A', E', R', x')|y'$, where (A', E', R', x'') has a full code with property P. We build a system with field $A \cup \{y\}$.

Let $E(c, c')$ if and only if $c, c' \in A \cup \{y\} \wedge (c = c' = y \vee$ the systems coded by, c, c' using $y, w, a_1, \ldots, a_7$ are isomorphic).

Let $R(c, c')$ if and only if $c, c' \in A \cup \{y\} \wedge ((c' = y \wedge P(c, y, w, a_1, \ldots, a_7))$ $\vee (c, c' \in A \wedge$ the systems coded by c, c' using $y, w, a_1, \ldots, a_7$ bear the relation $\in'$)).

The system (A, E, R, y) is as required.   □

Lemma 8.28   *The interpretation of Separation holds.*

PROOF   Let $(\exists a)(\forall b)(b \in a \leftrightarrow b \in c \wedge \varphi)$ be an instance of Separation, where a is not free in φ. Of course, parameters are allowed in φ. Let c be interpreted as the system (A, E, R, x). Let P be the property of full codes of systems $(A', E', R', x') \in' (A, E, R, x)$ that obey the interpretation of φ. Because the interpretation of φ involves only $\approx$ and $\in'$, we see that if two systems $(A', E', R', x'), (A'', E'', R'', x'') \in' (A, E, R, x)$ are isomorphic, then (A', E', R', x') has a full code with P if and only if (A'', E'', R'', x') has a full code with P.

Let P^* be P restricted to full codes of systems $(A, E, R, x)|z$ with $R(z, x)$. Let $y > A$. By Lemma 8.27, let (A^*, E^*, R^*, x^*) be a system such that every $(A, E, R, x)|z$ with $R(z, x)$ has $(A, E, R, x)|z \in' (A^*, E^*, R^*, x^*)$, and every $(A^*, E^*, R^*, x^*)|u$, $R^*(u, x^*)$ is isomorphic to some $(A, E, R, x)|z$,

$R(z, x)$. Then (A^*, E^*, R^*, x^*) witnesses the interpretation of this instance of Separation.   □

Lemma 8.29   *The interpretation of Pairing holds.*

PROOF   Let (A, E, R, x), (A', E', R', x') be systems. Let $y > A, A'$. Let P hold of the full codes of (A, E, R, x), (A', E', R', x'). Now apply Lemma 8.27. This creates a witness for Pairing applied to these two systems.   □

Lemma 8.30   *The interpretation of Union holds.*

PROOF   Let (A, E, R, x) be a system. Let P hold of the full codes of the $(A, E, R, x)|y$ such that for some $z, R(z, x) \wedge R(y, z)$. Let $A < y$, and apply Lemma 8.27. This creates a system that witnesses Union applied to (A, E, R, x).   □

Lemma 8.31   *Let (A, E, R, x), (A', E', R', x') be systems. Suppose that for all systems $(A^*, E^*, R^*, x^*) \in' (A, E, R, x)$, we have $(A^*, E^*, R^*, x^*) \in' (A', E', R', x')$. Then (A, E, R, x) is isomorphic to some system whose domain is a "subset" of A'.*

PROOF   Let $A, E, R, x, A', E', R', x'$ be as given. Let B be the "set" consisting of the y such that $R'(y, x')$, and $(A', E', R', x')|y$ is isomorphic to some $(A^*, E^*, R^*, x^*) \in' (A, E, R, x)$. Let C be the "set" consisting of the b that lies in the field of some $(A', E', R', x')|y, y \in B$. Let $u > A, E, R, x, A', E', R', x'$. Then $(C \cup \{x'\}, E^{**}, R^{**}, x')$ is the desired system, where E^{**} is E restricted to $C \cup \{z'\}$, and R^{**} is the binary "relation" given by

$$R**(c, d) \leftrightarrow (R'(c, d) \wedge c, d \in C) \vee (c \in B \wedge d = x').$$

By Lemma 8.25, $(A, E, R, x) \approx (C \cup \{x'\}, E^{**}, R^{**}, x')$.   □

Lemma 8.32   *The interpretation of Power Set holds.*

PROOF   Let (A, E, R, x) be a system. By Lemma 8.30, it suffices to form a system (A', E', R', x') such that for all systems α whose domain is a "subset" of A, we have $\alpha \in' (A', E', R', x')$. This is clear by Lemma 8.31.   □

The following will be used in Section 6.9.

Lemma 8.33   *The interpretation of Foundation holds. In fact, the interpretation of the schematic form of Foundation holds.*

PROOF   Let $(\exists x)(\varphi)$, where φ is a formula in $\in, =$. Let the variable x be interpreted as the system (A, E, R, x). Let z be R minimal such that the interpretation of φ holds at $(A, E, R, x)|z$. Note that the interpretation of φ involves only $\approx$ and $\in'$. Hence, $(A, E, R, x)|z$ witnesses foundation for $(\exists x)(\varphi)$.   □

The following is not needed but is of interest.

Lemma 8.34   *The interpretation of $(\forall x)(\exists y)(x \subseteq y \wedge y$ is transitive) holds.*

PROOF   Let (A, E, R, x) be given. Then (A, E, R', x) is as required, where $R'(a, b) \leftrightarrow R(a, b) \vee (a \in A \wedge b = x)$.   □

Lemma 8.35 *The interpretation of Infinity holds.*

PROOF By Lemma 8.17, let R be a prewellordering with no greatest element, where every nonempty "subset" of any $\{y: R(y,x)\}$ has an R greatest element. Because of Lemmas 8.6 and 8.9, we are in the standard situation with the second form of the Peano axioms for the natural numbers. Therefore, we can build arithmetic and definitions by induction. It is then straightforward to build a copy $(\mathit{fld}(R), S, =_R)$ of the set theorist's $(V(\omega), \in, =)$, where S is a binary "relation" on $\mathit{fld}(R)$ that respects $=_R$. We can put this in the form of a system $(\mathit{fld}(R), =_R, S, x)$, which will witness Infinity under our interpretation. □

Lemma 8.36 *The interpretation of all axioms of Z holds. In addition, the interpretations of Foundation (scheme) and Transitive hold.*

PROOF By Lemmas 8.26, 8.28, 8.29, 8.30, 8.32, 8.33, 8.34, and 8.35. □

Theorem 8.37 *Z is interpretable in B + SDE. In fact, Z + Foundation (scheme) + Transitive is interpretable in B + SDE.*

PROOF By Lemma 8.36 and 8.34. □

6.9 Interpretation of ZF in MBT

We now show that ZF is interpretable in B + SDE + UI.

By Lemma 8.36, we have only to prove the interpretation of Collection in B + SDE + UI, using the same interpretation of $L(>, \gg, =)$ in $L(\in, =)$ as we used in Section 6.8.

By Theorem 5.2, MBT proves B + SDE + UI. Hence, in this section we will have established that the interpretation of Section 6.8 is also an interpretation of ZF in MBT.

Before we prove the interpretation of Collection from ZF, we first show that B + SDE + UI proves a form of Collection in $L(>, =)$.

Lemma 9.1 *Let $n \geq 2$. Suppose $x_1, \ldots, x_n$ is coded by c using $y, a_1, \ldots, a_n$. Then $(\forall u)((\exists v)(u <^* v <^* c) \leftrightarrow u = x_1 \vee \ldots \vee u = x_n \vee u = a_1 \vee \ldots \vee u = a_n)$.*

PROOF Let $n, x_1, \ldots, x_n, c, y, a_1, \ldots, a_n$ be as given. Then $x_1, \ldots, x_n \leq y, A(y, a_1, \ldots, a_n)$. Also, let $u_1 >_{ex} \{x_1, a_1\} \wedge \ldots \wedge u_n >_{ex} \{x_n, a_n\} \wedge y >_{ex} \{u_1, \ldots, u_n\}$, where $A(y, a_1, \ldots, a_n)$.

We claim that the u's are incomparable under $\leq$. To see this, suppose $u_i \leq u_j$. Then $a_i < u_j, a_i \leq x_j \vee a_i \leq a_j, a_i \leq a_j, a_i = a_j$, which is a contradiction.

It follows that $v <^* c \leftrightarrow v = u_1 \vee \ldots \vee v = u_n$. Also, for each i, $u <^* u_i \leftrightarrow u = x_i \vee u = a_i$, using $u_i >_{ex} \{x_i, a_i\}$ and $A(y, a_1, \ldots, a_n)$. □

COLLECTION $(>, =)$. $(\forall y_1, \ldots, y_n < x)(\exists z)(\varphi) \rightarrow (\exists w)(\forall y_1, \ldots, y_n < x)(\exists z < w)(\varphi)$, where $n \geq 1$ and φ is a formula of $L(>, =)$ in which w is not free.

Lemma 9.2 *B + SDE + UI proves each instance of Collection $(>, =)$.*

PROOF Let n, φ be as given. We can assume that the free variables of φ are among $x_1, \ldots, x_m, y_1, \ldots, y_n, z$, which are distinct variables, and distinct from x, w.

Fix $x, x_1, \ldots, x_m$ such that the implication fails. Let $x' > x, x_1, \ldots, x_m$. Let $A(x' a_1, \ldots, a_{m+n+1})$. By Lemma 8.7, let v be such that

$$(\forall y_1, \ldots, y_n < x')(\exists c < v)$$
$$(c \text{ codes } x_1, \ldots, x_m, y_1, \ldots, y_n \text{ using } x', a_1, \ldots, a_{m+n}).$$

Let $w \gg v$. Given that the implication fails, let

$$y_1, \ldots, y_n < x.$$
$$(\exists z)(\varphi(x_1, \ldots, x_m, y_1, \ldots, y_n, z)).$$
$$\neg(\exists z < w)(\varphi(x_1, \ldots, x_m, y_1, \ldots, y_n, z)).$$

Now let c code $x_1, \ldots, x_m, y_1, \ldots, y_n$ using $x', a_1, \ldots, a_{m+n}$. By Lemma 9.1, we can define $x_1, \ldots, x_m, y_1, \ldots, y_n, a_1, \ldots, a_{m+n}$ from c, after repetitions are removed. However, we may not be able to define them in any particular order. So we say that we can define

$$\{x_1, \ldots, x_m, y_1, \ldots, y_n, a_1, \ldots, a_{m+n}\} \text{ from } c.$$

Now let t be the number of $m + n$ tuples $\alpha_1, \ldots, \alpha_{m+n}$ from $\{x_1, \ldots, x_m, y_1, \ldots, y_n, a_1, \ldots, a_{m+n}\}$ such that

$$\neg(\exists z < w)(\varphi(\alpha_1, \ldots, \alpha_{m+n}, z).$$

That is, the statement $\varphi(c, w) =$

$$\text{there are exactly } t \text{ tuples } \alpha_1, \ldots, \alpha_{m+n} \text{ from}$$
$$\{x_1, \ldots, x_m, y_1, \ldots, y_n, a_1, \ldots, a_{m+n}\} \text{ such that}$$
$$\neg(\exists z < w)(\varphi(\alpha_1, \ldots, \alpha_{m+n}, z)$$

holds. Note that t is a standard integer, and not a parameter in the above equation. Hence, we can apply Unlimited Improvement. Given that $w \gg y$, there are arbitrarily good $w' > w$ such that $\varphi(c, w')$. For $w' > w$, the count must be at most t.

In fact, for sufficiently good $w' > w$, there is a choice of $\alpha_1, \ldots, \alpha_{m+n}$, namely, $x_1, \ldots, x_m, y_1, \ldots, y_n$, such that

$$\neg(\exists z < w)(\varphi(\alpha_1, \ldots, \alpha_{m+n+1}, z)$$

goes from true to false for $w = w'$. This is because $(\exists z)(\varphi(x_1, \ldots, x_m, y_1, \ldots, y_n, z))$. Therefore, for sufficiently good $w' > w$, the count is at least $t + 1$. This is a contradiction. ☐

Lemma 9.3 *The interpretation of Collection* $(\in, =)$ *is provable in* $B + SDE + UI$.

PROOF Let (A, R, E, x) be a system. Assume that

(1) for all systems $(A', E', R', x') \in' (A, E, R, x)$, there exists a system (A^*, E^*, R^*, x^*) such that $P((A', E', R', x'), (A^*, E^*, R^*, x^*))$,

where P is expressed in terms of $\in'$, $\approx$ on systems, with some systems used as parameters.

Let $y > A$. Let $w \gg z \gg y, a_1, \ldots, a_6 \wedge v \gg u \gg w, a_7 \wedge A(y, a_1, \ldots, a_6) \wedge A(w, a_7)$. By Lemma 7.9, every 6-ary "relation" $< y$ is coded by some $c < v$ using $y, w, a_1, \ldots, a_7$. In particular, each 6-ary "relation" representing a system whose field is $< y$ is coded by some $c < v$ using $y, w, a_1, \ldots, a_7$. We have

(2) for each $c < v$ that codes a system $(A, E, R, x)|z, R(z, x)$,
   using $y, w, a_1, \ldots, a_7$, there exists a full code
   for a system (A^*, E^*, R^*, x^*) such that $P((A, E, R, x)|z, (A^*, E^*, R^*, x^*))$.

By Collection $(>, =)$, let w be such that

(3) for each $c < v$ that codes a system $(A, E, R, x)|z, R(z, x)$,
   using $y, w, a_1, \ldots, a_7$, there exists a full code $< w$
   for a system (A^*, E^*, R^*, x^*) such that $P((A, E, R, x)|z, (A^*, E^*, R^*, x^*))$.

Using the bound w and Lemma 8.27, we can construct a witness for the interpretation of this instance of Collection $(\in, =)$ with the given system parameters. $\square$

Theorem 9.4 *ZF is interpretable in* $B + SDE + UI$ *and in* $MBT = B + DE + SUI$.

PROOF From Theorem 8.37, Lemmas 9.2, 9.3, and Theorems 5.1, 5.2. $\square$

6.10 Some Further Results

We mention some further results that will appear in Friedman (in preparation).

We have shown that MBT does not prove VSDE, and does not prove SSDE. Furthermore, we have considered the following additional axiom:

STAR. There exists a star. In other words, something which is better than something, and much better than everything it is better than.

We have shown that MBT + STAR can be interpreted in some large cardinals compatible with $V = L$, and some large cardinals compatible with $V = L$ are interpretable in MBT + STAR.

Suppose we eliminate the diversity in VSDE and SSDE.

VERY STRONG EXACTNESS (VSE). Let x be much better than something better than a given range of things. Then x is much better than something exactly better than the given range of things.

SUPER STRONG EXACTNESS (SSE). Let x be much better than something and a given range of things. Then x is better than something exactly better than the given range of things.

We claim that $(\omega^{\omega+1}, >, \gg)$ is a model of $B + VSE + SSE + UI + SUI$, where we define $\alpha \gg \beta \leftrightarrow \alpha > \beta + \omega^\omega$. This is evident except for UI and SUI. But here we can use an elimination of quantifiers for small ordinals. This provides an interpretation of B + VSE + SSE + UI + SUI within EFA = exponential function arithmetic, or equivalently $I\Sigma_0(\exp)$. In fact, we obtain a consistency proof of B + VSE + SSE + UI + SUI within EFA.

To obtain a fragment that is mutually interpretable with P = Peano Arithmetic, we need only use $>$, $=$. Drop the last axiom of B, and use

DIVERSE EXACTNESS (in $>$, $=$). $y > \varphi \to (\exists z)(z >_{\text{ex}} \varphi \wedge \neg < y)$, where φ is a formula in $L(>, =)$ in which y, z are not free.

If we strengthen MBT by replacing $L(>, =)$ with $L(>, \gg, =)$ in SUI, then we obtain an inconsistent system. If we strengthen B + SDE + UI by replacing $L(>, =)$ with $L(>, \gg, =)$ in UI, then we also obtain an inconsistent system.

There are substantial connections between these systems and mereology. For background on mereology, see Simons (1987), Varzi (2007), and Hovda (2009).

Specifically, we can read $x > y$ as "y is a proper part of x" and read $x \gg y$ as "y is an infinitesimal part of x."

Acknowledgments

This research was partially supported by Templeton Grant #15400.

References

Enayat, A., Schmerl, J., and Visser, A. 2008. ω-models of finite set theory, www.phil.uu.nl/preprints/lgps/, #266, Logic Group Preprint Series, Department of Philosophy, Utrecht University.

Feferman, S. 1960. Arithemetization of metamathematics in a general setting. In *Fundamenta Mathematicae* 49, 35–92.

Friedman, H. 2007. Interpretations, according to Tarski. Lecture 1 of the 19th Annual Tarski Lectures, Department of Mathematics, University of California, Berkeley. www.math.ohio-state.edu/%7Efriedman/manuscripts.html, #60.

Friedman, H. In press. Forty years on his shoulders. In *Horizons of Truth: Gödel Centenary*. Cambridge University Press.

Friedman, H. In preparation. *Concept Calculus*, www.math.ohio-state.edu/%7Efriedman/manuscripts.html.

Friedman, H., and Visser, A. In preparation. *Interpretations between Theories*.

Hovda, P. 2009. What is classical mereology? *Journal of Philosophical Logic* 38: 55–82.

Kaye, R., and Wong, T. L. 2007. On interpretations of arithmetic and set theory. *The Notre Dame Journal of Formal Logic* 48 (4): 497–510.

Simons, P. 1987. *Parts: A Study in Ontology*. Oxford: Oxford University Press.

Varzi, A. C. 2007. Spatial reasoning and ontology: Parts, wholes, and locations. In *Handbook of Spatial Logics*, M. Aiello et al. (eds.) pp. 945–1038. Berlin: Springer-Verlag.

PART FOUR

Perspectives on Infinity from Physics and Cosmology

CHAPTER 7

Some Considerations on Infinity in Physics

Carlo Rovelli

> It is incumbent on the person who specializes in physics to discuss the infinite. And to inquire whether there is such a thing or not, and, if there is, what it is.
>
> Aristotle, *Physics* III, 202b34

I am a theoretical physicist, and, following Aristotle's injunction, I do consider my responsibility to discuss the problem of the notion of infinity in the world – in particular, to "inquire whether there is such a thing or not." I will do so here by illustrating some aspects of the notion of infinity in the natural sciences.

I focus on the recent evolution of our understanding of this notion with regard to two classical problems where infinity has long been discussed: the infinite *divisibility of space* and the infinite *extension of physical space*, that is, infinite in the small and infinite in the large. The two disciplines that currently deal with these two problems within physics are quantum gravity and cosmology. I briefly summarize the current understanding of these two problems within these two disciplines.

This discussion leads me to some general considerations on the notion of infinity. I point out the risk of confusion that such a notion generates, if we mistake limitations of our mind and our intelligence with properties of the natural world, or hints of the supernatural.

I point out the danger that is inherent when we try to fill up the mystery of the world with groundless solutions. In my opinion, uncertainty is preferable to unfounded certitudes.

7.1 Infinite Divisibility of Space: Quanta of Space

In the recent development of fundamental physics, the terms of the problem of the infinite divisibility of space have changed, owing to a number of theoretical and empirical results. These developments concern the discipline called *quantum gravity*, which is the domain of theoretical physics in which I work.

Quantum gravity is more the name of a problem than the name of a theory. The problem is viewed by many as the major open problem in fundamental physics. It originates from two momentous advances in the physics of the last century: *quantum theory* and *general relativity*. These two theories have proved extraordinarily effective and ground most of what we know today about the physical world. However, they are currently formulated on the basis of assumptions that largely contradict one another. The effort to find a coherent scheme in which these two theories can make sense together, and therefore the effort to get back to a coherent picture of the physical world, is the task of the research in quantum gravity.

The problem is not yet satisfactorily resolved, but there exist today well-developed theories that are possible tentative solutions. The two best developed of these theories are *string theory* and *loop quantum gravity* (see, for instance, Rovelli 2004a). Other directions of research include, among others, *noncommutative geometry* and *causal set theory*. All these theories have much to say about the infinite divisibility of space, and quite remarkably, they are consistent with one another in this regard. They open a novel perspective on the possibility of thinking about the problem of the infinite divisibility of space.

In the following section, I illustrate this new perspective, with particular emphasis on the *loop quantum gravity* (or *loop gravity*) approach, given that this is the development in which I have participated. In a later section, I discuss the relevance of a theory that is still tentative, in the context of the discussion of a general problem such as the one posed by the notion of infinity.

General relativity is Einstein's final theory of spacetime. It has successfully replaced the Newtonian conceptualization of space and time. The empirical success of general relativity has been triumphal, and today the theory is one of the best-confirmed scientific theories ever. In this theory, Newtonian space, which for Newton was a "continuous substratum" without dynamical properties, is reinterpreted as a continuous "substance" with dynamical properties, that is, a continuous substance that can be stretched and bent like a rubber sheet. Space stretches and bends following some equations that Einstein wrote, which are today called *Einstein's equations*.

In other words, general relativity is formulated in a background-independent way: it does not presuppose from the very beginning an arena for physical processes to develop, but rather it shows that such an arena is born out of a dynamical entity.

On the other hand, according to *quantum theory*, any entity with dynamical properties of this kind is always "quantized." This means, in particular, that, when observed at sufficiently small scale, this entity manifests itself in the form of "packets" or small "chunks," or "quanta." A well-known example of this "quantization" phenomenon is the fact that electromagnetic waves, when observed on a sufficiently small scale, turn out to be formed by a cloud of tiny particles: the photons. Quantum theory also has had spectacular confirmations. It is at the basis of much of the current technology, such as all the computer technology. It is one of the best theories of the world we have ever had, and it therefore encodes the basis of our present knowledge about the physical world.

Now, if we rely on quantum theory and general relativity, we are necessarily led to conclude that space, being an entity with dynamical properties like the electromagnetic field, is itself made by small "quanta" as well. In other words, to our best knowledge about the natural world, it is quite likely that space is *not* infinitely divisible.

Loop gravity reaches precisely this conclusion. Indeed, using the equations of this theory, it is possible to calculate with precision the size and the properties of the individual quanta of space. For instance, we can compute the number of such "atoms of space" that exist within, say, a centimeter cube. Needless to say, this number is very large (of the order of a one followed by a hundred zeros). It is a very large number, but it is not an infinite number.

The other tentative theories of quantum gravity arrive at similar conclusions. For instance, *string theory* was first formulated in a "background-dependent" manner, namely, it presupposes, from the very beginning, an arena for physical processes to develop. However, the recent developments in the theory are all in the direction of the background-independent formulation, namely, a formulation in which such an arena is born out of more primitive elements. But even if we start on a fixed background, it turns out that if we attempt to probe arbitrarily small regions of space using string theory, we fail necessarily, because any test particle we may think of using for this purpose would open up into an extended string when we try to confine it too much. In other words, we cannot test arbitrarily small spatial regions in string theory either. Thus, string theory leads to analogous indications that space is *not* infinitely divisible, although more indirectly.

Other approaches such as *noncommutative geometry* and *causal set theory* are even more radical in this regard. They take the existence of a finite granularity of space as one of their basic assumptions. That is, whereas loop gravity and string theory derive the granularity of space from our current basic knowledge of the physical world, these other approaches take it as an input for the definitions of the theory. More precisely, the divisibility problem is in some sense transcended in noncommutative geometry, given that topology can be generalized to such an extent that local concepts become meaningless.

Let us reflect on the meaning of this finite divisibility of space that appears in the research in quantum gravity. It is not difficult to imagine that *space* is not a continuous quantity. Space can be like a T-shirt, which is smooth and continuous if seen from a distance, but whose fabric reveals a small-scale discrete structure if observed close-up. Imagine billions and billions of extremely tiny chunks that are, themselves, indivisible.

Such a replacement of a continuous structure with a discrete structure is typical of the evolution of science in the last century. The most characteristic example is the discovery of the atomic structure of matter. Water in a glass, or air in a room, appears to us as an infinitely divisible continuum. We have learned in the last century that this apparent continuum is actually an aggregate of discrete atoms, and we are today well used to this idea. The amount of water in a glass is, ultimately, just the *number of atoms* in the glass.

Similarly, space itself can be thought of as an ensemble of *atoms of space*, each having finite volume. This should not be interpreted imagining that one individual atom of space has an extension, and this extension has a measure. Rather, the volume of a region of space, namely, its very "extension," is nothing else than the *number of atoms of space* in the region. Extension is our *approximate* conceptualization of the number of quanta of space.

This picture is not consistent with our common intuition about space. It is also not consistent with the Euclidean geometry that we use to formalize it. But this does not mean that the picture is inconsistent. It only means that our common intuition about space can be incorrect, and that Euclidean geometry may not be the correct mathematical tool for describing real physical space, below a certain scale.

It would not be the first time that science contradicted our common intuition. According to our intuition, after all, the earth is flat, and it does not move! In conclusion, contemporary research in fundamental physics is seriously considering the possibility that space is not infinitely divisible. There is no necessarily infinite toward the small.

7.2 Infinite Extension of Space: The Archytas Problem and the Solution of Dante Alighieri and Albert Einstein

Archytas of Tarentum was a fifth-century Greek thinker. An argument by Archytas has reached us, thanks to a fragment of Eudemus preserved by Simplicius. In this argument, Archytas claims that the universe must be infinitely extended because of the following: "If I arrived at the outermost edge of the heaven, could I extend my hand or staff into what is outside or not? It would be paradoxical not to be able to extend it." More in detail: if there is nothing blocking my hand, space continues, but if there is a "wall" that blocks my hand from being further extended, the wall is beyond the boundary; hence, space continues there. Archytas can advance to the new limit, the other side of the wall, and ask the same question again, so that there will always be something into which his hand can be extended, beyond the supposed limit. Hence, space is clearly unlimited. This argument was considered by Aristotle to be the "most important" reason why people believe in the existence of the infinite (Aristotle 1952, p. 203b22).

A solution of this problem was given by Albert Einstein (1996) in a celebrated paper written in 1917. This very solution was foreseen by the greatest of the Middle Age poets, Dante Alighieri. I do not know if Dante got this idea from somebody else or if it is his own idea. The idea, as expressed by Dante, is the following:

> In his poem, Dante describes his great trip across the universe and gives a grandiose vision of the entire universe. The universe described by Dante has no boundary: there is no place where space ends, and Archytas could attempt to push his hand ahead. Nevertheless, the universe described by Dante has a finite extension. It is not infinitely extended. This seems paradoxical at first, but it is not.

Let's see how this works. According to Dante, who followed Aristotelian cosmology to some extent, the universe has a spherical structure centered on the earth. The earth is a sphere. (The common idea that in the Middle Ages people thought that the earth was flat is completely false. Around the Mediterranean, the earth was universally known to be spherical since about five centuries before Christ.) Around the earth, according to Dante and Aristotle, there are several larger concentric spheres: the spheres of the Moon, Venus, and so on, all the way until the sphere of the fixed stars, or the "primus movens," the "first mover." Each of these spheres surrounds the previous one.

What next? What is around the sphere of the fixed stars? Here Dante departs from Aristotle. For Aristotle, the universe ends there, with all the difficulties raised by Archytas trying to push his hand further. For Dante, things are not so simple. In

Canto XXVIII of *Paradise*, Dante describes his vision, in which the sphere of the fixed stars is not the final boundary of the universe. Rather, around the sphere of the fixed stars there is *still* another sphere. However, this other sphere is not *larger* than the sphere of the fixed stars, but rather *smaller*. In this "external" sphere, there are singing angels. Around this "external" sphere, there is still another one, with more singing angels, and this is even *smaller*. In fact, there is a series of spheres, increasingly small, until they shrink to a single point, which is a point of light, where, Dante says, is God:

> *Un punto vidi che raggiava lume*
> *acuto sì, che 'l viso ch'elli affoca*
> *chiuder conviensi per lo forte acume.*
> [A point beheld I, that was raying out/Light so acute, the sight which it
>   enkindles/Must close perforce before such great acuteness.]

This peculiar geometry, which Dante attributes to the universe, has no boundary, but at the same time it is finite, that is, it has a finite extension. As beautifully noticed by Mark A. Peterson (1979), this is precisely a geometry that is today well known under the name of "three-sphere."

To understand what a three-sphere is, think about the geometry of a normal sphere, for instance, the geometry of the surface of the earth itself. The peculiar shape of the surface of the earth can be described as follows. Imagine you are at the North Pole. The North Pole is a point (at a latitude denoted 90° north). Let us (arbitrarily) choose the North Pole as our "center" for describing the surface of the earth. Around the North Pole we have a circle, formed, say, by the first geographical parallel (say, 85° north). Around this, we have another circle, which is a bit more south and is *bigger* than the first, namely, another geographical parallel (say, 80° north). We can continue increasing the size of these circles, each circle surrounding the previous one, until we arrive at a very large circle, which is the equator of the earth (latitude 0°). Does the surface of the earth stop there? Of course not: beyond the equator there is still another circle, which surrounds the equator. This is the parallel at latitude 5° *south*. Notice that in spite of the fact that this parallel surrounds the equator, it is nevertheless *smaller* than the equator, precisely as Dante's spheres! Around this parallel, there is still a smaller one that surrounds it, until we arrive, if we keep going south, to a single point, which is the South Pole.

This is precisely the description that Dante gives of the universe. The earth is like the North Pole, and the point where Dante sits God is like the South Pole. The difference is that the circles are replaced by spheres and we are talking about a three-dimensional space instead of a two-dimensional surface: in fact, a three-sphere instead of the (common) two-sphere.

The surface of the common two-sphere can be described as two disks (the northern and the southern hemispheres) that are attached along a circle: the equator. If you get to the boundary of the first disk and keep going, you simply enter the second disk. In the same manner, Dante describes the universe as two balls. When you get to the boundary of the first ball, namely, to the sphere of the fixed stars, you can keep going, and you simply enter the second ball. You never meet a real boundary, yet space is finite.

Of course, Dante gropes for a language to express an idea conceived intuitively and nonverbally, but in doing so he actually gives the essence of a well-defined mathematical construction. Although his conceptual tour de force went unnoticed, or was not understood, and so had no direct effect, as far as I know, on cosmological thinking, Dante had nevertheless expressed something entirely new, which today is simple mathematics taught in any university of the world.

This mathematical possibility could be seen just as the fantasy of a great poet, if it weren't that such a three-sphere and such a universe without boundary but with finite extension are a perfectly concrete possibility that is seriously contemplated by modern cosmology, as one of the possible shapes of our real universe. In fact, modern cosmology emerges from a remarkable paper by Albert Einstein, written in 1917, in which Einstein applies his equations to the entire universe and shows that a possible shape for the universe, according to these equations, is precisely such a three-sphere. In other words, it is very plausible that the universe could be without boundary but finite, precisely as the surface of the earth is finite but has no boundaries.

This implies that today we have a solution for the argument by Archytas of Tarentum: if you arrive at the end of the world and make one more step, you may still be within the universe, in the same manner in which if you walk all the way around the earth and then make one more step, you do not fall out of the surface of the earth: you just come back home. One of the strongest arguments for believing in the infinite extension of physical space has today vanished.

7.3 What Do We Learn from Theories that Are Not Even Sure?

The two physical theories that I have mentioned, quantum gravity and cosmology, are far from representing established scientific knowledge. That is, we are certainly not *sure* that space is made out of quanta or that the universe is a three-sphere. In fact, we do not know. The two theories represent tentative theoretical constructions for making sense of the world, both waiting for empirical corroboration. Can they nevertheless be relevant for discussing a general problem such as that of the notion of infinity?

I think that the main lesson of natural science is that we are not born with infused knowledge, nor are we born with the best possible conceptual framework for thinking about reality. Rather, observation, reflection, and discussion can help us discover newer and better understandings of the world around us. This does not only mean that we can learn more *facts* by observing nature with care. More importantly, it means that we can constantly refine the *concepts* that we use for understanding the world. We do so by enlarging, refining, and sharpening the ensemble of notions that we use for thinking about the world.

A specific physical theory is therefore not so much a set of new discovered facts, but rather, much more interestingly, a novel way of thinking: a new way of conceiving the world and understanding it.

Historically, each step ahead has built heavily on the previous ones. The old successful theories that have been later replaced by more general theories (such as classical quantum mechanics, which has been replaced by quantum mechanics) still maintain all their interest and value, not only because often (as is the case of classical mechanics) they are effective and commonly utilized within their respective domains, but, more importantly, because their conceptual legacy survives and nurtures the descendent theories, almost as a father does for his children.

In considering a novel hypothetical theory that is being proposed, one should therefore distinguish between two different levels. On the one hand, we must ask whether the novel theory is "physically correct" or not. A consistent theory may very well turn out to be physically wrong: the only reliable judge we know is empirical evidence. The strength of natural sciences has always been the use of empirical evidence as a means to discard theoretical hypothesis. From this point of view, we must take most of modern physics speculations with great care: nothing tells us that there really are multiple universes, strings, loops, or the like, and we should definitely not make hard deductions on the basis of the reality of these objects.

On the other hand, any new theoretical construction is also a possible new model for thinking about the world, and the very fact that a model of this sort *is possible* has deep consequences for our understanding of the universe. In particular, the existence of a coherent conceptual scheme can disprove facile deductions based on what we know, or what we think we know.

Back to our problem, the *possibility* that space is *not* infinitely divisible, realized by loop quantum gravity, bears strongly on all discussions on infinity in the small. The *possibility* that the universe is boundless but finite, implied by cosmology, bears strongly on all discussions on infinity in the large.

These possibilities indicate that the difficulty with infinity might very well be in our erroneous formulation of the problem, or in our erroneous hidden assumptions, and not in the *nature of things*. Irrespective of which quantum gravity will ultimately turn out to be confirmed by experiments, the very existence of the conceptual possibility that space is *not* infinitely divisible and *not* infinitely extended should deter us from taking our own "natural" intuitions about infinity as solid. Science has constantly taught us that what appears "obvious" or "intuitive" to us is often just a prejudice or, more precisely, an unwarranted extrapolation.

7.4 The Danger of Speculating about Infinity

The two cases illustrated in the preceding section cover the two major examples of infinities discussed in the history of thinking: infinity toward the small, namely, infinite divisibility of space; and infinity toward the large, namely, infinite extension of space. In both cases, the discussion shows that the traditional arguments put forward in support of the existence of these infinities can very well be wrong. In fact, they are wrong because they implicitly rely on intuitions and ideas that we have about physical space. To extend these intuitions and ideas too far is an illegitimate extrapolation.

And what if infinity was always nothing else than the place where our illegitimate extrapolations go wrong? My own impression is that it could very well be so. The two cases illustrated, indeed, undercut one of the sources of our confusion with infinity. They warn us that the source of the problems raised by our *intuition* about infinity might be in the limitation of our intuition, not in the nature of things.

I certainly do not want to suggest that these examples provide a general *solution* to the problem of the infinite, and I am not claiming to be a depository of the corresponding knowledge. Precisely to the contrary, I would like to express a warning against facile argumentations, based on our own certitudes or intuitions, that might be disproved by the acquisition of a slightly more extended knowledge.

I think that what is truly infinite may just be the abyss of our ignorance. We keep learning, but the little we already know is uncertain, limited in scope, and partial. Our mind has evolved on earth to solve problems about scales included between millimeters and kilometers, and about time spans between seconds and decades. When we try to use this limited and mortal intelligence for looking much farther, we probe our limits.

On the other hand, our brain contains more than a hundred thousand billion synapses. If each one of these can be activated or not, then the number of possible configurations of our brain is something like a one followed by billions and billions of zeros. That is an immensely large number. This means that the space of possible thoughts is far larger than anything we can even vaguely imagine. For all practical purposes, we can consider this number infinite. The space of the thinkable is effectively infinite, and we haven't explored more than an infinitesimal corner of it. This is fantastic complexity; indeed, the natural object around us that we understand the least is the very one we use for understanding – the few pounds of our own brain. But careful, "effectively infinite" does not mean infinite. It means "very large," in the same way in which the quanta of space may be "very small," but not "infinitely small," and the universe may be "very large," but not "infinitely large."

Thanks to this extraordinarily versatile tool that is our brain, we have found out that we can partly overcome some of our natural limits. We have evolved to think about meters and kilometers, but so effectively that we can now talk about atoms and galaxies. We keep overcoming our limits. We have been able to do so repeatedly, thanks to the complexity of the harmonious dance of the billions and billions of neurons inside our head. But the world remains far more extended and complex than the smallness of what this dance can grasp. The world remains far more extended and complex than the smallness of our spirit. Some limitations may very well be intrinsic. We may have so many neurons, and feel so smart and profound, but none of us can even multiply 379,887 times 88,699,673 in our mind, something that a tiny plastic machine can do in a beep. With all our neurons, we are very stupid indeed.

I suspect that what we feel as infinity is only the immensity of what we realize that we cannot reach. Our emotional sense of infinity, the one that grasps us when we look at the stars in the night, when we ponder the mathematics of the continuous, when we contemplate the eons of time before and after us, when we keep asking, with Archytas, "what's beyond?" – this is just the sense of our smallness, our frailty, our ignorance. We are ripples that last a moment over an immensity that we probably have no means to understand.

What science and the history of science teach us, I believe, is such a sense of the infinity of what we *do not* know. With this comes immediately a profound suspicion for all certitudes, especially those based on our own intuitions or direct personal experiences, or the intuitions and the experiences of our ancestors, including the most vivid of these. Science keeps telling us that it is so easy to nurture false conviction and beliefs. It is so easy to fill the unknown that the infinite represents with our empty fantasies and our desires.

Aristotle says in the passage about the argument by Archytas mentioned previously, "What is outside the heaven [is] supposed to be infinite because [it] never gives out in our thought" (Aristotle 1952, p. 203b22). However, *our thought* may very well be misled, regarding distant reality.

Infinity appears to me as a faraway point, like Dante's other center of the universe, where we can project so much of our dreams and our desires, without the risk of having to go there and finding out that there might be nothing at all.

References

Aristotle. 1952. *Physics*. Translated by R. P. Hardie and R. K. Gaye. In *The Works of Aristotle*, vol. 1. Chicago: R. P. Gwinn.

Einstein, A. 1996. Cosmological considerations in the general theory of relativity. *Preuss. Akad. Wiss. Berlin Sitzber* (1917), p. 142. In *The Collected Papers of Albert Einstein. Vol. 6: The Berlin Years: Writings, 1914–1917*. Princeton: Princeton University Press.

Peterson, M. A. 1979. Dante and the 3-sphere. *American Journal of Physics* 47 (12): 1031–35.

Rovelli, C. 2004a. *Quantum Gravity*. Cambridge: Cambridge University Press.

Rovelli, C. 2004b. *What Is Time? What Is Space?* Rome: Di Renzo Editore.

CHAPTER 8

Cosmological Intimations of Infinity

Anthony Aguirre

8.1 Introduction

The question of whether the universe is infinite or finite is an old and contentious one in philosophy. Einstein's general relativity (GR) has, however, brought the question largely into the domain of science by giving us mathematical models of just what an infinite (or finite) universe would look like. It has also provided genuinely new ways of conceptualizing old questions. For example, a seemingly insuperable objection to a finite universe would be to ask what is beyond the edge of the finite universe; however, GR teaches us that the universe may be finite but without boundary, in just the way that the surface of a sphere is. And, in contrast to Newtonian gravity, GR has no trouble dealing with an infinite spacetime.

In this chapter I explore the question of infinity in cosmology, in terms of infinite space, infinite time, and an infinite number of states.

8.2 Infinity in Classic Cosmological Models

8.2.1 Homogeneous and Isotropic Spacetimes

Cosmological models describing our observed universe have been based almost entirely on two ingredients: Einstein's theory of general relativity (GR) and the "cosmological principle." The latter takes a number of specific forms, but the general idea is that on very large spatial scales, the universe is homogeneous and isotropic. This leads, in GR (or any other metric theory of gravity), to a spacetime described by the "Friedmann-Lemaître-Robertson-Walker" (FLRW) metric:

$$ds^2 = -dt^2 + R^2(t)\left[\frac{dr^2}{1 - kr^2} + r^2(d\theta^2 + \sin^2\theta d\phi^2)\right],$$

where $k = \pm 1$ or 0, and $R(t)$ is a "scale factor" that essentially converts coordinate distances (such as radial distances in r) into physical distances; $R(t)$ increasing in time describes an expanding universe such as we see. At a given time ($dt = 0$), this metric describes a homogenous and isotropic space, and there are only three types: flat "Euclidean" space with $k = 0$, the positively curved $k = +1$ space of a three-sphere, and the space of constant negative curvature with $k = -1$. For $k = 0$ or $k = -1$ the universe is, at any time, spatially infinite. Moreover, Einstein's equations yield a relation between $H \equiv \frac{1}{R}\frac{dR}{dt}$, the energy density ρ, and k, so that by observationally measuring ρ and H, k can be determined.

This leads to the intoxicating possibility of determining observationally whether the universe is infinite or finite. But this is an illusion: it rests on the very strong assumption that the FLRW metric holds not just inside the observed region of the universe, but also everywhere outside it. Without this assumption, it is perfectly possible to be in a locally overdense region and determine $k = +1$ even while the universe is infinite, or to locally determine $k = -1$ in a very large but finite universe.

Despite this caveat, the FLRW metric has served as a workhorse in cosmology and underlies the two main classic cosmological models: the Big Bang, and the Steady-State. Both have interesting connections with the idea of a physically realized infinity.

8.2.2 The Big Bang Model

In the Big Bang model, it is assumed that the FLRW metric with $k = \pm1$ or 0 applies back to very early times. Under certain assumptions regarding the energy density and pressure of the cosmic "fluid," this implies that there exists a time $t = 0$, a finite duration to the past, at which $R(t) \to 0$. Cosmological attributes such as the value of k are simply prescribed as "initial" conditions shortly after $t = 0$. The cosmic evolution can then be evolved forward until today, when, for example, R_0 is an observational quantity related to the time-varying scale over which the universe is curved (for $k = 0$, this curvature scale is always infinite).

It is, however, important to distinguish the Big Bang theory (a very well-tested and observationally successful theory of the universe's evolution from an early, hot, dense epoch) from the idea that the FLRW metric applies at arbitrarily early times. Among other reasons, if we strictly maintain the FLRW metric, the finite age of the universe leads to a rather strange situation for the spatially infinite $k = 0$ and $k = -1$ versions: although the physical distance between any two objects of fixed coordinate separation goes to zero as $t \to 0$, nonetheless, there will always be objects, at any time, that are physically arbitrarily far apart; thus, there is no sense in which the universe is at any time "small" or "at a point." Rather, the whole infinite system springs into being all at once at $t = 0$. It is unclear to what degree this troubled various cosmologists over the years. Einstein had a strong preference for $k = +1$, but for reasons more connected with "Mach's principle"[1] than with the Big Bang.[2]

---

[1] Roughly speaking, Mach's principle asserts that nonaccelerated (inertial) frames should be determined by the matter distribution of the universe, as those that are moving at constant velocity with respect to the rest frame of that matter distribution.

[2] Einstein also initially preferred the idea of a static, eternal universe and introduced the cosmological constant to make this possible. Aside from being disproven observationally, I have often wondered how Einstein did not realize that this model is (a) unstable to small perturbations or (b) in violation of Olber's paradox.

Perhaps even more troubling, because of the finite speed of light, a finite age means a finite distance over which information can propagate. This means that such a model contains an infinite number of regions that can never have communicated with each other or exerted any causal influence over one another. It then seems exceedingly strange that they should, at any given time,[3] all have just the same properties, as demanded by the cosmological principle and FLRW metric. This strangeness (which afflicts the $k = +1$ model as well, but with only finitely many disconnected regions) is often termed the "horizon problem" and was part of the impetus behind the invention of inflation, which is discussed in Section 8.3.

8.2.3 The Steady-State Model

The Steady-State model of Hoyle, Bondi, Gold, Narlikar, and others (e.g. Bondi and Gold (1948), Hoyle and Narlikar (1996), Hoyle (1948)), took a somewhat different standpoint. There, the cosmological principle was extended to time as well as space, so that the universe *at all times* looks essentially the same on large enough scales. This is a very constraining symmetry: it leads both to $k = 0$ and to an exponential form for $R(t)$.

A fascinating aspect of the Steady-State model is that although the universe is expanding it does not get any bigger! That is, the scale factor in the metric grows exponentially, so the space between galaxies increases (as we observe). But in the model, new matter is created to fill in the gaps, so as to maintain a constant average density of matter. So (by construction) the universe is exactly the same (statistically, and on very large scales) at each successive time, and nothing has really happened as the universe has expanded.

It would certainly seem that at a later time the universe must have more matter in it than at an earlier time (since, after all, it is both expanding and acquiring more matter), but this is one of the very neat things about infinity: an infinite set can be a subset of itself! For example, we can take the set of integers, double each element (and thus have precisely as many as before), then add an additional (odd) integer to go with each element, and end up precisely with what we started. So also with the Steady-State.

Although beautiful in many ways, the Steady-State is simply not correct as a description of the observed universe; however, as we will see, it has returned in a different form through the idea of inflation.

8.3 Infinity in Inflationary Cosmology

8.3.1 The Idea of Inflation

Puzzles – like the horizon problem – concerning the "initial" conditions in the Big Bang model led a number of scientists in the early 1980s to develop the theory that

---

[3] "At a given time" is an ambiguous phrase in GR, but the point here is that even if we assume strict thermal statistical equilibrium so that the particle/field content is basically determined by bulk temperature and density, it is extremely special if there exists *any* coordinatization such that constant time surfaces are statistically homogeneous and isotropic.

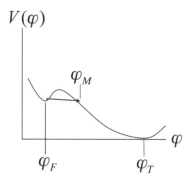

Figure 8.1. A double-well inflationary potential $V(\phi)$. For a homogeneous cosmological scalar field in such a potential, the field dynamics are much like a ball on a corresponding landscape: there is an acceleration proportional to the steepness of the slope and a friction term opposing the velocity. Thus, the field at a point will generally evolve from its initial value "down the hill" to oscillate around, then settle into, a minimum. With quantum effects, tunneling through a barrier can also occur in a potential like this one. In this model, $V(\phi_F)$ corresponds to an inflationary vacuum energy, and $V(\phi_T)$ corresponds to our observed vacuum energy. A bubble forms as the field tunnels through the barrier to the field value $\phi = \phi_M$. Inflation within the bubble can occur as the bubble rolls down the potential, ending when the field nears the true minimum at ϕ_t.

the early universe included a brief phase of exponential expansion, driven by vacuum energy (in the form of the potential energy $V(\phi)$ of a scalar field ϕ). Alan Guth showed that such a phase, which he dubbed "inflation," would lead to a universe that is large, homogeneous, of very nearly flat spatial geometry, and free of troublesome preinflationary relics such as magnetic monopoles Guth (1981).

In subsequently developed pictures of inflation, the exponential expansion ends when ϕ (the "inflaton" field) relaxes to near a minimum of its potential $V(\phi)$. This picture is slightly complicated by quantum fluctuations: as the inflaton classically "rolls" to the minimum (Fig. 8.1), it also undergoes quantum fluctuations up and down the potential. These are an important feature: they lead inflation to end in slightly different times in different nearby places. This translates into density perturbations of just the type (nearly scale-invariant, Gaussian, and adiabatic) that are required to seed galaxy formation later and that are observed in the temperature inhomogeneities of the cosmic microwave background (CMB).

The ability of inflation to give rise to just the sort of early state that explains our observed universe has led to fairly widespread acceptance of it as an indispensable ingredient of modern cosmology (see, e.g., Guth (2004), Linde (2007a) for recent reviews), and the evidence in its favor appears to keep improving. Most recently, the three- and five-year data from the *WMAP* observations of the CMB indicate a slight departure from scale invariance in the density perturbations, which is a generic prediction of inflation models (Komatsu et al. 2008). There are some candidate alternatives to inflation, but I think that it is safe to say that none are anywhere near as compelling, at least in the minds of the vast majority of cosmologists, as inflation.

But inflation, if accepted, comes with a certain amount of rather heavy baggage.

8.3.2 An Infinite Time of Inflation

Consider a region of the universe, with volume U, that is inflating. Now you suppose divide it into eight regions of equal volume $U/8$ and wait for a time Δt necessary for the scale factor to double. During this time, the classical tendency will be for ϕ to roll to lower values in all eight subvolumes, but the quantum fluctuations mean that in some, occasionally, the field will go up instead. Now, as it turns out, the quantum fluctuations become stronger at larger values of $V(\phi)$; thus, if the region starts with a high enough value, at least one of the eight volumes will, on average, fluctuate up during the time Δt. During the same time, that subvolume will grow into just the original volume U. Thus, *while seven-eighths of the universe has gone down in field value, the total volume that is inflating has stayed constant.* Now if you consider a region starting at a field value ϕ_i slightly above this, it can be shown (e.g., Linde (1986)) that the volume of the universe at $\phi > \phi_i$ will, in fact, increase exponentially in time. This phenomenon is often called "everlasting inflation" (or often "eternal inflation," but I will reserve that term for a different use) and means that while inflation may end locally to create a "Big Bang-like" region such as we observe, globally it will go on forever, continually spawning such regions. The universe is thus infinite in time, and at each time, the universe contains exponentially more volume of both inflating space and noninflating space than at the time before.

The back-reaction of quantum effects on the large-scale classical geometry makes it rather tricky to rigorously work out the structure of these "stochastic inflation" spacetimes. It is therefore useful to look at a related example in which these issues are less troublesome.

The model is a "double-well" potential as shown in Figure 8.1. The upper well at $\phi = \phi_F$ corresponds to an inflationary vacuum energy, and the lower well at $\phi = \phi_T$ to no vacuum energy (or perhaps the very small one that we appear to observe in our universe now). A region of the universe starting at ϕ_F would classically inflate forever. Because of quantum mechanics, however, the field can *tunnel* through the barrier to create a region of ϕ_T.

This leads to a first-order phase transition in which the universe would seem to convert from the ϕ_F phase into the ϕ_T phase. Because the space in the ϕ_F phase is inflating, however, it can be shown that even though the nucleated bubbles of the ϕ_T phase grow at the speed of light, they nonetheless fail to take over all of the inflating volume Aguirre and Gratton (2003), Guth and Weinberg (1983), Vilenkin (1992). Instead, the volume with $\phi = \phi_F$ increases exponentially with time, just as in the fluctuation-driven version described earlier.

The usefulness of this example is that the process of decay from ϕ_F to ϕ_T is well described mathematically (it was first treated by Coleman and Deluccia Coleman and Luccia (1980)), so (neglecting collisions between bubbles; see next section) we can assemble a fairly accurate global picture of the inflationary universe.

8.3.3 An Infinite Number of Pocket Universes Form at Each Time

The regions of the universe in which inflation ends (when the inflaton either rolls or tunnels to near the bottom of the well) might be (and often are) called "bubble

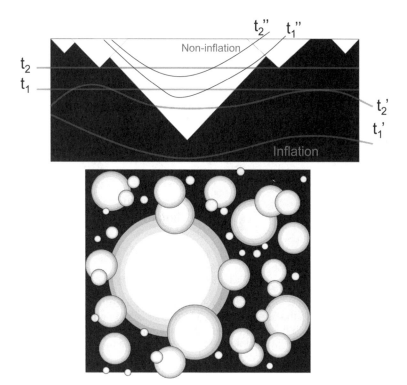

Figure 8.2. A depiction of inflation in the double-well potential. Black regions are at ϕ_F, white are at ϕ_T, and gray regions are at field values between the two. The top is a spacetime diagram drawn so that light travels on diagonal lines, depicting bubbles expanding at the speed of light. (Note that the required scaling implies that the upper boundary signifies $t \to \infty$, and that as it is neared, a finite distance on the diagram represents an infinite physical distance.) The spacetime can be decomposed into space and time in many different ways, for example, by surfaces like t_1, t_2, or by surfaces like t'_1 and t'_2, or even by t''_1 and t''_2. In the two former cases the spatial slices would be finite; in the latter they would be infinite. The bottom diagram shows the distribution in two spatial dimensions at a fixed t.

universes," or "pocket universes," because, to an observer inside, they would appear homogeneous and isotropic over large regions – and thus described by the FLRW metric above – even while this homogeneity would break down on very large scales.

To see in exactly what sense, however, as well as discuss infinity in this model, we must confront the somewhat thorny issue that in GR there is no unique way in which to decompose spacetime into space and time. Any given observer has a well-defined time, but this cannot be extended to define a globally defined time, because other observers will not agree. Thus, questions such as "What is the spatial volume of this region?" and even "Is the universe spatially infinite or finite?" become dependent on how surfaces of simultaneity are defined (Fig. 8.2).

With this in mind, let us examine the picture that develops from the double-well potential of Figure 8.1. Suppose that at some time we imagine the universe to be (a) spatially finite and (b) dominated by homogeneous inflaton vacuum energy with $\phi = \phi_F$. Then it is the case that Aguirre and Gratton (2003), (Komatsu et al. 2008), Vilenkin (1992):

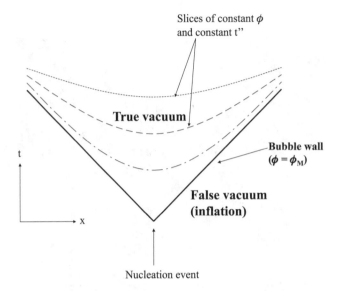

Figure 8.3. The geometry of a nucleated bubble. From the nucleation event, the bubble wall expands at (approximately) the speed of light. Nestled into this light cone are hyperboloids of constant ϕ, each of which has the geometry of an infinite negatively curved homogenous space. The sequence of nested hyperboloids corresponds to the time sequence during which the inflaton rolls down the hill toward ϕ_T in Figure 8.1.

1. There will be a spacetime decomposition such that our region's physical volume U expands exponentially with time t.
2. At each time interval dt, a number of bubbles of ϕ_T will nucleate, with the number given by $\lambda U f_{\text{inf}} dt$, where f_{inf} is the fraction of our region that has not thus far converted from ϕ_F to ϕ_T, and λ is a rate determined by the form of the potential in Figure 8.2.
3. Each such nucleated bubble will expand at the speed of light.
4. The fraction f_{inf} will approach zero as $t \to \infty$.
5. Nonetheless, the inflating volume $U f_{\text{inf}}$ will increase exponentially as $t \to \infty$, so that the number of bubbles nucleated in each time interval dt also increases exponentially with time.
6. There will exist other "slicings" of the spacetime in which the volume of our region increases more slowly, or even decreases with time Winitzki (2005).
7. Nonetheless, it will always be true that an infinite number of bubbles eventually form.

Note that it is also possible (as discussed in the next section) to choose the initial volume to be infinite, in which case infinitely many bubbles would form at each of the infinitely many times t Aguirre and Gratton (2002, 2003), Winitzki (2005).

8.3.4 Pocket Universes Are (Probably) Infinite

So far we have been discussing the inflating region outside of the bubbles. What happens inside? For this we must return to the issue of time slicing; but here there is a natural answer – not a unique way to decompose spacetime, but a particularly natural and physically meaningful way. Figure 8.3 shows the spacetime created by a bubble nucleation. The bubble wall accelerates away from the nucleation site, quickly approaching the speed of light so that the wall forms a "cone" in spacetime. Inside the cone are surfaces of $\phi = const.$, progressing from $\phi = \phi_M$ (which we can take to define the bubble wall) to $\phi = \phi_T$.

What is very interesting is that if we define $\phi = const.$ slices to be equal-time slices (like the slices t_1'', t_2'' in Fig. 8.2), then at each time the spatial sections are infinite, negatively curved, homogenous, and isotropic, that is, the interior of the bubble is described by the FLRW metric with $k = -1$ Coleman and Luccia (1980). Therefore, even if the exterior space can be described as always finite (although exponentially expanding), the interior – spawned by a local nucleation event – is at all times spatially infinite. The infinite time (and asymptotically infinite volume) of the exterior space has been converted into infinite space at all times inside.[4]

Now, the bubbles will often run into each other, but as it turns out, this does not change the infinite nature of the bubble interiors. As discussed in Aguirre et al. (2007a), Garriga et al. (2006a), in considering bubble collisions there is a center of the bubble. Far from this center, the probability to avoid being hit by another bubble approaches zero, but the physical volume at that distance diverges more quickly, so that the total physical volume is still infinite even if regions hit by bubbles are considered completely destroyed (which is itself unlikely; see Aguirre and Johnson (2008), Aguirre et al. (2007a), Bousso et al. (2006a)).

Although all of these aspects are clearly defined in the double-well case of Figure 8.2, they also appear to hold in the "single-well" case in which everlasting inflation is driven by quantum fluctuations. In this case, a surface in spacetime can be defined where the inflaton falls just below the critical value to sustain everlasting inflation. This region should then just classically roll down the potential. Once again we can define constant-time slices to be slices of constant ϕ, and analysis shows that these slices are spatially infinite Linde (1986).

In sum, then, we have an infinite number of times. At each of these, a large or infinite number of bubbles can form, so that an infinite number of bubbles are formed throughout the universe's evolution. Each of these bubbles is spatially infinite inside. This is a rather amazing level of physically realized infinity to contemplate.

8.4 Infinite Problems

8.4.1 The Inflationary Multiverse

This hierarchy of infinities not only is mind-boggling, but also causes enormous headaches in cosmology. The reason is that while the different bubble universes *may* be essentially identical, their properties may instead differ. In the past decade, theorists studying string/M theory have realized that the theory yields many different possible versions of low-energy physics Douglas (2003), Susskind (2003), some may have different values for the particle physics coupling constants. Others could have different

---

[4] It can be seen that this sort of model bridges the distinction between "actual" and "potential" infinity in the Aristotelian sense of an infinite set that is completed all at one time versus one that is given by a sequence of inclusions into the set. Here, these simply correspond to two different ways of decomposing spacetime, which in GR are equally valid.

particle physics symmetry groups or be supersymmetric.[5] With inflation, these possibilities may be brought into reality, realized as the post-inflation, low-energy physics of different bubble universes. Together these would comprise a true, physically realized, and quite infinite multiverse.[6]

The nightmare comes in trying to test this model: which of the universes should we compare to ours? The nicest possibility would be if these other universes somehow impacted our observations so that in some way we could actually see them. This may actually be possible at some level under some somewhat optimistic assumptions regarding the bubble collisions Aguirre and Johnson (2008), Aguirre et al. (2007a), Chang et al. (2007). But even if this pans out, it will not fully solve the problem, and much of what we can do will simply be to ask whether our universe is *possible* (i.e., whether our theory predicts that it exists somewhere in the multiverse) and if it is *likely* (i.e., if it is an abundant type of universe).

8.4.2 The Measure Problem

Asking for statistical predictions leads to a very thorny issue of both principle and implementation: how do we define a *measure* with which to define probability Aguirre and Tegmark (2005)? We might, for example, ask, "What properties does a randomly chosen universe have?" But what exactly is a universe? There is nothing to stop me from taking a single bubble and cutting it into infinitely many finite regions, or finitely many infinite ones, or perhaps even infinitely many infinite regions that I might call "universes."

Instead, I might ask something like, "What properties are seen at a randomly chosen point in space just after inflation has ended?" This sounds plausible but is also devilishly difficult to calculate, for two basic reasons. First, recall that volume depends on the slicing of spacetime and is nonunique. We can choose a nice slicing in which inflation ends at a single time, but then the equal-time slice will all be contained within a single bubble. So we need a "time" that encompasses multiple bubbles; there is no unique specification of this.

Second, even if we can choose such a thing, the volume seeing each set of properties will surely be infinite. How do we compare these infinities? The only hope, it would seem, would be to regularize the infinities somehow by considering finite regions, computing probabilities, and taking the regions' size to infinity. Is there a unique, or even generic way to do this?

Despite a good amount of work Garriga and Vilenkin (2001), Garriga et al. (2006b), Easther et al. (2006), Bousso (2006), Aguirre et al. (2007c, b), Vilenkin (2007), Linde (2007b), Bousso et al. (2007a, b), my view is that this problem remains fairly open. One might worry that there simply is no answer, as the measures are simply a few out of infinitely many measures one might assume. On the more optimistic side, many

---

[5] My understanding is that it is currently controversial whether there are finitely or infinitely many of these different solutions.

[6] Note that the term "multiverse" is used in multiple ways. Here, it is not used as a "set of all possible universe," but rather as the set of large, homogeneous spatial regions brought into being via the physical process of inflation as governed by a single fundamental theory of physics and cosmological boundary conditions.

of these measures do correspond to convergent regularizations of sensible-seeming quantities and thus ought to represent something meaningful. Moreover, many of the measures, while initially seeming distinct, have been shown to be either the same or closely related; perhaps there are only a small number of reasonable measures. The stakes are high, as this open question is a major hurdle currently blocking progress in understanding the universe on the largest scales, as well as whether string/M theory and inflation can plausibly reproduce the universe we see.

8.5 Is Infinite Qualitatively Different from "Really Really Big?"

In the preceding sections, I discussed how everlasting inflation "eventually" brings into being an infinite number of bubble universes, each of which is (probably) spatially infinite at each time (according to observers inside it, with their natural time slicing). In this section, I would like to discuss the question, "Are there issues in which the infinite is qualitatively or even observationally different from the finite but extremely large?" I have not proven to myself that this is so, but in the following section I discuss several instances in which the two cases seem, in some sense, truly different, and in which the distinction might even have observational import.

8.5.1 An Infinitely Old versus an "Almost Infinitely Old" Universe

Everlasting inflation completely alters the classical Big Bang picture of a universe springing into being a finite time ago, because the putative Big Bang (the time at which the universe is hot, dense, and homogeneous) results from reheating after an indefinitely long period of inflation. Indeed, inflation circumvents most of the classic singularity theorems indicating an initial singularity, because they are based on energy conditions violated by a field driving inflation. Nonetheless, it appears to be generally assumed that while inflation continues forever into the future, that it nonetheless started at some fixed time in a "Big Bang-like" event. But because everlasting inflation approaches a Steady-State mixture of inflating region and bubbles of noninflation, it seems reasonable to ask whether inflation might simply be in a steady state, so as to avoid any initial time or initial singularity and make inflation truly "eternal."

What would this look like? The idea would be to make the state approached by everlasting inflating into the exact state of the universe at any time Aguirre and Gratton (2002, 2003), Aguirre (2007). The universe would, therefore, be spatially flat (in terms of the inflating background), with a statistical distribution of bubbles that is given by the distribution of bubbles in everlasting inflation in the $t \to \infty$ limit.

This state has some rather interesting aspects:

- As described for the Steady-State model, the universe, while expanding and forming more bubbles, is always the same from one time to another (in the slicing of the inflating background that gives flat spatial sections). There is no sense in which there are more bubbles at one time than at the previous time.
- Although there is always an inflating region, the fraction of the volume at any time that is inflating (i.e., the inflating volume divided by the volume that would be inflating if

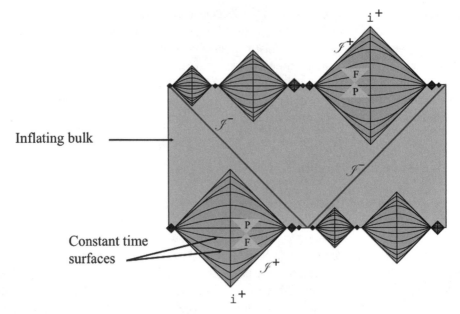

Figure 8.4. A conformal "slice" for false-vacuum driven past-eternal inflation as proposed by Aguirre and Gratton (2002). Bubbles are shown in a background de Sitter space, with $\mathcal{J}_-$ representing a (null) surface on which cosmological boundary conditions are placed by specifying that no bubbles exist there; it is the boundary of the region covered by $t > -\infty$, where t is the time variable in which the universe is statistically time translation invariant. The "arrow of time" points away from this surface, so that the entire universe is effectively time reversal invariant, while each bubble has a well-defined arrow of time pointing away from the inflating phase.

bubble nucleation were not accounted for) is zero. This is possible because the inflating region is an infinite fractal of dimension (very slightly) less than three. This is not terribly mysterious; it is just like an infinite rod in 3-space, which takes up a finite volume but a vanishing volume fraction because its dimensionality is less than three.[7]

- Although there is no initial time ($-\infty < t < \infty$) and all spatial sections are infinite, the spacetime so defined is nevertheless extendible; that is, the coordinates slicing de Sitter spacetime into geometrically flat equal-time surfaces only cover half of the full de Sitter spacetime (Fig. 8.4). This means that there is, in a sense, a past "boundary" to the universe, comprised of an infinite null surface on which boundary conditions can be placed that define the eternally inflating universe.[8] Interestingly, these same boundary conditions also define what happens if the manifold is extended: an identical copy of the eternally inflating universe must exist on the other side of the boundary!

---

[7] Perhaps a better analogy is a "Cantor dust," the three-dimensional version of the Cantor set, with dimension $\log 8/\log 3$. A similar idea, involving an infinite fractal distribution (of galaxies), was proposed by Mandelbrot (1982).

[8] This boundary is what is indicated by several singularity theorems proven over the years that purport to rule out past eternal inflation Brode and Vilenkin (1996), Borde et al. (2003). But they reveal just this boundary-value surface, not a singularity. Because the present model, which seems to me to completely undermine the singularity theorems, seems only to be at best accorded a footnote to oft-repeated claims that "inflation cannot be past eternal," I shall take my revenge here.

There are some rather strange aspects to everlasting (or "semi-eternal") inflation that may be resolved by this "truly eternal" sort of model.

Consider the expectation value of the time T since the putative initial singularity seen by an observer in semi-eternal inflation. At each successive time there are exponentially more (but still finitely many) observers than at the previous time (if we assume that observers are born at some rate per unit physical volume). Thus, the expectation value is infinite. But, of course, each observer was born at a finite time, so each observer is infinitely atypical, in the sense that the probability is zero of a randomly chosen observer being born earlier than the chosen observer. In the truly eternal case, this paradox goes away: there is no meaning, in the global sense, to any observer being older than any other.[9]

A similar paradox, sometimes called the "youngness paradox," arises when we ask about the time t since inflation ended found by a typical observer fulfilling some (non–age-related) criterion. Suppose, for example, that an "observer" must have some attribute A, and that A is correlated to t, but not in one-to-one relation with t (e.g., A might be "is gravitationally bound" or "has density ρ"). Now, consider some bubble and the observers fullfilling A. Most will have t_A (the time that is "naturally" linked with attribute A), with some scatter. Consider all the observers existing *at some* T and fulfilling A. Some will see a time t_A, as expected. But because of the exponential expansion, exponentially more of them will have inflated until just prior to T, then stopped inflating just in time to, by some unusual set of circumstances, attain attribute A in a time span shorter than t_A. So almost all observers at a given T will see a time $t \ll t_A$, and the connection – based on theory – that A implies t_A, will be broken. This can be made fairly exact; for example, if we take A to be "contains a baryon," then only an incredibly tiny fraction of all observers like us that exist at a given time T would see a microwave background temperature as low as we do Tegmark (2005).

As far as I can ascertain, this paradox does not arise in the fully eternal model, because there is no meaning to T, or to saying that one T should contain exponentially more observers than another. But this leads to a strange dilemma: should the statistical predictions made in eternal inflation be continuous or discontinuous when the universe is changed from semi-eternal inflation to truly eternal inflation? If continuous, it seems clear that no measure that explicitly makes use of the time T can make any sense. If discontinuous, this would imply a categorical distinction between the infinite and the "indefinitely large" that would be quite interesting.

If we take the former view, it suggests searching for measures that correctly encode the time translation invariance of the universe. One possibility would be to look for a global foliation of spacetime (both inside and outside of the bubbles) that accomplishes this. Another choice would be to abandon the importance of a foliation and compare regions at different "times," as is apparently advocated by Linde (2007b), the

---

[9] Interestingly, this reasoning very closely parallels that used in the "doomsday" argument that it is unnatural to imagine that our civilization will go on much, much longer with exponential growth because this would mean that we see extremely atypical values of the observable "people born before us." Thus, I hereby propose a new resolution of the doomsday paradox: the universe is infinite at all times, and we must consider not just the population of people in "our" civilization but all of the infinitely many others. Considering this whole set, there is no sense in which the population is growing, and we are just as likely to live at any time as at any other.

prescription of which removes the youngness paradox by giving a head start to certain bubble times.

8.5.2 Infinite versus Finite Number of States

Are the number of states accessible to the entire universe infinite or finite? In everlasting or eternal inflation, it would seem that the number is infinite, because the spacetime volume is, but this has been disputed (e.g., Dyson et al. (2002), Banks and Fischler (2004), Bousso et al. (2006a)). Quite independent of inflation or everlasting inflation, however, an interesting set of paradoxes – going back to Boltzmann – arises if we imagine the universe to be a physical system with a finite number N of microstates (i.e., states that completely characterize the physical system) that evolve according to a fixed Hamiltonian (i.e., a fixed deterministic evolution from one microstate to the next). In this context, imagine two (macro) states of the universe, s_i and s_o with entropies S_i and $S_o > S_i$ (i.e., $\exp(S_o)$ different microstates are "coarse-grained" into the macrostate s_o that is identified by some observables that are indistinguishable for that set of microstates, and likewise for S_i and s_i.) Here, the "i" and "0" can be considered to stand for the "initial" and "observed" states. Suppose further that the microstates in s_i nearly all evolve into microstates within s_o; thus, if the universe is in state s_i at some time, it will naturally evolve into s_o later. Now let the system evolve forever. After some time, the universe will reach equilibrium (attaining entropy of roughly $\ln N$), and thereafter (given some assumptions about ergodicity) states s_i and s_o, which are nonequilibrium states, will be realized only as ultra-rare downward fluctuations in entropy.

Now suppose we analyze all times at which the universe is in state s_o. What are the most typical histories of the universe just prior to this? Because, according to statistical mechanics, the probability of fluctuating a state decreases exponentially with the magnitude of the downward entropy fluctuation, analysis quickly reveals that in general, these histories will not include the "precursor" state s_i. Rather, they will almost certainly be a direct fluctuation from the equilibrium state to s_o, or (with somewhat smaller probability) a fluctuation to a state s_o' that is only very slightly lower entropy than s_o and then evolves into s_o.

As one final piece of groundwork, consider some set of attributes A, which may or may not describe a state s. Next, denote by A_o some attributes that describe our universe. (These might, for illustration, be "there is a cosmologist named Anthony on a planet around a star in a galaxy.") If s_o is the state our universe is in, then clearly A_o describes s_o, but unless A_o is so detailed as to uniquely fix the macrostate, there may be plenty of other states also described by A_o.

So far so good, and now we can discuss the paradoxes. Imagine that we ask the question, "Why do we see a particular direction of time, as defined by an increasing entropy?" In particular, why did the universe "begin" at low entropy, and what brought such a highly improbable state into being? The possibility that Boltzmann suggested (and attributed to his lab assistant Schuetz[30]) is that the universe is generally in equilibrium, but occasionally fluctuates to a nonequilibrium state like s_i, which then naturally evolves to s_o, thus accounting for the history that we see.

The immediate problem, however, is the fact stated above. Supposing that we are in state s_o now, the most probable precursor to our state was not s_i. A much more probable history is a direct "fluctuation" into the universe 5 minutes ago, complete with incoming photons, carefully arranged neurons, and so on, that fool us into thinking that the universe is much older than it appears. This is weird, but it gets worse. In reality, we do not know the current state of the universe – we know certain attributes A_o. The reasonable question we could ask is, "Given A_o, what was the history leading up to the current situation, and what will be observed in the future?" The answer will be totally contrary to the "natural" evolution of the system because of the laws of physics.

Imagine, for example, that A_o is indeed "a cosmologist named Anthony on a planet around a star in a galaxy."

Then the most probable[10] sequence of events consistent with these observations is the following:

1. The universe is in equilibrium.
2. A precisely galaxy-size region replete with stars and planets fluctuates into existence,[11] leaving the rest of the universe in equilibrium.
3. This fluctuated region evolves back to equilibrium.

How do we know this did not occur? By step 3. Because the galaxy did not evolve, but simply fluctuated, its dynamics will almost certainly lead to gross violations of "normal" physics. There would be no reason, for example, for the planets to be bound to the stars, and, in general, any data not included in A_o would appear incomprehensible. By specifying more and more in A_o, we could force (via this conditionalization to a more and more specific set of states) the universe to be "cohesive" for longer, but ultimately we are limited by our ability to observe an A_o, which is far from sufficient to specify the macrostate of the universe. In fact, we can take this to the logical extreme and realize that all that is really necessary for (say) me to account for my experience is the fluctuation of one disembodied brain, with the correct memories, and so forth. Of course, after an instant this brain would decompress in the empty equilibrium universe. Because I repeatedly do not observe this, I can quickly rule out the whole picture, as can the reader.

This *reductio ad absurdum* indicates that we cannot account for the coherence of the world around us if we imagine that the low entropy in our past was the result of a fluctuation from equilibrium, and this reasoning seems hard to deny. But if we apply the very same reasoning to any universe with finite entropy, we run into an almost identical problem. That is, even if such a universe starts at low entropy and spawns some "normal" observers soon thereafter, it will nonetheless attain equilibrium eventually and thereafter spawn an infinite number of "fluctuated" observers that will, statistically, infinitely outweigh the normal ones. The paradox persists.

---

[10] By most probable, I am speaking in the frequentist sense of the number of instances relative to the total number of instances for which A_o are realized. See Hartle and Srednicki (2007) for an extended argument against this view.

[11] While this conventional phrase gives a feeling of discontinuity, in this scenario the region is, in fact, naturally evolving according to its ordinary time evolution in a rather particular microstate. In fact, I strongly suspect that the "fluctuation" process would be the time reverse of the "evolve back to equilibrium" process of the next step. That is, the fluctuation would begin with a sort of reversal of the local arrow of time that would go on for "long enough" to reach the macroscopic configuration described by A_o, but no longer.

Thus, it appears that if we ask, "Given some set of conditions A_o, what should we observe next?" the theory described by a finite-state system described by a time-independent Hamiltonian cannot give the right answer: most realizations of the conditions A_o are preceded by nothing like the "natural" history $s_i \cdots s_o$, nor will they further precede anything like what we expect our universe to do. (For more discussion, see Dyson et al. (2002), which makes this old argument in detail in the modern context.)

Now, this might seem an artificial setup, but as noted previously, a number of people have suggested that if there is a fundamental positive cosmological constant (i.e., if the global minimum of potential energy for all fields is positive), then the universe should have a finite number of states. It also seems plausible that, at least after a long period of evolution, the effective Hamiltonian would become time-independent. From this standpoint, thermodynamics seems to lead to the conclusion that for us to observe what we do cosmologically, the universe must have an infinite number of states, rather than any finite number, no matter how stupendously large. With an infinite number of states one can imagine also an indefinite increase in entropy, and that the universe never achieves equilibrium, potentially Carroll and Chen (2004) (although not necessarily Bousso et al. (2006b), Page (2006) resolving the paradox.

What are we to make of this? One possibility is that the frequentist reasoning that leads to the result is simply incorrect, or that we should not be asking the questions we are asking. But then we are forced to accept (as the Bayesian reasoning of, e.g., Hartle and Srednicki (2007), would seem to imply) that there is no way to rule out that we are simply statistical fluctuations from equilibrium. A second possibility would be that frequentist reasoning is correct in principle, but that the *measure* (or some other element of reasoning) that we are using is incorrect, and that if done correctly, there would be a smooth transition from a huge number of states to an infinite set.[12] Third, the reasoning may indeed be telling us something profound: that the very coherence of our experience means that the universe has infinite possibilities.

8.6 Conclusions

It seems inescapable that, as finite beings, we can never prove that the universe is physically infinite: we cannot travel through infinite spaces or times or experience an infinite number of states. Nonetheless, I have argued that in modern cosmology, we may face the fascinating situation that the theories (particularly inflation) devised to explain the finite observed region of the universe also *naturally produce an infinite universe*, through a process called everlasting inflation. This can be the case even if the

---

[12] My current suspicion is that a better lens through which to view this paradox is the amount of information necessary to select out the subensemble of systems satisfying A_o. Imagine, for example, a closed and impenetrable box with a book of Shakespearean sonnets in it. Now wait forever, keeping the box at constant temperature. The book will slowly disintegrate and eventually attain equilibrium, but if we wait long enough, the macrostate will evolve back to the state of a book; this will disintegrate and eventually form another book (as well as all manner of other objects.) Suppose we ask, "Given that we find a book, which one will it be?" Well, the amount of information we would have to specify (and entropy generate) to monitor the box for the vast eons of time necessary for it to cycle into various books would be stupendous, and there would be no reasonable way to separate the box from the nonequilibrium measuring environment.

universe is initially finite, because the dynamics of inflation, played out over an infinite available duration, allow the creation of an infinite universe – even, in a sense, many such universes. Thus, although we cannot prove that the universe is infinite, strong evidence for inflation, along with the (strong but imperfect) theoretical link between inflation and "everlasting" inflation, leads to a strong inference of an infinite universe.

On even more speculative ground, I have discussed the possibility that in cosmology, *finity* might be problematic in certain ways, so that the very coherence and comprehensibility of our physical world is pointing to an infinite duration, or infinite number of states of the universe. Even if this conclusion is overreaching, however, the analysis of the paradoxes *or solutions to paradoxes* that infinity can generate in cosmology can bring novel perspectives on some ancient riddles.

Acknowledgments

I am grateful to the organizers of the "Infinity" workshop, Ellipsis Enterprises, in connection with which this paper was prepared. I also am grateful to Matt Johnson, Steven Gratton, and Max Tegmark for their company in banging our heads against some of the riddles of cosmological infinities. I thank Michael Heller for helpful comments on the manuscript.

References

Aguirre, A., and Gratton, S. 2002. Phys. Rev. **D65**, 083507.

Aguirre, A., and Gratton, S. 2003. Phys. Rev. **D67**, 083515.

Aguirre, A. 2007. arXiv **hep-th**, URL http://arxiv.org/abs/0712.0571v1.

Aguirre, A., Johnson, M. C., and Shomer, A. 2007a. Phys. Rev. **D76**, 063509.

Aguirre, A., Gratton, S., and Johnson, M. C. 2007b. Phys. Rev. **D75**, 123501.

Aguirre, A., Gratton, S., and Johnson, M. C. 2007c. Phys. Rev. Lett. **98**, 131301.

Aguirre, A., and Johnson, M. C. 2008. Physical Review **D77**, 123536, (c) 2008: The American Physical Society.

Aguirre, A., and Tegmark, M. 2005. JCAP **0501**, 003.

Banks, T., and Fischler, W. 2004. arXiv **hep-th**, URL http://arxiv.org/abs/hep-th/0412097v1.

Bondi, H., and Gold, T. 1948. Monthly Notices of the Royal Astronomical Society **108**, 252.

Borde, A., and Vilenkin, A. 1996. Int. J. Mod. Phys. **D5**, 813.

Borde, A., Guth, A. H., and Vilenkin, A. 2003. Phys. Rev. Lett. **90**, 151301.

Bousso, R. 2006. arXiv **hep-th**, URL http://arxiv.org/abs/hep-th/0605263v4.

Bousso, R., Freivogel, B., and Yang, I.-S. 2006a. Phys. Rev. **D74**, 103516.

Bousso, R., Freivogel, B., and Lippert, M. 2006b. Phys. Rev. **D74**, 046008.

Bousso, R., Freivogel, B., and Yang, I.-S. 2007a. arXiv **hep-th**, URL http://arxiv.org/abs/0712.3324v3.

Bousso, R., Harnik, R., Kribs, G. D., and Perez, G. 2007b. arXiv **hep-th**, 38 pages, 9 figures, minor correction in Figure 8.2, URL http://arxiv.org/abs/hep-th/0702115v3.

Carroll, S. M., and Chen, J. 2004. arXiv **hep-th**, URL http://arxiv.org/abs/hep-th/0410270v1.

Chang, S., Kleban, M., and Levi, T. S. 2007. arXiv **hep-th**, URL http://arxiv.org/abs/0712.2261v1.

Coleman, S. R., and Luccia, F. D. 1980. Phys. Rev. **D21**, 3305.

Douglas, M. R. 2003. JHEP **05**, 046.

Dyson, L., Kleban, M., and Susskind, L. 2002. JHEP **10**, 011.

Easther, R., Lim, E. A., and Martin, M. R. 2006. JCAP **0603**, 016.

Garriga, J., Guth, A. H., and Vilenkin, A. 2006a. arXiv **hep-th**, URL http://arxiv.org/abs/hep-th/0612242v1.

Garriga, J., Schwartz-Perlov, D., Vilenkin, A., and Winitzki, S. 2006b. JCAP **0601**, 017.

Garriga, J., and Vilenkin, A. 2001. Phys. Rev. **D64**, 023507.

Guth, A. H. 1981. Phys. Rev. **D23**, 347.

Guth, A. H. 2004. arXiv **astro-ph**, URL http://arxiv.org/abs/astro-ph/0404546v1.

Guth, A. H., and Weinberg, E. J. 1983. Nucl. Phys. **B212**, 321.

Hartle, J. B., and Srednicki, M. 2007. Phys. Rev. **D75**, 123523.

Hoyle, F. 1948. Monthly Notices of the Royal Astronomical Society **108**, 372.

Hoyle, F., and Narlikar, J. V. 1966. Proceedings of the Royal Society of London. Series A **290**, 162.

Komatsu, E., Dunkley, J., Nolta, M. R., Bennett, C. L., Gold, B., Hinshaw, G., Jarosik, N., Larson, D., Limon, M., Page, L., et al. 2008. arXiv **astro-ph**, URL http://arxiv.org/abs/0803.0547v1.

Linde, A. 2007a. arXiv **hep-th**, URL http://arxiv.org/abs/0705.0164v2.

Linde, A. 2007b. arXiv **hep-th**, URL http://arxiv.org/abs/0705.1160v2.

Linde, A. D. 1986. Phys. Lett. **B175**, 395.

Mandelbrot, B. B. 1982. p. 468.

Page, D. N. 2006. arXiv **hep-th**, URL http://arxiv.org/abs/hep-th/0612137v1.

Susskind, L. 2003. arXiv **hep-th**, URL http://arxiv.org/abs/hep-th/0302219v1.

Tegmark, M. 2005. JCAP **0504**, 001.

Vilenkin, A. 1992. Phys. Rev. **D46**, 2355.

Vilenkin, A. 2007. J. Phys. **A40**, 6777.

Winitzki, S. 2005. Phys. Rev. **D71**, 123507.

CHAPTER 9

Infinity and the Nostalgia of the Stars

Marco Bersanelli

The word "infinity" immediately evokes a sense of vastness. Since ancient times, human imagination has been invited to the idea of infinity by the immense landscape of the universe. The beauty of the firmament and the mystery of what lies beyond have always attracted the most brilliant minds. Generations of philosophers, theologians, and scientists have been challenged by the simple and inevitable question: is the universe finite or infinite? In recent times, scientific cosmology has translated this age-old problem into a rigorous mathematical language and has pointed to a set of astrophysical observations that are useful in addressing this issue. Hence, today we ask, what experimental data can help us to approach – if not answer – the question of the finiteness or infinity of the universe? If an answer has not been obtained, do we have a hope that deeper observations and more advanced theories may help us reach a conclusion in the future?

The meaning of the word "infinity" needs to be carefully understood in the different contexts in which it is used. The infinity of mathematicians[1] is related to, but not the same as, the infinity of physicists and cosmologists. When we refer to infinity in a metaphysical discourse, we normally mean the object of human longing for what is ultimately true, beautiful, just – a reality quite distinct from the notion of a physical or mathematical quantity that takes on an infinite value.

This is not necessarily to say that the infinity of the natural sciences has nothing to do with an existential or metaphysical infinity. In several ancient cultures, the perception

---

[1] The concept of infinity in mathematics (not discussed in this paper) is rich in itself and capable of illuminating the relationship with a theological concept of Infinity. As Russel (2008) poined out, "developments in mathematics since the 19th century may shed provocative new light on this issue. Georg Cantor in particular has given us a new conception of infinity which is much more complex than previously thought, with layers of infinities leading out endlessly to an unreachable Absolute. In effect, we now can say a lot more about infinity than merely that it contrasts with the finite – indeed, an infinite amount more! These revolutionary discoveries in mathematics might, then, lead to new insights into the ways we can use the concept of infinity in thinking theologically about God."

This chapter is adapted from a presentation given at the JTF-funded STARS conferences in Cancun in January 2007.

of the expanse of the night sky has been a central clue to the philosophical quest, suggesting in different ways the ultimate mystery that lies at the root of all things. In many religious traditions, the immensity of the universe has been perceived as a sign of the infinite divine power and, by contrast, of the fragile, yet prodigious, character of human life. Indeed, the stars have always mirrored our inextinguishable need for infinity as a distinctive feature of human existence. Dante Alighieri ends each of the three parts of his *Divine Comedy* with the same word, "stars," clearly a privileged image of our destiny. Interestingly, the etymology of the word "desire" is from the Latin "de-sidera" and may be translated as "nostalgia of the stars."

Thus, there are two issues at stake. First, I would like to discuss what modern cosmology can tell (if anything) about the infinity of the universe. A second related question is whether scientific knowledge may provide new opportunities to think and express infinity as understood in its metaphysical sense. In the first part of this chapter, I discuss how contemporary cosmology deals with the question of whether the spatial extension of the universe is finite or infinite. We will see that this question requires us to clarify what we mean by "the universe," and that the issue can be answered only partially. Then I consider some aspects of a metaphysical notion of infinity. The language of arts and poetry, in this case, will be better suited than that of mathematics. Next, I suggest a relationship we may recognize between the structure of the physical universe as emerging from science and the experience of infinity as an ultimate reality. Finally, I exemplify how the image of the universe we have today illuminates some of the ways in which the Judeo-Christian tradition has expressed the infinity of the Creator through the sign of the created universe.

9.1 Vast Universe and Physical Infinity

Scientific cosmology has the objective of unveiling the structure, geometry, evolution, and underlying physical laws of the universe through its own powerful and highly selective methods. Among the key issues on the nature of the universe is the possibility that some form of infinity is actually realized.[2] Is infinity something that belongs to the real world? In particular, is the extension of cosmic space infinite or finite?

Historically, a key precondition to speak comfortably of "the universe" has been to assume that the average characteristics we see from our particular location are typical of what other potential observers view from anywhere in the universe. This apparently innocuous assumption, known as the "cosmological principle," has far-reaching consequences. It implies spatial isotropy and homogeneity on large scales and that the laws of physics, the fundamental constants, and the types of elementary particles and forces are the same everywhere in the universe. This is a prerequisite for scientists to be able to treat the cosmological problem with forgiving simplicity and effectiveness through general relativity. Deviations from uniformity, such as those

---

[2] In relativistic cosmology infinity of space is to be regarded as a particular case of the infinity of the spacetime manifold. Other possible infinities, referred to as "singularities," may occur in nearby spacetime regions where some physical quantity diverges and the structure of spacetime breaks down. In this work, however, I shall restrict discussion to the infinity of spatial sections of the universe.

making you and me, our planet, stars, single galaxies, and even clusters of galaxies, are only second-order perturbations to an otherwise uniform distribution of matter and energy. The cosmological principle is an advanced form of Copernican outlook: not only does the universe have no physical *center*, but it also lacks any *structure* if we look at it as a whole.

Perhaps owing to its immense benefit to theoreticians, the cosmological principle was widely accepted well before observations could effectively verify its validity. Remarkably, recent data confirm the gradual tendency toward uniformity at large dimensions.[3] If we look at portions of the universe of sizes >100 Mpc or so,[4] we find that the general statistical distribution is maintained, whereas the details change from region to region, suggesting that the cosmological principle is indeed a good approximation of the real universe – at least within the limits of our current data. But how far can we verify its validity with observation?

Since the discovery of cosmic expansion in the late 1920s, our view of the universe has changed drastically (Hubble 1929; Hubble and Humason 1931).[5] Change and evolution are not only characteristics of biological life, of our planet, of stars and galaxies, but also of the universe as a whole. We live in a historical, contingent universe: no instant of time, at cosmic scale, is the same as any another. Expansion also means that back in the past the universe was smaller, hotter, and denser than it is today. The present estimates of the age of the universe[6] tell us that back in cosmic history some 13.7 billion years ago the temperature and energy density reached fantastic values everywhere in space.

The Hubble Ultra Deep Field image (Beckwith et al. 2006), taken with the Advanced Camera for Surveys on board the *Hubble Space Telescope* (*HST*), shows some of the farthest known galaxies, whose light has traveled for about 13 billion years before hitting the *HST* mirrors. Those galaxies belong to our universe when it had less than 10 percent of its present age. Even at those distances, if we account for evolution, the data are consistent with cosmic uniformity. Can we look further back? Remarkably, the answer is yes. In 1965 Arno Penzias and Robert Wilson serendipitously discovered a prodigious fossil light (Penzias and Wilson 1965), named cosmic microwave background (CMB),[7] a sea of photons pervading the universe and released in an early phase (redshift $z \cong 1100$) of cosmic history. The great isotropy of the CMB, better than one part in 10,000, provides further strong support for the cosmological principle. The CMB photons were released when the expansion cooled the temperature below 3000 K and allowed the first atoms to form from electrons and light nuclei. This took place

---

[3] The current most precise data come from large galaxy surveys such as the Sloan Digital Sky Survey (SDSS) and the Two Degree Field Galaxy Redshift Survey (2dFGRS). In addition, on very large scales we have compelling evidence of high isotropy from CMB observations.

[4] 1 Mpc (megaparsec) $= 10^6$ pc (parsec); 1 pc $= 3.26$ light-years.

[5] Vesto Slipher, working at Lowell Observatory, back in 1914 announced the discovery of "nebular redshifts," and he contributed the radial velocity data used by Hubble in his 1929 discovery paper.

[6] The *WMAP* data in the standard ΛCDM model give a cosmic age of $13.73 + 0.13/ - 0.17$ billion years (Spergel et al. 2007). This is in agreement with estimates based on globular clusters (Chaboyer and Krauss 2002) and white dwarfs (Richer et al. 2004).

[7] The word "microwave" reflects the fact that the radiation spectral range, as observed today, is in the millimeter-microwave range.

when the universe was 380,000 years old, only 0.003 percent of its present age. Despite the enormous number of galaxies, the voids between them are huge and the universe is essentially empty. Therefore, the CMB traveled rather undisturbed and brings to us a remarkably faithful view of the early universe, back to the time when the photons last interacted with matter. Before then, the universe was opaque to light. This ultimate curtain – the so-called last scattering surface – represents a physical obstacle to direct observation into earlier cosmic epochs: beyond that limit our view of the universe is blocked.[8]

Are there ways to get direct information from what lies beyond the last scattering surface? We could go a bit further, at least in principle, but not by much. In the future (well beyond what is predictable today), we might be able to detect cosmic neutrinos, which easily cross the hot plasma unimpeded and reach us directly from a universe only one-second old.[9] If primordial gravitational waves will some day be detected, they would get us a direct signature from the inflation era, some 10^{-35} seconds of the Big Bang.[10] Eventually, however, we approach the ultimate barrier of our cosmic horizon: we cannot get information from regions farther away than the distance traveled by light in the entire lifetime of the universe. It is not a matter of improving our instruments and observing strategy; rather, it is a fundamental limit set by the finite speed of light and the finite cosmic age. As a consequence, we can observationally test the validity of the cosmological principle only within a limited region of space. The extension of isotropy and uniformity to the *entire universe* – something we may call a "strong cosmological principle" – is ultimately unverifiable. That would be the case for the extrapolation to the universe as a whole of any other physical property observed within our cosmic horizon. The expanding hot Big Bang universe brings with it the notion of a horizon that delimits the part of cosmic space we can probe: the observable universe is definitely finite.

9.2 Hints of Infinity in the Primordial Music

If we assume a strong cosmological principle, then the metric of spacetime can be represented in the Friedmann-Lemaître-Robertson-Walker (FLRW) form, and general relativity describes beautifully the evolution and geometry of the universe under the action of gravity. The theory leads to differential equations whose solutions depend on adimensional density parameters (normalized to a critical density ρ_C) that quantify the

---

[8] However, by observing in detail the properties of the CMB (its frequency spectrum, its angular distribution or anisotropy, and its polarization), we can learn a great deal of what happened in the first 380,000 years of cosmic infancy.

[9] The cosmic neutrino background is expected to have a present equivalent temperature of 1.9 K.

[10] High-precision polarization measurements of the CMB may also lead indirectly to such a result (see, e.g., Boyle, Steinhardt, and Turok 2006). The *Planck* satellite, launched in May 2009, should be able to explore this possibility. The ultimate conceivable early phase is the Planck era, at times 10^{-43} seconds from the beginning of the universe, when space and time become subject to quantum fluctuations and Einstein's theory of gravity fails. This frontier corresponds to energies exceeding 10^{19} GeV and sizes smaller than 10^{-35} cm. Currently we have no way to describe the universe in this state. To extend physics beyond this limit, we would need a quantum theory of gravity. Superstring theory might offer an attractive approach, but it is not known whether it would lead to successful synthesis.

relative contribution of different types of matter-energy components: $\Omega_M = \rho_M/\rho_C$ for matter, $\Omega_R = \rho_R/\rho_C$ for radiation, $\Omega_\Lambda = \rho_\Lambda/\rho_C$ for dark energy.[11] These parameters are not fixed by any known theory, and only experiments may be able to pin down their values. They are of great importance to our discussion of infinity, because they govern the dynamics and shape of cosmic space. In particular, the total energy density parameter $\Omega_0 = (\rho_M + \rho_R + \rho_\Lambda)/\rho_C$ is directly related to the adimensional constant k, which controls the global curvature of space: $\Omega_0 = 1$ corresponds to the familiar flat, Euclidean[12] space, $\Omega_0 < 1$ to a hyperbolic space with negative curvature, and $\Omega_0 > 1$ to a positively curved spherical space. If we assume that the FLRW metric holds over the entire spacetime and that the topology of space is simply connected, then the value of the curvature constant determines finiteness ($k = +1$) or infinity ($k = 0$, $k = -1$) of spatial sections. Under these assumptions, measurements of the total energy density parameter Ω_0 provide a clear answer to our problem.

It is remarkable that modern science has allowed us to speak of the finiteness and infinity of the universe in an elegant, self-consistent, and concise way and, even more amazingly, to point at observable parameters that may discriminate between the different scenarios. In the past few years, observations of the CMB (Bennett et al. 2003) and of distant supernovae (Riess et al. 2005) and systematic studies of the large-scale distribution of galaxies (for a review, see Schindler 2002) have yielded quantitative estimates of the cosmic density parameters with unprecedented precision. Even more precise measurements are expected in the near future. We live in a golden age for cosmology!

A most direct way to measure spatial curvature is to measure very large triangles: depending on how the sum of the internal angles compares to the Euclidean value, 180 degrees, we can infer the curvature of the underlying space. In particular, a standard length seen from a given distance will be subtended by different angles depending on the space curvature. In a two-dimensional analogy, if a given standard rod is subtended by an angle θ_0 on a flat surface (Euclidean space), it will be subtended by an angle $\theta > \theta_0$ on a spherical surface (positive curvature) and by an angle $\theta < \theta_0$ on a hyperbolic surface (negative curvature). If we have an a priori estimate of the length of a source, then a precise measure of the angular diameter (i.e., a measure of θ) carries information on the curvature of space. Of course, the larger the triangles, the more precise the measurement we can hope to achieve. Traditionally, a method to measure large triangles in the universe has been to exploit gigantic sources like radio galaxies, whose radio lobes extend at megaparsec distances from each other: the triangle is made by the two lobes and our observing point. Unfortunately, this yields a poor measure of the curvature of space because the distance between the two lobes is too small and not sufficiently known.

---

[11] In the late 1990s two independent research groups (Perlmutter et al. 1999; Garnavich et al. 1998), using observations of Type Ia supernovae, concluded that, contrary to every expectation, the rate of cosmic expansion in recent cosmic epochs has undergone a new era of acceleration. This requires some unknown form of energy characterized by negative pressure to act as an anti-gravity component. Calculations showed that in order to explain the observed acceleration, the so-called dark energy must contribute as much as $\sim 2/3$ of the total energy density of the universe.

[12] Hereafter "Euclidean space" refers to the three spatial dimensions of the pseudo-Euclidean Minkowski four-dimensional spacetime.

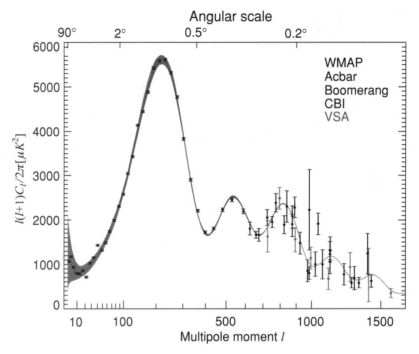

Figure 9.1. Measurements of the angular power spectrum of the CMB radiation. The oscillating pattern results from acoustic oscillations in the primordial plasma. The best fit yields accurate measurements of cosmological density parameters. The angular scale of the main peak, about 1 degree, is directly related to the total energy density, and its value indicates that cosmic space is very nearly Euclidean. The thickness of the model line represents the cosmic variance limit. (From Hinshaw et al. 2007, *courtesy of the WMAP Science Team*)

Are there larger and better-known triangles in the universe? Yes, the CMB provides a tool to measure giant triangles in the sky. The ancient radiation exhibits tiny deviations from perfect uniformity produced by density and velocity perturbations in the primeval plasma. Since the first detection of these anisotropies, observed at a level $\delta T / T \cong 10^{-5}$ by the differential microwave radiometer on board the *Cosmic Background Explorer* (*COBE*) satellite (Smoot et al. 1992), a number of experiments from ground and balloon have been obtained, culminating with the *Wilkinson Microwave Anisotropy Probe* (*WMAP*) satellite mission.[13] The statistics of the anisotropies are described by a power spectrum that shows an oscillating pattern as a function of angular scale (Fig. 9.1). The so-called acoustic peaks in the power spectrum are the result of oscillating sound waves in the photon-baryon fluid, a sort of primordial music of the forming cosmos (Hu and White 2004). This behavior is predicted by theory and provides evidence of a coherent onset of the initial fluctuations as expected in inflationary cosmologies (Guth 1997). The angular scale of the first peak corresponds to the angular size of the causal horizon at the last scattering surface. The giant cosmic triangles we need to measure space curvature are defined by our distance to the last scattering surface and by the length of the causal horizon at last scattering. Indeed, these are the largest causally connected triangles in the universe!

---

[13] The three-year results of *WMAP* were released in March 2006 (see Hinshaw et al. 2007).

In this case, however, we do not measure single triangles, as with radio galaxies; rather, we use a statistical measure as defined by the CMB anisotropy power spectrum. The angular scale θ_{peak} of the main peak, or the corresponding multipole $\ell_{peak} \approx \pi/\theta_{peak}$, is directly related to the total energy density parameter, $\ell_{peak} \cong 220/\sqrt{\Omega_0}$, so that CMB anisotropy data yield a remarkably direct evaluation of the spatial curvature (for a review see Bersanelli, Maino, and Mennella 2002). As shown in Figure 9.1, the peak amplitude of the anisotropy distribution occurs at $\theta_{peak} \approx 1°$, or $\ell_{peak} \cong 220$, corresponding to the critical density $\Omega_0 \cong 1$ with only a few percent uncertainty.

Cosmic background anisotropy data set particularly tight constraints on spatial curvature when combined with observations at lower redshifts. In particular, supernova data and statistical studies of matter distribution (the so-called baryonic acoustic oscillations [BAOs][14]) help us break the degeneracy with the effects of the changing rate of expansion during cosmic history. Recent results from the five-year *WMAP* data (Komatsu et al. 2009) combined with BAO results (Percival et al. 2007) and with supernova data (Astier et al. 2006; Wood-Vasey et al. 2007) yield an estimate of $0.9915 < \Omega_0 < 1.0175$. This is indeed remarkable: even with measurements at $\sim$1 percent precision, we still fail to detect a curvature of cosmic space. A very low curvature is an expectation of inflation, as the initial exponential expansion would have stretched the original curvature to a very low level, just as a given portion of the surface of a balloon flattens as the balloon is inflated. This may just mean that the observable universe is far too small compared to the overall curvature radius. However, some analyses (Dunkley et al. 2005) based on generalized assumptions on the modes of the primordial perturbations indicate a slight preference for a closed (finite) universe. Although even the latest *WMAP* results do not completely dissipate this tendency, these hints are weak and highly tentative. In any case, given that we are so close to a critical universe, high-precision measurements such as those expected from *Planck*[15] are crucial to evaluate whether a measurable deviation from flatness can be detected.

We seem to remain persistently on the verge between a finite and an infinite world. This can be translated in lower limits to the curvature radius of the universe, which has to be greater than 32 Gpc for a positive curvature and 46 Gpc for a negative curvature.[16] It is as if Eratosthenes in his famous measurement of the radius of the earth in 250 BC was not able to measure any curvature: then his conclusion would have been that the earth might be flat and infinite, or that its radius is greater than a given size compatible with the accuracy of his observation.

---

[14] Baryons generate an oscillatory signature that is visible not only in the CMB power spectrum (Fig. 2) but also in the power spectrum of the large-scale structure of the universe at moderate redshifts. These BAOs have been seen recently in large galaxy redshift surveys (e.g., Eisenstein et al. 2005; Cole et al. 2005). Just as the oscillations producing CMB anisotropies, BAOs in late-time structures provide a constant comoving length scale (a standard ruler) that defines another useful set of "cosmic triangles" to measure global curvature of space.

[15] The ESA *Planck* satellite was launched on May 14, 2009, and it is designed to obtain a full-sky map of the CMB with an unprecedented combination of sensitivity, angular resolution, and spectral coverage. See http://planck.esa.int.

[16] These curvature radii are derived for a Hubble constant of 72 km s^{-1} Mpc^{-1}, as measured by the *HST* Key Project (see Freedman et al. 2001). Note also that in a negative curvature space with simply connected topology, space is infinite for any value of the radius of curvature.

9.3 Neverending Territory

These results imply that space is very close to a flat, or Euclidean, geometry. A spatially flat universe is, of course, spatially infinite. If this is really the case, and if we take seriously a "strong cosmological principle," we conclude that we live in a limitless distribution of matter and radiation extending to infinity in all directions with a high degree of isotropy and homogeneity on large scales. The universe is characterized everywhere by the same basic ingredients and physical laws that we find in our cosmic surroundings. Therefore, the only changes from one region to another in the boundless depth of space would be due to the random local outcomes of the same basic processes. Because the spatial extension is infinite, one should also conclude that any system occurring with a nonzero probability, no matter how improbable, must be realized again and again an infinite number of times. The very fact that we see a given object, say, the Matterhorn, means that that particular object has nonzero probability to be formed. Therefore, the Matterhorn must have infinite identical copies existing somewhere in the universe. The formation process of DNA-based life forms, even if exceedingly unlikely, should be expected to take place indefinitely often in the infinite universe. If we consider a large enough portion of space, the same exact history that life has undergone on our planet, its evolution to complex organisms up to the emergence of consciousness, should be found to occur identically in some remote planets.[17] If, in addition, one is prepared to think that conscious beings, such as ourselves, are completely defined by the physical systems supporting them, then deeper paradoxes arise (Tegmark 2003). Somewhere, very far away, right now there would be an infinity of beings *indistinguishable* from yourself, reading exactly these same words about the infinity of the universe. . . .

The degree of weirdness one can fantasize about by playing with probabilities in an infinite universe is rather arbitrary. This situation reflects unsolved problems in defining a meaningful *measure*, that is, how to compute what is common and rare within an infinite set (Bousso 2006). Recently, a discussion has developed around one such extreme case, the so-called Boltzmann brain paradox. We know that quantum field fluctuations can materialize particles with a very small, but nonzero, probability. If instead of a single particle we now want, say, the quantum formation of a full carbon atom complete with its twelve nucleons and electrons perfectly arranged, then the probability will be vastly smaller – but still nonzero. In principle, we also have a tiny chance of a sudden quantum appearance of rocks, Matterhorns, complex systems, cells, and so on. "Boltzmann's brains" are conscious observers – naked brains in empty space! – popping out of quantum vacuum for a very short time, perhaps endowed with a full array of virtual memories of a fictitious past personal history. Some recent calculations in connection with inflationary models (Dyson, Kleban, and Susskind 2002;

---

[17] The situation is further exacerbated in models invoking a "perfect cosmological principle," which postulates that the average properties are maintained not only in spatial extension but also in time. In such a universe anything that is happening around you right now not only is happening elsewhere infinitely often, but it has always happened and will happen endlessly, in the past and in the future. Such a scenario was popular in the early 1960s through the Steady-State model of Fred Hoyle, Tommy Gold, and Hermann Bondi. The model was then discarded under observational evidence against it, most notably after the discovery of the CMB in 1965 by Arno Penzias and Robert Wilson.

Albrecht and Sorbo 2004; Linde 2007) seem to indicate that Boltzmann's brain events would be far more probable than the quantum appearance of a life-supporting universe, suggesting that right now you are more likely to be a Boltzmann's brain than what you think you are. . . .

Most cosmologists see these nonsenses as pathological symptoms of some flaws somewhere in the model: these situations are just too antiaesthetic to be taken seriously. Even the infinite repetition paradox, arising for the flat universe model preferred by current data, may be indicative of one such situation. It is worthwhile, therefore, to look carefully at various hidden assumptions that we make when going from the observational results that Ω_0 is very close to unity to the conclusion that we actually live in an infinite, flat universe.

9.4 The Impossible Proof

A fundamental problem with the $k = 0$ universe is that it is unprovable. In fact, assuming a global FLRW metric, one would require an infinitely precise measurement of $\Omega_0 = 1$ (or equivalent quantity) to demonstrate the flatness of space: any uncertainty around the (true) critical value can accommodate an infinity of solutions with positive or negative curvatures. Every physicist knows that no experiment, however precise, can give results without a finite error bar.

There are deeper uncertainties built in by nature itself that make a spatially flat universe unprovable. Generally, these are related to limitations in the *observability* of the universe by us. A remarkable example applies to CMB data. Even supposing (unrealistically!) an *infinite precision* in the measurements of the CMB fluctuations, the power spectrum would still be limited in accuracy by a "cosmic variance" because of the finite statistics of CMB samples that we can observe from a single location in the universe. This ineliminable uncertainty becomes more important when we probe large angular scales ($\propto \sqrt{2/(2\ell + 1)}$), where only a few independent sky regions ($2\ell + 1$) may be compared to each other to yield the power spectrum coefficients. Current data at low ℓ's are already limited by cosmic variance (see Fig. 9.1). To overcome this limitation, it would be necessary to gather observations of the CMB from several observers distributed at cosmological distances from each other – a possibility that appears unthinkable. It is possible, therefore, that the answer to our big question on the finiteness or infinity of the universe is hidden forever inside this kind of fundamental cosmic uncertainty.

In principle, this limitation might be partly overcome with extremely precise measurements of the Sunyaev-Zel'dovich (SZ) effect[18] on large samples of clusters of galaxies at high redshift (Kamionkowski and Loeb 1997). The basic idea is that SZ polarized scattering on distant clusters is sensitive to the CMB field as seen by the cluster; therefore, in principle it can give information on the properties of the surface

---

[18] The Sunyaev-Zel'dovich effect is inverse Compton scattering of CMB photons off hot electrons in the gas of clusters of galaxies. Scattered photons are boosted in energy and their spectrum is distorted so that, in the solid angle subtended by a cluster, we observe a decrease of the CMB temperature at low frequencies and an increment at high frequencies (for a review see, e.g., Rephaeli 1995).

of last scatter as seen by a source at cosmological distance from us. The polarization of the CMB photons scattered by the cluster's hot electron gas provides a measure of the CMB quadrupole moment as seen by the cluster. Therefore, CMB polarization measurements toward several clusters would probe the anisotropy on a variety of realizations of primordial fluctuations, thus reducing the cosmic variance uncertainty. However, the feasibility of such observations to the required accuracy is questionable, certainly far from what is currently achievable. More fundamentally, this approach would lead to a *reduction* of cosmic variance, not to its suppression. Even assuming that we can reach the limits of just-formed clusters ($z \cong 10$), we would still explore a limited portion of space. Perhaps future methods, independent of the CMB or BAOs or supernovae, will measure the curvature to very high precision. However, cosmic variance limitations will necessarily surface as a consequence of the limited information we can obtain from a single location in the universe. We are bound within our cosmic horizon, set by the distance covered by light in the finite age of the universe.

Let's now assume that somehow, in spite of the fundamental limits just described, we *know* that space is *exactly* Euclidean, or it has a negative curvature ($k = 0$ or $k = -1$). A common misconception in cosmology is to state that this directly leads to infinite spatial sections. However, this is true only if two conditions are met. First, we must assume that the FLRW metrics – a good approximation in our past light cone – is maintained indefinitely beyond our cosmic horizon. This is far from obvious, and it is in competition with other conjectural scenarios. Some inflationary models, for example, postulate a highly inhomogeneous superhorizon distribution with a variety of domains, possibly with different local metrics and sets of fundamental parameters. However, both the multidomain inflation picture and the all-encompassing FLRW universe ultimately elude observational verification.

In addition, spaces with null or negative curvature are infinite *only if* they have a simply connected topology. In more complex topologies, such as a 3-torus, perfectly Euclidean spaces have spatially finite solutions (Ellis 1971). This can happen in a variety of situations: in three dimensions there are eighteen locally homogeneous and isotropic Euclidean spaces; for negative curvatures the number of possibilities is infinite. No known theory provides a prediction of the topology of the universe; thus, only observations may be able to decide. A finite dodecahedral Euclidean space was proposed to explain the apparent lack of power at low multipoles in the CMB anisotropy spectrum (Luminet et al. 2003).[19] Given that the scale of perturbations cannot exceed the size of the polyhedron itself, a cutoff in wavelengths of density fluctuations is expected, which would explain the observed low power at large scales. Observational opportunities to test cosmic topology include the observation of multiple images of high-redshift objects, prediction of total density parameters associated with particular models (e.g., $\Omega_0 = 1.013$ for the Poincaré dodecahedral space), and matching features in the CMB sky. The latter has not been confirmed by *WMAP* and will be tested by the *Planck* survey. Of course, observational verification may happen only if we are lucky and the fundamental domain has an appropriate size: if such a domain is much larger than our Hubble volume, then the (whole) universe might be perfectly Euclidean (or hyperbolic) and yet finite, but we would never know.

---

[19] The low power at low multipoles (first point at low multipoles in Fig. 9.2) is confirmed also in the most recent *WMAP* data.

Let's now assume that we do live in a simply connected universe. Let's suppose that future experiments will be sufficiently precise to detect a slight deviation, either positive or negative, from the Euclidean space ($k \neq 0$, $\Omega_0 \neq 1$). Indeed, that would be a great discovery. However, even in that case we would not be able to draw strong conclusions on the universe as a whole. A small curvature might be a "local" effect rather than a global feature – just as a valley is a local concave shape on the overall spherical surface of a planet. Our view of the universe is similar to that of a sailor in the middle of the ocean with a visibility limited by mist at large distance (resembling the last scattering surface). Waves in the real ocean are typically at a given length; this preferential scale may correspond to the size of galaxies in the real universe. However, CMB observations tell us that in the early cosmic plasma, waves were present on all scales with nearly constant amplitude (the so-called Harrison-Zel'dovich scale-invariant spectrum). This means that even on the largest dimensions we do expect deviations from the unperturbed scenario.[20] It's as if our sailor saw waves of all possible lengths, including some so extended that they encompass the entire horizon and beyond. If the sailor happened to be in the valley of a very long cosmic wave, a precision measurement of the curvature of the water surface would lead him to the erroneous conclusion that the earth is negatively curved or flat and infinite in extent.

It is clear that strong claims on the infinity of space based on the apparent spatial flatness of the observable universe might be as naive as those ancient ideas on the flatness of the earth based on local and inaccurate observation, but with an important difference: in the case of cosmic space, the ambiguity would remain even with infinitely accurate data. This is because our cosmic horizon, unlike the visibility horizon on the earth surface, is a *fundamental* boundary (Barrow 1999; Ellis 2006) not surmountable with improved instruments or more refined theory.

9.5 Infinity Is Not Enough

Even if we assume the flat infinite universe, we are still confronted with the question of why something like life (here and now at least) is actually happening. Even though any system with nonzero probability would occur infinitely often in such a universe, this doesn't mean that anything can happen: the possibility of something happening depends on the ingredients and the physical laws of that universe. A lattice structure repeating itself in all directions, such as that pictured by Maurits C. Escher in his famous *Cubic Space Division* (Escher 1952), makes an infinite universe with a very sharp – and low – limit to complexity. In an infinite FLRW universe, complexity and life would not appear unless the physics has highly specialized characteristics. In fact, it turns out that many basic parameters in our universe, such as the value of coupling constants, the mass and charge of elementary particles, the rate of cosmic expansion, the amplitude of density perturbations in the early universe, the number of space and time dimensions, and the form of physical laws, are extremely sensitive to life-supporting conditions. In fact, their measured values turn out to be precisely tuned for complex structures and life to emerge (Barrow and Tipler 1986).

---

[20] These largest fluctuations have not been processed during the history of the universe because they are outside the causal horizon.

Generalized versions of physical infinities have been invoked to mitigate this impression of a cosmic interconnectedness oriented toward life (Rees 2003). Various scenarios have been proposed in which selection effects are a key to explaining the apparent peculiarity of our cosmic setup. Interestingly, in this attempt many have been willing to give up the long-lived cosmological principle and to go nearly to its opposite: from infinite uniformity to limitless diversity. Starting from different theoretical standpoints (Linde 1994; Giulini et al. 1996; Kallosh and Linde 2003), recent speculations have proposed that our universe could be regarded as one of an infinity of parallel universes, causally disconnected from each other, characterized by different realizations of those parameters and properties that locally we perceive as life-encouraging. Thus, the fine-tuning issue would be reduced to an anthropic selection effect by observers in the multiverse. Although interesting, these ideas are problematic from several points of view (Ellis, Kirchner, and Stoeger 2004), including their intrinsic difficulty to undergo empirical verification.

If taken too far, infinite multiverse scenarios lead to rather embarrassing paradoxical situations (Bersanelli 2005). It has been argued that the ultimate form of multiverse is one in which every subuniverse is identified as a mathematical structure possessing an actual physical existence (Tegmark 2004). In accepting this view, one should realize that, even in this most general case, a particular criterion has been assumed to define existing universes within the infinite multiverse, that is, the requirement of being a "mathematical structure." This would mean that a particular capability that the brain of our species has achieved through biological evolution, such as the development of the mathematical language, is believed to define what does or does not exist at the multiverse level. We end up with a picture of reality that is dangerously similar to a materialized projection of the set of logical possibilities of our human mind. This may appear even more rigidly *homo-centered* than the anthropic flavor that multiverse speculations seek to remove.

How would we judge these ideas from their aesthetic angle? Science, of course, is not driven by our own taste, and eventually it is the persistent reality of the facts, gathered by careful and repeated observations,that prevails and defeats any undue prejudice we may have tacitly nursed. On the other hand, it is also clear that physicists are motivated and guided in their work in some important way by their aesthetic perception. Cosmologist Mario Livio quite correctly notes (Livio 2000) that although the central role of aesthetics in fundamental science is normally not explicitly recognized, in practice it is adopted wholeheartedly by physicists, suggesting an underlying "cosmological aesthetic principle." Aesthetic guidance may be particularly relevant in scarcely constrained situations that leave ample room for conjectures and speculation, such as the one we are discussing here. Although the evidence of a deep aesthetic component in scientific research is compelling, we are, of course, left with the challenge of clarifying what we mean by beauty in this context. Most physicists identify symmetry, effectiveness, and simplicity as some of the key words defining the canons of scientific beauty.[21] It is also rather conventional to include various forms of the

---

[21] See, for example, Chandrasekhar (1990). A number of works have documented the importance of aesthetics in the scientific creativity of most of the great scientists (see, e.g., Beveridge 1950; Bersanelli and Gargantini 2009).

generalized Copernican or mediocrity principle as an aesthetic element that a credible scientific theory is expected, or even required, to exhibit. However, I think that this aspect of the debate deserves some attention. Pushing too much on the mediocrity ideal of nature is questionable from the aesthetics point of view, and it may hide a pitfall for a sound attitude toward knowledge. The ultimate mediocre universe is indeed maximally symmetric and infinitely simple, but it is also absolutely uninteresting. Things don't get better in plenitude multiverses, such as those described earlier, in which "every thing that can exist, does exist"; although at first sight they may appear to offer a most rich and diverse reality, they can also be seen as ultimately boring and featureless. Nothing really happens in a world where everything always goes on infinitely often.

Both the strong cosmological principle and the plenitude universe paradigms seem to lead to a rather poor aesthetic appeal. Perhaps the reason is that, although for opposite reasons, such models are ill-assorted with the concepts of rareness and uniqueness, which have deep aesthetic significance on their own. Aesthetics has its requirements. It seems that spatial infinity, in order to be perceived as a fascinating concept, has to maintain some kind of element of selected variety and genuine surprise. Perhaps a new theory will turn out to include some of these aspects, apparently lost in our present attempts to describe a global vision of the universe. However, as we shall see, things become more clear and interesting when we look at the aesthetic content of the observable universe.

9.6 Infinity and the Heart of Human Nature

Throughout human history, the debate on whether the physical universe is finite or infinite has been intense and changing from epoch to epoch (Barrow 2005; Luminet and Lachièze-Rey 2006). Today, we are still in trouble when we are asked what we mean by "the entire universe," and we may be left with an ineliminable uncertainty about its finite or infinite character. However, modern cosmology has unambiguously answered part of the question, showing beyond doubt that the *observable* universe is finite in a fundamental sense. We also learned that our accessible universe, although finite, has astonishing size. Space is spangled with galaxies, disposed in rather regular structures of filaments and voids. At the center (by definition) of the observable spacetime is ourselves, an infinitesimal and mysterious fragment of the whole. We have all reasons to believe that any other cosmic observer, located anywhere in space, would have roughly the same cosmic view. At the present cosmic epoch, from here or from any other cosmic location, about 100 billion galaxies are visible.

The vastness of cosmic space, in addition to being investigated by scientific cosmology, is suggestive of a notion of infinity that is beyond the physical sciences and pertains to the heart of human existence. The fact that our deepest desire is made for "something infinite" is deeply rooted in human nature. Fyodor Dostoevsky expressed the profound need of an "*infinitely great*" as a very condition for human existence: "The mere presence of the everlasting idea of the existence of something infinitely more just and happy than I, already fills me with abiding tenderness and – glory – oh, whoever I may be whatever I may have done! To know every moment, and to believe that somewhere there exists perfect peace and happiness for everyone and for everything, is much more important to a man than his own happiness. The whole law of human existence consists merely of making it possible for every man to bow down before what is infinitely great. If man were to be deprived of the infinitely great, he would refuse to go on living, and die of despair" (Dostoevsky 1872).

Clearly, the "infinitely great" that Dostoevsky mentions is not merely an endless expanse of space or material. We all realize that Dostoevsky's "infinitely great" is something of a different nature than any measurable physical quantity. It is not a very large number, say, the number of elementary particles in the universe, or even the infinite set of elementary particles in the infinite universe; nor is it a mathematical infinity larger than any physical counterpart. This "infinitely great" speaks of our deepest longing for happiness, of our ultimate need for forgiveness and peace. In fact, even those infinite arrays of infinite worlds postulated by multiverse theories would be way too narrow to satisfy the extent of human aspiration. The "Infinity" that the human heart longs for cannot be filled by any endless amount of space, time, or matter. In the words of Italian poet Giacomo Leopardi,

> ... the inability to be satisfied by any worldly thing or, so to speak, by the entire world. To consider the inestimable amplitude of space, the number of worlds and their astonishing size – then to discover that all this is small and insignificant compared to the capacity of one's own mind; to imagine the infinite number of worlds, the infinite universe, then feel that our mind and aspirations might be even greater than such a universe; to accuse things always of being inadequate and meaningless; to suffer want, emptiness, and hence ennui – this seems to me the chief sign of the grandeur and nobility of human nature.
>
> (Leopardi 1981)

The physical universe, however infinite and diverse, does not seem to fulfill the yearning of a single fragile human being. Again Dostoevsky commented that the bee knows the secret of its beehive, the ant knows the secret of its anthill, but man does not know his own secret. The fact that we understand each other when we speak of the "infinitely great" signals that indeed all human beings are structured with a fundamental capacity for such infinity. If relationship with the infinite is indeed part of the very structure of human beings, then we expect that there will be aspects of human existence that cannot be constrained within a limited perimeter. It is interesting to explore the dynamism of various levels of human experience in which infinity appears as a clear feature.

Human desire, first of all, cannot be locked up within any finite measure. It "bursts through the walls of any place within which one would want to restrain it" (Giussani 1997, p. 79). The human heart is an aspiration to Infinity, as it is apparent when considering the impossibility of our desire to be completely fulfilled by any finite achievement. Similarly, the human mind is continuously open to an ultimate meaning. No partial answer will really satisfy the quest for deeper explanation. Scientific inquiry is a wonderful example of how human reason, after every finite achievement, is relentlessly struggling to reach beyond the acquired knowledge. But perhaps the personal experience in which reference to an Infinity is most undisputable is love:[22] the boundless intergalactic space becomes like nothing compared to the true love for a single

---

[22] The recent Encyclical Deus Caritas Est by Pope Benedictus XVI describes with unprecedented depth the dimensions of love.

person. A human act of free donation or forgiveness stands in front of the infinity of the universe. Blaise Pascal has acutely expressed the presence of dimensions in human nature (mind, charity) that are not reducible to any physical reality: "All bodies, the firmament, the stars, the earth and its kingdoms, are not equal to the lowest mind; for mind knows all these and itself; and these bodies nothing. All bodies together, and all minds together, and all their products, are not equal to the least feeling of charity. This is of an order infinitely more exalted. From all bodies together, we cannot obtain one little thought; this is impossible and of another order. From all bodies and minds, we cannot produce a feeling of true charity; this is impossible and of another and supernatural order" (Pascal 1660).

Now we ask, what is the most adequate language to speak of the kind of infinity that we have named "metaphysical"? In the first part of this chapter, we have used the language of physics: we spoke of density, curvature, triangles, and so forth, to discuss the infinity of physical space. That was a powerful language in that context. However, when dealing with infinity as a component of human experience, mathematics and physics are not of great help. Poetry and art seem to provide a more adequate language. The aesthetic perception of infinity has inspired a great amount of literature and artworks. The beauty of many natural phenomena and the vastness of the cosmic scenery are evocative of a metaphysical infinity, so that artists often take inspiration from them. What are the features adopted in artistic expressions that have proved most effective in conveying a sense of infinity?

9.7 The Art of Immensity

The most elementary representation of infinity is probably in the idea of repetition. Even in prehistorical rock carving and paintings we find remarkable examples of repeated symbols and patterns, sometimes representing stars or lunar phases.[23] This testifies to the early emergence of the idea of "something" that may be reproduced indefinitely, perhaps foreshadowing the concept of number. Architectural designs worldwide have played with the idea of repetition as an invitation to infinity, as it can be admired in buildings as diverse as the Parthenon in Athens or the Registan in Samarkand, as Michelangelo's dome of St. Peter in Rome or the Shinto Shrine in Miyajima Island in Japan. The geometric sophistication of Islamic art has probably reached the height of richness of self-reproducing patterns. Western modern art has also exploited the same idea. Several paintings by Escher, such as *Depth*, for example, (Escher 1955), or some of the works by Andy Warhol, representing interminable ranks of Coca Cola bottles (Warhol 1962a) or of Marilyn Monroe's lips (Warhol 1962b), seem to propose an unscrupulous technique to evoke a moment of infinity. However, although limitless reproduction of a pattern does introduce some idea of infinity, its communication potential is rather limited. From an aesthetics standpoint, the infinite homogeneous universe, reproducing countless copies of anything you like or dislike, may suffer from the same weakness.

---

[23] Some of the most striking examples are the drawings discovered in 1940 in the caves of Lascaux, France, painted around 18,000 BC by the Cro-Magnon man. See also Barrow (1995).

Interestingly, although the thought of an infinite uniform physical space is rather mute to our imagination, the view of a vast natural landscape (the sight of the night sky, but also an ocean, a mountain range, a desert) may be capable of powerfully evoking a sense of metaphysical infinity. That's where art finds inspiration. German painter, sculptor, and poet Caspar David Friedrich, in his famous *The Wanderer above a Sea of Mists* (Friedrich 1818), expresses in an imposing way a perception of a limitless and attractive ultimate reality. The gaze of the human figure, at the center of the painting, is invisible but perceived as irresistibly aimed at and beyond the horizon. The farthest mountains and peaks are degrading with an element of vagueness, crossed by distant clouds and haze. In his *Monk by the Sea* (Friedrich 1809), Friedrich exacerbates the same situation, proposing an undefined horizon that paradoxically adds force to the presence of an impending and unreachable mystery. Just as in Dostoevsky's and Leopardi's words, in these paintings the experience of immensity that is evoked is not reducible to the extension of more land beyond the reach of our view, but rather suggests a reality that surpasses the very notion of a measure.

In many works by Giovanni Segantini, such as *Natura* (Segantini 1896–1899), we feel dominated by a vast horizon; the human presence is disproportionately small in the landscape, and yet somehow central in the scene. In Vincent van Gogh's *Avenue of Poplars at Sunset* (van Gogh 1884), we see the effect of a combination of the repetitive double line of trees, the marked line of the horizon backlit by the sun, and the human figure, once again seen from the back, whose shadow renders his direct relation with the sky. The famous *Angelus* by Jean François Millet (1860) is a powerful demonstration that the Infinite to whom the prayer of the countrymen is addressed is not the physical horizon surrounding the vast fields, highlighted by the pink-orange shading of the sunset; however, that physical horizon clearly represents a sign of that Infinite Reality to whom the prayer is directed.

If from painting we move to poetry, we should consider again Giacomo Leopardi, known as "the poet of infinity." Perhaps his best known piece of poetry, precisely entitled *Infinity*, introduces the intriguing aesthetic invention of the "hedge":

> Always dear to me was this solitary hill,
> And this hedge, which from so great a part
> Of the farthest horizon excludes the gaze.
> But sitting and gazing, boundless
> Spaces beyond that, and superhuman
> Silences, and profoundest quiet
> I in my mind imagine (create); wherefore
> The heart is almost filled with fear. And as
> I hear the wind rustle through these plants, that
> Infinite silence to this voice
> I go on comparing: and I recall to mind the eternal,
> And the dead seasons, and the present
> And living one, and the sound of it. So in this
> Immensity my thought is drowned
> And the shipwreck is sweet to me in this sea
>
>   (see Brose 1983)

In the brilliant poetic intuition of Leopardi, the presence of a hedge partly impeding a direct view of the ultimate landscape, does not diminish but rather enhances the impression of an underlying vastness. Perhaps the presence of a hedge somehow helps

Figure 9.2. Edward Hopper (1882–1967), *Cape Cod Morning* (1950), 86.7 × 101.9 cm. Natural Museum of American Art, Washington, D.C., USA.

in moving our perception from a physical infinity (just endless space) to a metaphysical infinity (a mysterious but real presence to which our being is outstretched). Because we are more profoundly attracted by a metaphysical than a physical infinity, Leopardi's hedge adds considerable aesthetic power to his verse. In fact, the poetic force of this image does not depend on whether we assume that what lies beyond the hedge is an immense finite territory or an actually infinite space: now our attention has been pointed to an Infinite of a deeper nature.

We can see the same concept at work in some modern paintings. In Edward Hopper's *Cape Cod Morning* (Hopper 1950), for example, we find almost literally depicted a hedge that "from so great a part of the farthest horizon excludes the gaze" (Fig. 9.2). We do not see the horizon, hidden by the plants in the background, but this conveys an even stronger sense of the vastness through the presence of the human figure, behind the windows, looking at infinity into a limpid sunrise. We reflect ourselves in the gaze of that woman: perhaps we see more clearly infinity through the eyes of another than through our own eyes. Equally impressive is Hopper's *Rooms by the Sea* (Hopper 1951), in which the "hedge effect" reaches an apex: the portion of unconfined sea just glimpsed through the open door attracts all our attention and imagination, with a force that is incomparable to what we would have with an open view on the sea as a whole.

Now we can go back to our initial question: what contribution can our scientific vision of the universe give to our perception of infinity as an ultimate reality? Our quick surveys on infinity from the scientific and artistic angles may suggest a clue.

We have seen some examples of how artistic language can be effective in expressing metaphysical infinity. Repetition of patterns is a most basic feature alluding to infinity, a technique largely exploited in architecture and in ancient as well as modern art. Representations of a vast horizon, with a purposely introduced element of indefiniteness, are a potent rendering of infinity. The idea of a "hedge," of an obstruction that partly veils the direct vision of what lies beyond, can be sapiently used to magnify, rather than to inhibit, the perception of an ineffable, ultimate mystery. In many cases, the presence of a human figure, small and apparently insignificant, incarnates the paradoxical nature of the relationship between the ephemeral conscious creature and the "infinitely great." How does the structure of our physical universe appear to us according to the present scientific understanding? I claim that the salient characteristics of the physical universe as unveiled by science foreshadow all of the aforementioned aesthetic elements used in artworks to express an Infinity.

Cosmology has shown to us the vertiginous immensity of cosmic space – similar to but much larger than Segantini's alpine views. The depths of the universe are filled with rich structures in a rather regular pattern resembling a carefully manufactured Persian carpet. These structures repeat themselves through billions of light-years, most significantly on the scales of stars, galaxies, and clusters of galaxies. Nonetheless, we have a finite horizon, a cosmic hedge in spacetime, an ultimate veil that "from so great a part of the farthest horizon excludes the gaze." This last curtain of the cosmic theater also has an element of indefiniteness: it is not sharp and fixed, but rather expands in time, and it treasures the secret of whether the entire picture is physically finite or infinite – a secret that we may have to accept as forever hidden to our inquiry. Then we find, in our physical world, the last decisive element: the human presence, so fragile and apparently insignificant in the cosmic picture, and yet a crucial term of a universe that thinks itself. Man is the realization of a reality that is capable of the infinite. The vast, perhaps infinite universe does not identify itself with the "infinitely great"; rather, it is perceived as a sign of it. The physical universe thus may be seen as an icon of that ultimate Infinite that we call God.

9.8 What Is Man?

Different religious traditions have express in a variety of ways their images of the divine. In the Judeo-Christian tradition, infinity is a key attribute of God. As all creatures, human beings are finite and fleeting, yet God created man "in his own image, in the image of God he created him,"[24] so that he shares something of the infinity of God. This direct relationship of every single human being with the infinity of God provides a foundation for the unlimited value of each human person, irrespective of his/her status, power, health, or intellectual capability: "In only one case is this point . . . , [the] single human being, free from the entire world, free, so that the world together and even the total universe cannot force him into anything. In only one instance can this image of a free man be explained. This is when we assume that this point is not totally the fruit of the biology of the mother and father, not strictly derived from the biological tradition

---

[24] Genesis 1:27.

of mechanical antecedents, but rather when it possesses a direct relationship with the *infinite*, the origin of all of the flux of the world . . . that is to say, it is endowed with something derived from God" (Giussani 1997, p. 91).

In a Judeo-Christian understanding there are various manifestations of God's infinity. In the beginning,[25] His infinite power is expressed in his being the Creator of the universe ("maker of heaven and earth . . . of all things visible and invisible"[26]). Creation is not to be limited to a "start-up" action, taking place in some remote time in the past, in a Big Bang or some other way. As St. Augustine (354–430 AD) suggested, we should rather think of time itself as a creature, that is, as part of the creation of the universe: "[I]f any excursive brain rove over the images of forepassed times, and wonder that Thou the God Almighty [. . .] didst for innumerable ages forbear from so great a work, before Thou wouldest make it; let him awake and consider, that he wonders at false conceits. For whence could innumerable ages pass by, which Thou madest not? [. . .] or what times should there be, which were not made by Thee? or how should they pass by, if they never were?" (O'Donnell 1992).

Thus, creation is primarily the ontological, radical dependence of all creatures from the Creator. He pulls into being all things, as well as all instants of time and of cosmic and human history. His power fills the infinite gap between being and nothingness.

Creation is God's free participation of existence to every creature: "I have loved you with an everlasting love."[27] The same power that draws us out of nonexistence reveals itself in His compassion for our imperfection and meanness. His love is stronger than our betrayal: "Can a mother forget the baby at her breast, and have no compassion on the child she has borne? Though she may forget, I will not forget you!"[28] Here we find another situation in which the infinity of God is directly at work: forgiveness. Whereas creation is the infinite power of calling things out of nothingness, mercy and forgiveness turn something-that-is (evil, sin) into nothingness (Giussani, Alberto, and Prades 1998). Only an infinite power can do this.[29]

Perhaps the most extreme manifestation of the infinite power of the Judeo-Christian God is incarnation. The Christian event is the claim that the Infinite God has entered the finite human history. The unreachable, hidden God has taken the form of a human presence, within space and time: Jesus of Nazareth, "born of a woman."[30] In a sense, incarnation might be seen as the most unexpected realization of infinity in a physical reality: the highest of all infinities, God, coincides with a finite being, the man Jesus.

As far as we know, the Jewish were not particularly advanced in systematic observations of the sky, and their astronomical knowledge was mainly inherited from he Babylonians. However, they were deeply touched by the immensity and beauty of the firmament, and often, in the Bible, the sky is mentioned as a privileged sign of the infinite power of God. It is interesting to revisit some of these examples, showing how

---

[25] Genesis 1:1.
[26] Nicene Creed (325 AD).
[27] Jeremiah 31:3.
[28] Isaiah 49:15.
[29] Luke 5:23–24.
[30] Galatians, 4:4.

our modern scientific understanding of the universe provides some resonances with the biblical language that amplify our appreciation of the infinity of God.

The evidence of a disproportion between our human nature and the cosmic realm was vividly recognized. Hence, in the words of the prophet, the "height" of the universe is used to express the incommensurable depth of God's vision compared to our own: "As the heavens are higher than the earth, so are my ways higher than your ways, and my thoughts than your thoughts."[31] Our modern awareness of the size of the universe amplifies the power of this image and may consolidate an awareness of our humble condition in front of the mystery of being.

One is often reminded of the fragility and transience of human existence in the Old Testament.[32] There is a clear awareness that man is almost nothing in the universe: "When I consider your heavens, the work of your fingers, the moon and the stars, which you have set in place, what is man that you are mindful of him, the son of man that you care for him?" Yet this apparently marginal creature shares something of the infinity of the creator: "Yet you made him little lower than God and crowned him with glory and honor."[33] Small as we are in the cosmic scene, each human being is a realization of the self-awareness of the universe. We are given the "glory and honor" to be placed in a relation with everything that exists. Fleeting as he is, man was made able to grasp and enjoy the panorama of creation, with its beauty and vertigo, and to search and host the meaning of all things.

The God of the Bible is the creator of a purposeful universe. Every creature is part of a great design in which life and human existence have a central role. Interestingly, in the Jewish tradition we find hints of an intuition that the entire universe, not just the earth and its benign environment, is beneficial for life and our existence: "He stretches out the heavens like a canopy, and spreads them out like a tent to live in."[34] It is a remarkable anticipation of the modern discovery of deep anthropic interconnections in nature, necessary for life to exist, not only in the direct local environment but also at an astronomical and fundamental level.

A characteristic trait of the Jewish religious heritage is a clear distinction between *Creator* and *creation*. It has been pointed out (Hodgson 2002) that this notion, incorporated in Christian theology and developed in medieval Western Europe, is probably a key intellectual prerequisite for the birth of modern science. God's transcendent and yet caring infinity is often expressed making use of the comparison with the physical universe, but always without confusion between the two. The Bible insists that man does not belong to any finite creature, but only to the Infinite who made them: "When you look up to the sky and see the sun, the moon and the stars – all the heavenly array – do not be enticed into bowing down to them."[35] In the book of Wisdom the confusion between the finite creature and the infinite Creator is pointed out as a sad mistake: "Surely vain are all men by nature, who are ignorant of God, and could not out of the good things that are seen know him that is: neither by considering the works did they

---

[31] Isaiah 55:9.
[32] Genesis 3:19.
[33] Psalms 8:3–5.
[34] Isaiah 40:22.
[35] Deuteronomy 4:18.

acknowledge the workmaster; but deemed either fire, or wind, or the swift air, or the circle of the stars, or the violent water, or the lights of heaven, to be the gods which govern the world. With whose beauty if they being delighted took them to be gods; let them know how much better the Lord of them is: for the first author of beauty hath created them. But if they were astonished at their power and virtue, let them understand by them, how much mightier he is that made them. For by the greatness and beauty of the creatures proportionably the maker of them is seen."[36]

In the Psalms, the stability of the heavens is used to signify the infinite faithfulness and eternity of God.[37] The vastness and beauty of the firmament are a sign of the infinite power and kindness of the Creator.[38] Yet the universe in its majesty is perceived as non-absolute reality: cosmic ages are short-lived,[39] no space is hidden to God.[40] Here we most clearly encounter the unique cosmological contribution introduced by the Jewish culture: their unprecedented vision of a contingent universe created by a unique, personal God. Probably no other ancient culture, before modern science, has had such a clear conception of the universe as a contingent reality. Any natural phenomena, including the firmament, are metaphoric images of God's infinite power: all creatures are ephemeral so that nothing, not even the whole universe, is self-standing, absolute, and unchangeable: "The heavens will vanish like smoke, the Earth will wear out like a garment . . . But my salvation will last forever."[41] It is most impressive to consider that Jesus, suggesting his own divine nature, used an echo of the same humanly impossible statement: "Heaven and earth will pass away, but my words will never pass away."[42]

9.9 Epilogue

The impression of vastness of the night sky is probably our most immediate and fascinating image of infinity. Since ancient times the question of whether physical space is infinite or finite has stirred up human intelligence. One of the most ambitious objectives of contemporary cosmology is to address this question in terms of measurable properties of the universe. A first remarkable and conclusive answer supported by strong empirical and theoretical evidence is that our observable universe is finite and expanding in time. When we try to say something about the "universe as a whole," we look at spatial curvature, which is determined by the average energy density in the universe. Recent data from the CMB, distant supernovae, and large-scale structures indicate that the energy density parameter is within ∼1 percent of the critical density that characterizes a Euclidean geometry. This is in agreement with the expectations of inflation theories, to be tested with future CMB experiments, that postulate an early rapid expansion that would have stretched any initial curvature of space.

[36] Wisdom 13:1–3.
[37] Psalms 136.
[38] Isaiah 55:9.
[39] Psalms 90:4.
[40] Psalms 139:8.
[41] Isaiah 51:6.
[42] Mark 13:31.

A Euclidean space, for simple topological realizations, is infinite. New observations expected in the near future may improve our understanding and possibly provide new breakthroughs. Several reasons, however, lead us to conclude that questions regarding the "entire universe" are ultimately undecidable. In particular, we have shown that the statement that the universe is spatially flat is unprovable.

Even assuming an ideal measurement, fundamental uncertainties (cosmic variance, superhorizon fluctuations) arise. Furthermore, flatness does not necessarily imply an infinite space, if the topology is not simply connected.

Sometimes, aesthetic reasons are advocated in support of various forms of Copernican principles in cosmology. Similar arguments are used, more or less implicitly, in connection with multiverse visions that saturate the space of all physical possibilities (plenitude). On the same aesthetic grounds, however, we note that these scenarios require us to abandon concepts such as rareness, uniqueness, and unexpectedness, whose aesthetic significance appears undisputable and perhaps necessary. We need a concept of infinity in which genuine novelty can happen.

The notion of infinity of spacetime, or of other physical quantities, does not coincide with the Infinite as the ultimate object of human aspiration and desire. We have identified a number of features used in modern artistic works to express metaphysical infinity. We have pointed out that our view of the universe, as it emerges from current astrophysical observation, is characterized by many of those features that convey aesthetic force to infinity in artworks. In this view, our physical universe may be seen as an ultimate metaphor of the Infinite.

The Judeo-Christian tradition suggests several expressions of infinity as an attribute of God and as a capacity of human nature. The belief that each human being has a direct relationship with the Infinite provides a solid foundation to the irreducible dignity of the individual, a foundation that is hard to maintain otherwise. This is reflected in the fact that infinity is an ineradicable element of the most significant dimensions of human experience, such as desire, reason, and love. Modern science provides a view of the physical universe that may enrich our appreciation of the infinity of God as expressed in the biblical language. The Christian image of the divine is the fascinating reality of a God who becomes involved within the finite horizon of space and time, sharing the drama of human existence.

References

Albrecht, A., and Sorbo, L. 2004. Can the universe afford inflation? *Physical Review D* 70: 06352.

Astier, P., et al. 2006. The Supernova Legacy Survey: measurement of ΩM, $\Omega \Lambda$ and w from the first year data set. *Astronomy and Astrophysics* 447: 31.

Barrow, J. D. 1995. *The Artful Universe*. Oxford: Clarendon.

Barrow, J. D. 1999. *Impossibility. The Limits of Science and the Science of Limits*. London: Vintage.

Barrow, J. D. 2005. *The Infinite Book*. London: Jonathan Cape.

Barrow, J. D., and Tipler, F. J. 1986. *The Anthropic Cosmological Principle*. Oxford: Clarendon.

Beckwith, S. V. W., et al. 2006. The Hubble Ultra Deep Field. *Astronomical Journal* 132: 1729–55.

Bennett, C. L., et al. 2003. First-year Wilkinson Microwave Anisotropy Probe (WMAP) observations: Foreground emission. *Astrophysical Journal Supplement* 148: 97.

Bersanelli, M. 2005. An echo of ancient questions from contemporary cosmology. In *100 Perspectives on Science and Religion*, Charles L. Harper, Jr. (ed.), pp. 121–26. Philadelphia and London: Templeton Foundation Press. Publication in honor of Sir John Templeton.

Bersanelli, M., and Gargantini, M. 2009. *From Galileo to Gelmann: The Wonder that Inspired the Greatest Scientists of All Time In Their Own Words*. Philadelphia and London: Templeton Foundation Press.

Bersanelli, M., Maino, D., and Mennella, A. 2002. Anisotropies of the cosmic microwave background. *La Rivista del Nuovo Cimento* 25: 1.

Beveridge, W. I. B. 1950. *The Art of Scientific Investigation*. London: William Heinemann.

Bousso, R. 2006. Holographic probabilities in eternal inflation. *Physical Review Letters* 97: 191302.

Boyle, L. A., Steinhardt, P. J., and Turok, N. 2006. Inflationary predictions for scalar and tensor fluctuations reconsidered. *Physical Review Letters* 96: 111301.

Brose, M. 1983. Leopardi's "L'infinito" and the language of the romantic sublime. *Poetics Today* 4:1, 47–71.

Chaboyer, B., and Krauss, L. M. 2002. Theoretical uncertainties in the subgiant mass-age relation and the absolute age of ω Centauri. *Astrophysical Journal* 567: L45.

Chandrasekhar, S. 1990. *Truth and Beauty. Aesthetics and Motivation in Science*. Chicago: University of Chicago Press.

Cole, S., et al. 2005. The 2dF Galaxy Redshift Survey: Power-spectrum analysis of the final data set and cosmological implications. *Monthly Notices of the Royal Astronomical Society* 362: 505.

Dante Alighieri. *Inferno* XXXIV 139; *Purgatorio* XXXIII 145; *Paradiso* XXXIII 145. Commented by Attilio Momigliano (1974). Firenze: Sansoni Editore.

Dostoevsky, F. 1872. *The Devils*, p. 656. *Translated by David Magarshack* (1971). Middlesex: Penguin Books.

Dunkley, J., et al. 2005. Measuring the geometry of the universe in presence of isocurvature modes. *Physical Review Letters* 95: 261303.

Dyson, L., Kleban, M., and Susskind, L. 2002. Disturbing implications of a cosmological constant. *Journal of High Energy Physics* 0210, 011., vol. 10, n. 11.

Eisenstein, D. J., et al. 2005. Detection of the baryon acoustic peak in the large-scale correlation function of SDSS luminous red galaxies. *Astrophysical Journal* 633: 560.

Ellis, G. F. R. 1971. Relativistic cosmology. In *General Relativity and Cosmology, Proceedings of the XLVII Enrico Fermi Summer School*, R. K. Sachs (ed.). New York: Academic Press. Pag. 104.

Ellis, G. F. R. 2006. *Issues in the Philosophy of Cosmology, Handbook in Philosophy of Physics*, J. Butterfield and J. Earman (eds.). Amsterdam: Elsevier.

Ellis, G. F. R., Kirchner, U., and Stoeger, W. R. 2004. Multiverses and physical cosmology. *Monthly Notices of the Royal Astronomical Society* 347: 921.

Escher, M.C. 1952. *Cubic Space Division*, 27 × 26.5 cm, *Cordon Art*, Baarn, Holland.

Escher, M. C. 1955. *Depth*, 32 × 23 cm. *Cordon Art*, Baarn, Holland.

Freedman, W. L., et al. 2001. Final results from the Hubble Space Telescope Key Project to measure the Hubble constant. *Astrophysical Journal* 553: 47.

Friedrich, C. D. 1809. *Monk by the Sea*, 110 × 171.5 cm. Gemäldegalerie, Berlín, Germany.

Friedrich, C. D. 1818. *The Wanderer above a Sea of Mists*, 74.8 × 94.8 cm. Hamburger Kunsthalle, Hamburg, Germany.

Garnavich, P. M., et al. 1998. Constraints on cosmological models from Hubble Space Telescope observations of high-z supernovae. *Astrophysical Journal* 493: L53.

Giulini, D., Joos, E., Kiefer, C., et al. 1996. *Decoherence and the Appearance of a Classical World in Quantum Theory*. Berlin: Springer.

Giussani, L. 1997. *The Religious Sense*. Translated by J. Zucchi. Montreal: McGill-Queen's University Press.

Giussani, L., Alberto, S., and Prades, J. 1998. "Simone mi ami tu?" *Tracce* 10 (November), pages 5–6.

Guth, A. H. 1997. *The Inflationary Universe. The Quest for a New Theory of Cosmic Origins*. Reading, MA: Addison-Wesley.

Hinshaw, G., et al. 2007. Three-year Wilkinson Microwave Anisotropy Probe (WMAP) observations: Temperature analysis. *Astrophysical Journal Supplement* 170: 288.

Hodgson, P. 2002. *The Roots of Science and Its Fruits*. London: Saint Austin Press.

Hopper, E. 1950. *Cape Cod Morning*, 86.7 × 101.9 cm. Natural Museum of American Art, Washington, D.C., USA.

Hopper, E. 1951. *Rooms by the Sea*, 73.7 × 101.6 cm. Yale University Art Gallery, New Haven, Connecticut, USA.

Hu, W., and White, M. 2004. The cosmic symphony. *Scientific American* 290: 44.

Hubble, E. 1929. A relation between distance and radial velocity among extra-galactic nebulae. *Proceedings of the National Academy of Sciences* 15: 168.

Hubble, E., and Humason, M. L. 1931. The velocity-distance relation among extra-galactic nebulae. *Astrophysical Journal* 74: 43.

Kallosh, R., and Linde, A. 2003. M-theory, cosmological constant and anthropic principle. *Physical Review D* 67: 023510.

Kamionkowski, M., and Loeb, A. 1997. Getting around cosmic variance. *Physical Review D* 56: 4511.

Komatsu, E., et al. 2009. Five-year Wilkinson Microwave Anisotropy Probe observations: Cosmological interpretation. *Astrophysical Journal Supplement* 180: 330.

Leopardi, G. 1981. *Pensieri*, LXVIII, p. 113. Translated by W. S. Di Piero. Baton Rouge and London: Louisiana University Press.

Linde, A. 1994. The self-reproducing inflationary universe. *Scientific American* 271: 32.

Linde, A. 2007. Sinks in the landscape, Boltzmann brains, and the cosmological constant problem. *Journal of Cosmology and Astroparticle Physics* 0701: 022.

Livio, M. 2000. *The Accelerating Universe*. New York: Wiley.

Luminet, J.-P., and Lachièze-Rey, M. 2006. *Finito o infinito? Limiti ed enigmi dell'universo*. Cortina Raffaello, Milano (Italy).

Luminet, J.-P., et al. 2003. Dodecahedral space topology as an explanation for weak wide-angle temperature correlations in the cosmic microwave background. *Nature* 425: 593.

Millet, J. F. 1860. *Angelus*, 55.5 × 66 cm. *Orsay Museum*, Paris, France.

O'Donnell, James J. 1992. *Augustine: Confessions. Text and Commentary*. 3 vols. Oxford: Oxford University Press.

Pascal, B. 1660. *Pensées*, XII, p. 793. Translated by W. F. Trotter (1938). New York: P. F. Collier & Son.

Penzias, A., and Wilson, R. 1965. A measurement of excess antenna temperature at 4080 Mc/s. *Astrophysical Journal* 142: 419.

Percival, W. J., Cole, S., Eisenstein, D. J., et al. 2007. Measuring the baryon acoustic oscillation scale using the Sloan Digital Sky Survey and 2dF Galaxy Redshift Survey. *Monthly Notices of the Royal Astronomical Society* 381: 1053.

Perlmutter, S., et al. 1999. Measurements of Omega and Lambda from 42 high-redshift supernovae. *Astrophysical Journal* 517: 565.

Rees, M. J. 2003. Numerical coincidences and "tuning" in cosmology. *Astrophyics and Space Science* 285: 375.

Rephaeli, Y., 1995. Comptonization of the cosmic microwave background: The Sunyaev-Zeldovich effect. *Annual Review of Astronomy and Astrophysics* 33: 541.

Richer, H. B., et al. 2004. Concerning the white dwarf cooling age of M4: A reply to De Marchi et al. on "a different interpretation of recent deep HST observations." *Astronomical Journal* 127: 2904.

Riess, A. G., et al. 2005. Cepheid calibrations from the Hubble Space Telescope of the luminosity of two recent Type Ia supernovae and a redetermination of the Hubble constant. *Astrophysical Journal* 627: 579.

Russell, R. J. 2008. The God who infinitely transcends infinity: Insights from cosmology and mathematics. In *Cosmology from Alpha to Omega: Theology and Science in Creative Mutual Interaction*, pp. 56–76. Philadelphia: Fortress Press.

Schindler, S. 2002. Ωm: Different ways to determine the matter density of the universe. *Space Science Reviews* 100 (1/4): 299–309.

Segantini, G. 1896–1899. *Natura*, 234×400 cm. Gottfried Keller-Stiftung, Saint-Moritz, Switzerland.

Smoot, G. F., et al. 1992. Structure in the COBE differential microwave radiometer first-year maps. *Astrophysical Journal* 396: L1.

Spergel, D. N., et al. 2007. Three-year Wilkinson Microwave Anisotropy Probe (WMAP) observations: Implications for cosmology. *Astrophysical Journal Supplement* 170: 377.

Tegmark, M. 2003. Parallel universes. *Scientific American* 288 (5): 30–41.

Tegmark, M. 2004. Parallel universes. In *Science and Ultimate Reality from Quantum to Cosmos*, J. D. Barrow., P. C. W. Davies, and C. L. Harper (eds.). Cambridge: Cambridge University Press. pp. 459–491.

van Gogh, V. 1884. *Avenue of Poplars at Sunset*. Kröller-Müller Museum, Otterlo, Netherlands.

Warhol, A. 1962a. *Coca-Cola Bottles*. New York, USA.

Warhol, A. 1962b. *Marlyn Monroe's Lips*. Smithsonian Institution, Hirshhorn Museum and Sculpture Garden, Washington, D.C., USA.

Wood-Vasey, W. M., et al. 2007. Observational constraints on the nature of dark energy: First cosmological results from the ESSENCE supernova survey. *Astrophysical Journal* 666: 694.

CHAPTER 10

Infinities in Cosmology

Michael Heller

10.1 Introduction

Traditionally, the problem of infinity in cosmology appeared in the form of questions concerning space and time. These questions were notoriously difficult and led to paradoxes. Does space extend to infinity? If so, how can something by definition beyond empirical control be a subject matter of scientific inquiry? If not, what is beyond the edges of space? Similar questions were involved in the time problem: If time had a beginning, what was "before" the beginning? If the age of the universe is infinite, what supplies infinite amounts of energy indispensable to compensate for dissipative losses? Relativistic cosmology did not solve these questions; it only tamed them by elaborating strict mathematical tools, in terms of which they could be expressed while awaiting more empirical data that would help to further elucidate them.

Quite recently, the problem of infinity has emerged in speculations in cosmology on the, so to speak, metalevel. Wanting to neutralize difficult problems related to the initial conditions of the temporal beginning of the universe, some people started to speculate about "all possible universes." If all possibilities are implemented somewhere in the huge collection of "everything," then even the most intricate questions lose their disturbing force. The strangest things just happen (an infinite number of times) someplace. If we live in this extremely strange world of ours, this is because "more usual" worlds do not admit the existence of the "intelligent observer." In this approach, the old problems with space and time infinities of our world have been superseded by the many-faceted infinity problem of the "multiverse" (the name recently coined for the collection of all possible universes). However, the point is that this approach will remain a vague ideology unless we are able to at least tame it with the help of strict mathematical concepts, as was the case with space and time infinities in relativistic cosmology.[1]

---

[1] There are a great variety of interpretations of the term "multiverse." See, for instance, a "list" composed by Tegmark (2007); some items on his list are rather vague. There are attempts to introduce some rigor in this area; see, for instance, Ellis (2007) or McCabe (2005).

In Sections 10.2 through 10.5, we consider the problem of infinities in relativistic cosmology and focus on mathematical methods to work with them. Essentially, one meets two kinds of infinities in cosmology: regions "infinitely distant" in spacetime, and "regions," called singularities, at which the standard structure of spacetime breaks down. When one approaches such regions, some physical magnitudes tend to infinity. The strategy one adopts in cosmology is to learn something about these "forbidden regions" by collecting information from regular domains of spacetime as one approaches the "forbidden domains." In this way, one constructs various spacetime boundaries, some elements of which represent singularities, and some other "infinitely distant" regions.

In Section 10.6 we deal with various aspects of the infinity problem as it arises within the multiverse ideology. We present some attempts to subdue this concept to a more demanding analysis and indicate limitations in which all such attempts are involved.

In Section 10.7 we try to draw a philosophical and theological lesson from the previous considerations.

10.2 Conformal Infinity

If we want to discuss the problems of space and time infinities in relativistic cosmology, we should keep in mind that these two kinds of infinities are but two components of the one spacetime infinity problem. Quite often, for both technical and philosophical reasons, we prefer to discuss this problem in its "two-component" form (space and time separately) rather than in its spacetime invariant form, but then we are immediately confronted with additional computational and conceptual questions that have to be solved. In the standard cosmological models, owing to their high geometric symmetries, these questions are solved quite naturally, but in more general cases they often require skill and ingenuity.

As mentioned in the introduction, in relativistic cosmology we meet two kinds of spacetime infinities: points at spacetime infinity (i.e., points infinitely distant from the observer) and singularities (such as the one in the Big Bang). Both of these kinds of infinities are a challenge to our *horror of boundaries*: we hate everything that limits the field of our potential knowledge.

Points at infinity are inaccessible, by definition. In general relativity there are various categories of "points at infinity." For instance, an infinite time could be needed to reach such a point, or a signal of infinite velocity would be required to send a message to such a point. Can we say something meaningful about such infinitely distant regions? In some cases, and to a certain extent, the answer is positive. The so-called conformal transformations, well known in modern geometry, rescale distances in spacetime in such a way that points at infinity are brought to finite distances; consequently, we are able to investigate these "infinities" by studying various patterns that they form after being conformally rescaled. However, the price we must pay for this is that any conformal transformation distorts various geometric properties of a given spacetime, with one exception: the structure of light propagation remains unaffected. In general relativity, light beams (photons) propagate along curves called *null* or *light-like geodesics*, and the geometry of these curves is the same before and after any conformal transformation of spacetime. In this sense, we can say that light propagation is a skeleton of the entire spacetime geometry.

In some cases, after the conformal rescaling and an additional procedure, called compactification, one can drawn a nice picture of a given spacetime. The compactification attaches to the picture clear boundaries that represent "points at infinity." Such pictures are called Penrose diagrams and are often used to study the "infinity structure" of different relativistic spacetimes. The boundaries of Penrose diagrams perspicuously display various properties of "conformal infinities." Some parts of them are formed by the end points of curves that are histories of freely moving massive particles (the so-called time-like geodesics); they are said to represent "time-like infinities" of a given spacetime. Some parts of them are formed by the end points of curves that are histories of photons (null geodesics); they represent "null" or "light-like infinities." Other parts of them are formed by the end points of curves that represent no physical objects (particles moving along such curves, called space-like geodesics, would have to move with velocities greater than that of light); they form "space-like infinities." Taking into account the fact that causal influences can be propagated along time-like and light-like curves, Penrose diagrams are said to represent the *causal structure* of various spacetimes.

From this short reminder of well-known facts about spacetime architecture[2] we can learn an interesting philosophical lesson. In general relativity "infinitely distant" regions of spacetime are not mere shadowy horizons. They make their presence in the regular domains of spacetime through their causal properties. Geometry of light is invariant with respect to conformal transformations, and it creates a skeleton supporting the causal structure of spacetime, which, in turn, serves as a groundwork for the entire physics. The structure of conformal infinity is an essential ingredient of global physics.

10.3 Infinitely Divergent

Other barriers for our knowledge are situated at finite distances from us in spacetime. They are formed by singularities of various kinds. There are mild singularities that can be removed by simple prolongations of a given spacetime (the so-called regular singularities); there are singularities similar to stringlike inhomogeneities in spacetime (quasi-regular singularities); and there are malicious singularities, called strong curvature singularities, having the property that on approaching them the spacetime curvature and some other physical magnitudes blow up to infinity. To the latter kind belong the initial singularity in the Big Bang cosmological models and the Big Crunch singularity in the closed standard cosmological model. The initial singularity is situated in our finite past, and the final singularity in our finite future. These expressions are meaningful only with respect to cosmological models in which, owing to their high symmetries, spacetime naturally decomposes into one-dimensional time and the family of three-dimensional instantaneous spaces. In such cosmological models we can meaningfully speak about the unique world's history, and possibly about its abrupt

---

[2] The fundamental reference is Hawking and Ellis (1973). The reader could find a good introduction in Geroch and Horowitz (1979) and, on the more popular level, Geroch (1978).

beginning and end. However, this very particular situation cannot serve as a model for a general definition, or even as an adequate description, of singularities.[3]

In the theory of relativity, decomposition of spacetime into space and time separately depends on the choice of the local reference frame; consequently, the concept of the world history, as measured by time, has no invariant meaning. On the other hand, the history of a single observer, or of a massive particle, or of a photon has a well-defined meaning because it is represented by a suitable curve in spacetime: by a time-like curve in the case of an observer or of a massive particle, and by a null geodesic curve in the case of a photon. The concept of the prolongation of a curve in the theory of relativity is a tricky one (given that length depends on the reference frame); however, in the case of geodesic curves there is a sense in which one can meaningfully speak about its prolongation.[4] Suppose that all time-like and light-like geodesic curves in a given spacetime can be indefinitely prolonged (in the above sense).[5] This would mean that histories of all physical objects never meet any obstacles – they can happen indefinitely. In such a case, spacetime is said to be singularity-free. In contrast, if at least one such geodesic curve cannot be indefinitely prolonged, this would mean that it has encountered a singularity.[6] In the standard cosmological models, all time-like and light-like geodesic curves break down at the initial or final singularities.

With the help of this construction it was possible to prove, in the 1960s and 1970s, several theorems on the existence of classical singularities (i.e., without taking into account possible quantum gravity effects). It has turned out that singularities are not by-products of some simplifying assumptions, as was so far commonly believed, but rather that they are deeply rooted in the structure of the present theories of gravity.

Let us notice that in this approach singularities are not "points" or some "pathological regions" in spacetime but are just a name announcing the fact that the structure of spacetime somewhere breaks down. Indeed, if we know only that a curve cannot be prolonged, we know that something is going awry, but we do not know the nature of the obstacle. To borrow the analogy from theology, we could say that singularities are defined *apophatically*,[7] that is, purely negatively, by denying to them the structure of spacetime. What can we do to overcome this *docta ignorantia* situation? We try to organize the end points of geodesics that break down at singularities into some meaningful totality that would inform us, at least to some extent, about the structure of the spacetime edge representing something called singularity. To this end we investigate the behavior of time-like and light-like geodesics, as they are more and more prolonged, in order to decide which of them have the same end points. The class of curves having the same end point defines this end point. Instead of investigating the end points, we investigate the corresponding classes of curves to which we have access from within

---

[3] References cited in footnote 2 are also recommended for this section.

[4] We say that a geodesic curve can be indefinitely prolonged (technically, it is said to be complete) if the affine parameter along it can assume arbitrarily large values.

[5] We do not take into consideration space-like geodesics because they are not histories of physical objects.

[6] It goes without saying that this is a very rough, intuitive description. To change it into the working definition would require a host of technical details. The reader interested in them should consult the classical monograph by Hawking and Ellis (1973). Earman (1995), a more philosophically oriented book (but also highly technical), is also worth reading.

[7] For more about apophatic theology see Section 10.7.

spacetime. The collection of all such classes of curves (i.e., the collection of end points defined by them) defines the *geodesic boundary* (or *g-boundary*, for short) of spacetime. It was Robert Geroch who elaborated all details of this construction.

Let us notice the philosophy behind the g-boundary construction. We collect information from inside a given spacetime (by following the behavior of geodesics in it) to learn something about the way its structure breaks down. The *apophatic* character of our knowledge is mitigated by tracing vestiges of what we do not know in the domain open for our investigation. In this way we have learned that there are time-like, light-like, and space-like g-boundaries, similar to "conformal infinities."[8] However, this is only the beginning of the story.

10.4 The b-Boundary Crisis

In any spacetime, besides geodesic curves, there are plenty of other curves, and it is evident that the g-boundary method ignores them. Examples of spacetimes are known in which all causal geodesic curves can be prolonged indefinitely (such spacetimes are said to be *geodesically complete,* or g-complete, for short) but which are "singular in other curves." The problem is that in the case of nongeodesic curves no "prolongation procedure" is known. A new method has to be elaborated.

The job was undertaken by Bernd Schmidt (1971). His principal idea was to construct a "singular boundary" of spacetime as a collection of suitably defined end points of all curves in this spacetime (not only of geodesic curves) that can be interpreted as histories of physical objects. The main problem was how to define a length or a prolongation of such curves.[9] Schmidt has succeeded in doing so by using, as an auxiliary tool, a space of all local reference frames attached to all points in spacetime. Such a space, well known in differential geometry, is a part of the structure called *frame bundle* over spacetime. This is why Schmidt's boundary of spacetime has been called the *b-boundary*. It was soon acclaimed to be the best available definition of singularities in general relativity.

It was fairly complicated, however. Schmidt himself was able to provide a few "toy examples" that showed that in very simple cases his b-boundary construction works well (in agreement with natural intuitions). However, when Bosshardt (1976) and Johnson (1977), almost simultaneously but independently, computed the b-boundary for the closed Friedmann world model, the disaster became manifest. It has turned out that the b-boundary of this model consists of a single point. This means that the initial and final singularities, present in the Friedmann world, coincide. The beginning and the end of the world are the same! Moreover, from the topological point of view, all points of spacetime, together with the single b-boundary point, are "close" to each other.[10]

The reaction to this disaster was immediate. It became rather obvious that Schmidt's definition was bad and that one needed to look for a better one. Several constructions were proposed, but they either did not work or were not general enough. The so-called

---

[8] In fact, the method of conformal transformation is also helpful in investigating the structure of singularities.

[9] The definition had to be such that in the case of geodesic curves it should coincide with the previous one.

[10] More strictly, no spacetime point can be Hausdorff separated from the b-boundary.

causal boundary of spacetime, proposed by Kronheimer and Penrose,[11] gained some popularity, but without some hybrid combination with the g-boundary construction it was unable to distinguish between singularities and points at infinity. However, it has turned out that there is an aspect in which Schmidt's construction is useful. It shows what can happen to geometry when some physically meaningful magnitudes (like curvature) go to infinity. Let us look at the singularity problem from this point of view.

10.5 Structure of Singularities

Let us try to understand what is going wrong with the Friedmann singularities when an attempt is made to describe them with the help of Schmidt's construction. The standard method of working with differential manifolds (every spacetime is a differential manifold) is to use coordinates (like in elementary geometry). There is, however, an equivalent method: instead of coordinates, one can consider all smooth functions on a given manifold and define all relevant geometric magnitudes on this manifold in terms of these functions. Th... ...ized in the following way: If we have a ... , cusps, boundaries, or some other kind ... ly of functions that could play the role ... rms of it, develop the geometry of this ... ethod has been elaborated by, among ... to see what happens if we apply this ... ɔ initial and final singularities as they ... e time being, we disregard singulari- ... ith the collection of smooth functions ... y. However, if we try to extend this ... nt functions survive this attempt. In ... ngularities, only constant functions ... he the same value everywhere, that ... ɔses to a single "neighborhood."[13] ... ngularities are strict mathematical re... ...e behavior of infinities quite independently of whether they apply to the real universe or not. However, there are strong indications that they do refer to the real world. As mentioned in Section 10.3, in the 1960s and 1970s several theorems were proved demonstrating that singularities are not artifacts of simplifying assumptions made in the process of model constructions (as was at that time commonly believed), but rather, inherent structural properties of a broad class of gravity theories, including general relativity (Hawking and Ellis 1973). The important point is that these theorems refer to classical singularities, that is, they do not take into account quantum gravity effects. This is a serious limitation because there are strong reasons to believe that the very early stages of the cosmic evolution were governed by so-far-unknown quantum gravity laws. According to the prevailing view, the quantum gravity theory, when finally discovered, will eliminate singularities from

---

[11] This construction is explained in Kronheimer and Penrose (1967); see also Penrose (1979).
[12] Functions belonging to this family need not be smooth in the usual sense.
[13] This phenomenon was studied in detail in Heller and Sasin (1995).

the scientific image of the universe. Even if this is true, the classical singularity theorems tell us something important about the structure of the universe. Every quantum gravity theory must, in the limit of low energies, recover classical geometric theory of spacetime (general relativity). This means that it must contain enough powerful mechanisms that would be able to smooth out singularities with all associated infinities, but the mechanisms must not be so powerful as to erase them completely. They have to be reproduced in the limiting case. In other words, the looked-for quantum gravity theory must be "singular in the limit." In this sense, "infinities" should potentially be present even in "finite theories" of our universe (see Heller 1993).

Let us try to collect our results. First, in the standard cosmology the problem of infinity appears in two forms: in the form of an "infinitely distant," and in the form of an "infinitely divergent" (singularities). For both of these cases, mathematical methods have been elaborated to deal with infinities. Although we are unable to work directly with infinities, we apply to them various versions of the "strategy of limits." The standard "going to the limit" in a convergent series is the simplest and best-known method of this kind. Conformal transformations and attaching different kinds of boundaries to spacetime are more sophisticated ones. We can say that although infinities are beyond our direct reach, they can be mathematically "tamed," and only when they have been suitably tamed can they be incorporated into a sound physical theory or model.

This is a warning against considerations, not quite rare in popular books, on "cosmological infinities" that are based on a naive concept of infinity. Infinity should not be regarded as something "very, very, very . . . big," but rather as something irreducible to all "big things" we know. Without the aforementioned mathematical "taming" of infinities, it is better to refrain from speculations than to pronounce uncontrollable utterances.

Second, from the way we have tamed infinities in classical cosmology (i.e., without taking into account quantum gravity effects), we learn an important lesson: infinities are not "local effects." They are not strictly localizable at "infinitely distant" regions of spacetime or within the singularities in the beginning and end of the universe. Their presence can be "felt" everywhere in spacetime. For instance, the structure of the conformal boundary or of the b-boundary determines various causal properties of spacetime, which in turn have direct influence on local physics.[14]

The "nonlocality" of singularities can best be seen in a "pathology" they produce in the Friedmann model (when they are understood as b-boundary points): as soon as we try to incorporate singularities into the geometric description of the world, everything collapses to a single point.

10.6 Infinity of Universes

One of the reasons the multiverse idea gained popularity was purely ideological. Its aim was to neutralize teleological interpretations seemingly implied by the fine-tuning

---

[14] For instance, when the past conformal boundary is space-like, there are particle horizons in the model; when the past conformal boundary is null, there are no particle horizons in the model (see Hawking and Ellis 1973, p. 128).

of the initial conditions of the universe.[15] "The physical laws 'laid down' in the big bang seem to apply everywhere we can now observe. But though they are unchanging (or almost so), they seem rather specially adjusted" (Rees 1998, p. 250). This could be either a coincidence or an "intelligent design." To avoid this conclusion, one could adopt the following argument: "There may be other universes – uncountable many of them – of which ours is just one. In the others, the laws and constants are different. But ours is not randomly selected. It belongs to the unusual subset that allows complexity and consciousness to develop. Once we accept this, the seemingly 'designed' or 'fine-tuned' features of our universe need occasion no surprise" (Rees 1998, p. 250).

Many people support the multiverse idea because of some conclusions following from the chaotic scenario of inflationary cosmology and from cosmological models based on the superstring theory and the so-called M-theory, which they have independent reasons for believing. The details of these highly hypothetical models need not concern us here.[16] For our present point of view, the important fact is that almost all considerations regarding the multiverse are based on probabilistic arguments. We claim that some physical properties are *randomly distributed* in the multiverse, that our position in the multiverse is not *random*, and so forth. At the same time, at least in some formulations, the multiverse is claimed to be an infinite – and perhaps even uncountably infinite – collection of universes. The point is that we do not know whether the concept of probability can legitimately be defined on such a collection. Mathematically speaking, we do not know whether there is a probability measure on it (or even whether this collection is a set), and if such a measure does not exist, any probabilistic utterance about the multiverse is meaningless. If we want to improve the situation, we must impose quite demanding constraints on our understanding of multiverse. However, even if this is the case, the problem is utterly difficult (see Ellis 2007). Tegmark has put this bluntly: "As multiverse theories gain credence, the sticky issue of how to compute probabilities in physics is growing from a minor nuisance into a major embarrassment" (Tegmark 2003).

I do not claim that we should stop talking about the multiverse (even the most audacious speculations can have positive influence on science), but rather that we should be aware of our conceptual and linguistic constraints. Unfortunately, this is not always the case. Especially, when infinity and probability are combined together, unexpected situations are ready to be born. The same Tegmark (2003) does not hesitate to write the following: "If space is infinite and the distribution of matter is sufficiently uniform on large scales, then even the most unlikely events must take place somewhere. In particular, there are infinitely many other inhabited planets, including not just one but infinitely many with people with the same appearance, name and memories as you."[17] He claims that, according to his "extremely conservative estimate," in the infinite universe "the closest identical copy of you is about $10^{10^{29}}$ m away" (Tegmark 2003).

---

[15] For a debate concerning the multiverse idea see Carr (2007).

[16] See the corresponding chapters in Carr (2007).

[17] Tegmark justifies his conclusion by quoting the ergodic theorem (which, in its essence, is a probabilistic theorem). To be more precise, Tegmark, based on this theorem, claims that the probability distribution of outcomes in a given volume in the multiverse is the same as the probability distribution of outcomes in different volumes in a single universe, provided that each universe within the multiverse evolves from random initial conditions.

To this remark concerning the existence of the probability measure on an infinite collection of universes we should add the following caveat: It cannot be excluded that on higher levels of complexity, such as intelligent life, an "individualization principle" could be operating. Philosophers, for a long time, postulated its existence (for instance, Leibniz's *indiscernibility principle*), and we can find its traces even in elementary mathematics. For example, in the set of real numbers, each number is an individual entity that is never repeated in the entire uncountable infinite set of reals. The "individualization principle," in this case, consists in both peculiar properties of a given real number and in the ordering properties of the whole set of reals. If such a principle works with respect to such apparently simple entities as real numbers, should we not expect that something analogous could be at work at much higher levels of complexity?

The general conclusion of the present section is that, in contrast with "infinitely distant" and "infinitely divergent" in the standard cosmology, the concept of infinity, as it is involved in speculations around the multiverse idea, has not yet been mathematically tamed;[18] consequently, it must be looked on with extreme caution. In these matters, as long as good mathematics is absent, we do not know what we are talking about.

10.7 Theological Lesson

When the First Vatican Council, at its third session, declared that God is "almighty, eternal, immeasurable, incomprehensible, infinite in will, understanding and every perfection,"[19] it only articulated a long tradition existing in Christian theology and philosophy. In this tradition infinity was usually understood as that which has no limit, boundary, or end and which surpasses any measure or magnitude. It is obvious that such an understanding of Divine Infinity cannot be "mathematically tamed," which is why, when we apply the concept of infinity to God, we are relegated to its analogous or metaphorical use. The awareness of this fact in Christian theology often assumed the form of an *apophatic* or *negative* theology (*via negativa*) – a conviction that everything we assert about God is, in fact, about what God is not, rather than about what God is. Such an approach played an important role in the early Christian theology (Pseudo-Dionysius, the Cappadocian Fathers, etc.), but also in the later theological thinking (e.g., Meister Eckhart, St. John of the Cross).[20]

It is natural that apophatic tendencies can be found in the writings of those mathematically inspired authors who speculated about theological implications of the concept of infinity. It was Descartes, inspired by Nicholas of Cusa, who said that "there is more reality in the infinite substance than in the finite" (Descartes, 1641);[21] consequently, he regarded infinity as a primary mathematical concept and all other concepts

[18] There are attempts to put some order into the multiverse concept; see Ellis (2007) and McCabe (2005).

[19] Session 3, April 24, 1870: *Dogmatic Constitution on the Catholic Faith*, chap. 1. See: *The Scripture Documents. An Anthology of Official Catholic Teachings*, ed. by D. B. Bechard, The Order of Saint Benedict, Collegeville, Minnesota, 2002, pp. 14–18.

[20] An interesting study of apophatic theology can be found in Hadot (1987).

[21] This does not prevent Descartes from treating infinity in an apophatic way: "And I must not imagine that I do not apprehend the infinite by a true idea, but only by the negation of the finite, in the same way that I comprehend repose and darkness by the negation of motion and light" (Descartes 1641).

as derivative ones. Modern thinking about infinity was influenced by John Locke, who, in his "Enquiry Concerning Human Understanding," noticed that one can compare finite things, whereas no comparison is possible between the finite and the infinite (the same idea was much earlier expressed by Boethius in a theological context). This "lack of proportion" between the finite and the infinite is the essence of the apophatic theology. God is "totally different" from anything else to such an extent that even the noncontradiction law should be applied to Him with the utmost caution. For instance, Nicholas of Cusa thought that even in mathematical infinity the "opposites" can be identified: "the curved" and "the straight" are opposites with respect to each other, but the circumference of a circle, having an infinite radius, is both curved and straight (Nicholas of Cusa 1990).

Modern mathematics owes the rationalization of the infinity concept to Georg Cantor and Richard Dedekind. Cantor's theory of sets and Dedekind's theory of numbers became cornerstones of mathematics and paradigmatic models in thinking about infinity. Cantor considered his work as providing a great contribution to Christian theology by offering to it, for the first time in history, the true theory of the infinite.[22]

The "taming of infinity" in modern cosmology, which we briefly reviewed in Sections 10.2 through 10.5, is just an application of highly elaborated ideas, seminally present in Cantor's work, to the field of differential geometry. The question arises whether these purely scientific ideas can have any theological significance. Supposing the positive answer to this question, we can approach it in a naive or in a less naive way.

The naive approach would be by drawing an analogy between infinities in cosmology and the infinity of God. We start with certain properties of the world model (with some of its geometric aspects referring to infinity), and we argue that these properties can illuminate, or help us to imagine, the relationship between God and the universe. Both "infinitely distant" and "infinitely divergent" transcend the regular parts of space-time and at the same time are, as nonlocal elements of the model, somehow present everywhere in the model. Analogously, God transcends the world and at the same time is present within it. This approach is qualified as naive, given that the term "transcends" has a drastically different meaning in the cosmological and theological contexts.

A less naive approach starts with a theological or philosophical doctrine of God (it should be remembered that no doctrine of God can be devoid of naive elements) and tries, in light of it, to interpret some aspects of the cosmological model. One could, for example, start with a theological (or philosophical) doctrine of God, known under the name of panentheism. Etymologically, the term means "everything-in-God."[23] It goes beyond pantheism (identifying God with the world) in claiming that God "is larger" than the world; the world "is contained" in God. The world should be regarded as "godly," but the Divine infinitely surpasses the world. In some stronger versions, panentheism asserts that the Absolute, transcending the world, does not include it dualistically, but rather monistically, as an aspect in a whole. Panentheism could also be interpreted "in the Einstein style," by asserting that the mathematical structure of

---

[22] On Cantor's concept of infinity see Dadaczyński (2002, pp. 98–142).

[23] The term "panentheism" was devised by Karl C. F. Krause in 1828. The doctrine itself gained popularity owing to the process philosophy of Alfred North Whitehead and Charles Hartshorne.

the world is but a "mind of God" implemented in the work of creation (or rather that the mathematical structure of the world is our approximation to the "mind of God"). If so, then every mathematical model, which has a reference to the real world, also has a theological significance in that it gives us a hint of what a certain aspect of the "mind of God" could look like. In this approach, no "infinite properties" of a given world model have more major theological significance than its other properties, except that, from the psychological point of view, they can, more than the other ones, orient or elevate our thinking to the horizon of transcendence.

As mentioned previously, one of the motives to create the multiverse idea was to neutralize its philosophical significance: if there exists an infinite number of universes in which all possible initial conditions (and other parameters characterizing a given world) are somewhere realized, then no intelligent design is necessary, and God, as the Great Designer, seems superfluous. Indeed, such a reasoning makes the idea of design irrelevant as far as cosmological considerations are concerned, but it does not eliminate from theological (or philosophical) considerations the need of the ultimate understanding of reality. In the case of the multiverse, the metaphysical question "Why does there exist something rather than nothing?" seems to be even more relevant (or at least more appealing) than in the case of a single world. If the Christian idea of creation contains the answer to this question in the case of a single universe, it does the same in the case of the multiverse. To put this simply: for God it is equally easy to create an infinite number of universes as it is to create a single universe. And even if this is a little more difficult, God could not be interested in anything less than infinity.

Acknowledgments

I express my gratitude to Anthony Aguirre for his careful reading of the manuscript and penetrating remarks.

References

Bosshardt, B. 1976. On the b-boundary of the closed Friedmann model. *Communications in Mathematical Physics* 46: 263–68.

Carr, B. (ed). 2007. *Universe or Multiverse?* Cambridge: Cambridge University Press.

Dadaczyński, J. 2002. *Mathematics in the Eyes of the Philosopher*. Kraków: OBI; Tarnów: Bibls (in Polish).

Descartes, R. 1641. *Meditations on First Philosophy*. Translated by J. Veitch. Third Meditation, n. 24. http://oregonstate.edu/instruct/phl302/texts/descartes/meditations/meditations.html.

Earman, J. 1995. *Bangs, Crunches, Whimpers, and Shrieks*. New York: Oxford University Press.

Ellis, G. 2007. Multiverses, description, uniqueness and testing. In *Universe or Multiverse?* B. Carr (ed.), pp. 387–409. Cambridge: Cambridge University Press.

Geroch, R. 1978. *General Relativity from A to B*. Chicago: University of Chicago Press.

Geroch, R., and Horowitz, G. T. 1979. Global structure of spacetime. In *General Relativity: An Einstein Centenary Survey*, S. W. Hawking and W. Israel (ed.), pp. 212–93. Cambridge: Cambridge University Press.

Hadot, P. 1987. *Exercices spirituals et philosophie antique*. Paris: Études Augustinnienes.

Hawking, S. W., and Ellis, G. F. R. 1973. *The Large Scale Structure of Space-Time*. Cambridge: Cambridge University Press.

Heller, M. 1993. Classical singularities and the quantum origin of the universe. *European Journal of Physics* 14: 7–13.

Heller, M., and Sasin, W. 1995. Structured spaces and their application to relativistic physics. *Journal of Mathematical Physics* 36: 3644–62.

Johnson, R. A. 2005. The bundle boundary in some special cases. *Journal of Mathematical Physics* 18: 898–902.

Kronheimer, E. H., and Penrose, R. 1967. On the structure of causal spaces. *Mathematical Proceedings of the Cambridge Philosophical Society* 63: 481–501.

McCabe, G. 2005. Possible physical universes. In *Zagadnienia Filozoficzne w Nauce (Philosophical Questions in Science)* 37: 73–97 (in English).

Nicholas of Cusa. 1990. *On Learned Ignorance*. Translated by J. Hopkins. Minneapolis: Arthur J. Banning Press.

Penrose, R. 1979. Singularities and time asymmetry. In *General Relativity: An Einstein Centenary Survey*, S. W. Hawking and W. Israel (eds.), pp. 581–638. Cambridge: Cambridge University Press.

Rees, M. 1998. *Before the Beginning – Our Universe and Others*. New York.

Schmidt, B. 1971. A new definition of singular points in general relativity. *General Relativity and Gravitation* 1: 269–80.

Sikorski, R. 1972. *Introduction to Differential Geometry*. Warszawa: PWN (in Polish).

Tegmark, M. 2003. Parallel universes. *Scientific American* (May): 40–51.

Tegmark, M. 2007. The multiverse hierarchy. In *Universe or Multiverse?* B. Carr (ed.), pp. 99–125. Cambridge: Cambridge University Press.

PART FIVE

Perspectives on Infinity from Philosophy and Theology

CHAPTER 11

God and Infinity: Directions for Future Research

Graham Oppy

Philosophical investigation – in particular, metaphysical investigation – is rarely advanced through the consultation of dictionaries. In the present case, however, it will repay us to begin by considering the entries for "infinite," "infinity," and "the infinite" that are found in the OED (*The Oxford English Dictionary, Vol. V, H-K*):

Infinite adj. (omitting obsolete and archaic uses)

1. Having no limit or end (real or assignable); boundless, unlimited, endless; immeasurably great in extent, duration, or other respect. Chiefly of God or His attributes; also of space, time, etc., in which it passes into the mathematical use.
2. In loose or hyperbolical sense: Indefinitely or exceedingly great; exceeding measurement or calculation; immense, vast.
3. *Math.* Of a quantity or magnitude: Having no limit; greater than any assignable quantity or magnitude (opp. to finite). Of a line or surface: Extending indefinitely without limit, and not returning to itself in any finite distance (opp. to closed).
4. *Infinite series*: a series of quantities or expressions which may be indefinitely continued without ever coming to an end (but may or may not have a finite value or "limit" to which it approaches as more and more terms are taken).
5. *Gram.* Applied to those parts of the verb which are not limited by person or number.

Infinite absol. or as sb.

1. That which is infinite or has no limit; an infinite being, thing, quantity, extent, etc. Now almost always in the sing. with *the*; esp. as a designation of the Deity or the absolute Being.
2. *Math.* An infinite quantity.

Infinity n.

1. The quality or attribute of being infinite or having no limit; boundlessness, illimitableness (esp. as an attribute of Deity)
2. Something that is infinite; infinite extent, amount, duration, etc.; a boundless space or expanse; an endless or unlimited time.

3. *Math.* Infinite quantity: denoted by the symbol ∞. Also, an infinite number (of something).

4. *Geom.* Infinite distance, or that portion or region of space which is infinitely distant: usually in phr. *at infinity*.

As these entries make clear, the words "infinite" and "infinity" have a number of overlapping uses and meanings. This overlapping of uses and meanings can, it seems, be explained, at least in part, by appeal to historical considerations. However, it is a matter for investigation whether this overlapping constitutes an impediment to certain kinds of inquiries and theoretical activities. We might think of this investigation as a prolegomenon to serious discussion of the proper uses of the words "infinite" and "infinity" – and the concepts that these words express – in theological contexts.

11.1 A Question for Investigation

According to reliable authorities,[1] the origins of our words "infinite" and "infinity" can be traced back to the Greek word *peras* (πέρας), which can be translated as "limit," "bound," "frontier," or "border," and which has connotations of being "clear" or "definite." The Greek word *to apeiron* (ἄπειρον) – the "negation" or "opposite" of peras – thus can be understood to refer to that which is unlimited or boundless, or, in some cases, unclear and indefinite.

When the word *to apeiron* makes its first significant recorded appearance – in the work of Anaximander of Miletus – it is used to refer to "the boundless, imperishable, ultimate source of everything that is" (Moore 1998, p. 772). Thus, in this early usage, the word *to apeiron* has connotations ("imperishable," "ultimate source of everything") that are quite separate – or, at any rate, separable – from considerations about the absence of "limits," "bounds," "frontiers," "borders," "clarity," or "definiteness."

As Moore (1998, p. 773) points out, most of the Greeks associated much more negative connotations with *to apeiron* than are evident in the early usage of Anaximander: for the Pythagoreans – and, at least to some extent, for Plato – *to apeiron* "subsumed . . . all that was bad . . . it was the imposition of limits on the unlimited that accounted for all the numerically definite phenomena that surround us." Again, on this kind of usage of the term *to apeiron* has connotations ("chaotic," "irrational," "disorderly") that are quite separate – or, at any rate, separable – from considerations about the absence of "limits," "bounds," "frontiers," "borders," "clarity," or "definiteness."

In current English, we have the adjective "infinite," the noun "infinity," and the substantive "the Infinite." The standard use of the substantive form is "as a designation of the Deity or the absolute Being"; and so, of course, there is one standard use of the adjectival and noun forms that rides piggyback on this standard use of the substantive form. It seems to me that it is plausible to see the current use of the substantive "the Infinite" as a direct descendent of Anaximander's use of the word *to apeiron* with more or less the same connotations ("imperishable," "ultimate source of

---

[1] See, for example, Benardete (1964), Owen (1967), Rucker (1982), Moore (1990), and Barrow (2005).

everything"), except, of course, that *to apeiron* is personalized, that is, taken to have personal attributes and attitudes, in Christian theology.

In current English, however, we also have uses of the adjective and noun forms that are not obviously related to the standard use of the substantive form. In particular, there are uses of these terms in mathematics, including geometry, and applications of these terms to space and time, in which most of the connotations associated with the substantive form seem to play no role at all. While these uses of the term do have more or less clear connections to the absence of "limits" or "bounds," they have very little to do with considerations about the absence of "clarity" or "distinctness," and nothing at all to do with considerations about "the ultimate, imperishable, source of everything."

It is not clear to me whether this separation of considerations was achieved by the Pythagoreans. Given their metaphysical belief that the positive integers are the ultimate constituents of the world, it is a plausible conjecture that they did not recognize the discussion of "limits" or "bounds" – and the application of these terms to, say, space and time – as a separate topic for investigation in its own right. But, whatever the truth about this matter may be, it seems that some of the contemporaries and immediate successors of the Pythagoreans *did* come to see the discussion of these topics as an independent subject matter. It is, I think, plausible to view Zeno's paradoxes as a contribution to such a discussion, and even if that is controversial, it is surely right to see Aristotle's treatment of infinity as an investigation of "limits" and "bounds" – in the context of space, time, and matter – in their own right. (In *Physics*, book III, Aristotle makes mention of Anaximander's views about "the ultimate source of everything," but those views are entirely incidental to the theory of "limits" and "bounds" that Aristotle proceeds to elaborate and defend.)

However, once it is recognized that the investigation of "limits" and "bounds" – in the context of space, time, and matter – is a legitimate subject matter in its own right, then various questions arise about the application of the results of *that* investigation to the subject matter with which Anaximander was primarily concerned: "the ultimate source of everything." Even if it is true – as I think it is – that the historical entanglement of talk about "limits" and "bounds" with talk about "the ultimate source of everything" persists into the present, it is important to ask whether this entanglement has any *essential* significance for either the investigation of "limits" and "bounds" as a subject matter in its own right or for the investigation of "the ultimate source of everything" (as a subject matter in its own right).

Prima facie – at least! – there seems to be good reason to think that the investigation of "the ultimate source of everything" has no *essential* or *ineliminable* significance for the investigation of "limits" and "bounds" as a subject matter in its own right. Modern logical, mathematical, and physical theories depend on no substantive theological assumptions. No serious standard text in logic, mathematics, or the physical sciences begins with a chapter on "theological preliminaries" or "theological assumptions."[2]

---

[2] Meyer (1987) offers a "proof" that the claim that God exists is logically equivalent to the Axiom of Choice. However – even setting aside the evidence of tongue placed securely in cheek – it is clear that Meyer offers no more than a patch for *one* of the holes in the argument of Aquinas's second way. Lewis (1991) is typical of much technical literature in logic, mathematics, and the physical sciences: it contains a range of references to God, but none that are *essential* to the theory of parts of classes that Lewis elaborates and defends.

Moreover, the same point holds true for serious textbook discussions of infinities *in* logic, mathematics, and the physical sciences: there is no theological prolegomenon that is *required* for examinations of Conway numbers, renormalization in quantum field theory, Kripke models for intuitionistic logic, or any other particular topic in this domain.

But what about the other direction? Does the investigation of "limits" and "bounds" as a subject matter in its own right have some *essential* or *ineliminable* significance for the investigation of "the ultimate source of everything"? It is, of course, well known that some of those who have investigated "limits" and "bounds" as a subject matter in its own right have supposed that this investigation does have *important* consequences for the investigation of "the ultimate source of everything." (This is true, for example, of Cantor.) But the question that I wish to take up, in the remainder of this chapter, is whether it is true – and, if so, in what ways it is true – that those who wish to investigate "the ultimate source of everything" need to equip themselves with the fruits of an investigation of "limits" and "bounds" as a subject matter in its own right.

11.2 Predicates and Properties

Those who believe that there is a unique "ultimate source of everything" exhibit a range of different views concerning the language that they use when talking about this ultimate source. We can illustrate some of the range of views by considering simple subject-predicate sentences of the form "God is F," where "F" is a relatively simple and unstructured predicate.

There are, of course, questions about the interpretation of the word "God." Given that I don't wish to focus on *those* questions here, I shall simply assume that we can take it for granted that "God" is a proper name, and that the reference of this name is fixed by the description "the ultimate source of everything." (Others who accept the general account that is suggested here will prefer different reference-fixing descriptions: e.g., "the omnipotent, omniscient, perfectly good creator of the world *ex nihilo*" [Swinburne 1979, p. 8] or "the thing that is, in an objectively normative manner, the proper object for religious attitudes" [Sobel 2004, p. 10]. For the purposes of the present discussion, nothing turns on the exact phrasing of the reference-fixing description.) Of course, this account leaves it open that there is nothing that satisfies the reference-fixing description; if there is nothing that satisfies the reference-fixing description, then the name is empty.

Among those who suppose that "God" is not an empty name, there is a wide range of views about the *understanding* that it is possible for people to have of the properties that are possessed by the being who bears the name. Some suppose that we cannot grasp (apprehend, understand) *any* of the properties of God. Others suppose that we cannot *fully* grasp (apprehend, understand) any of the properties of God, but that we can have a *partial* or *incomplete* grasp (apprehension, understanding) of *some* of the properties of God. Yet others suppose that, while we can fully grasp (apprehend, understand) *some* of the properties of God, there are other properties of God of which we can – as a matter of logical or metaphysical necessity – have no more than a partial or incomplete grasp. And, perhaps, there are some who suppose that, while there are properties of God of which we remain – and will always remain – ignorant, there is no *logical* or

metaphysical barrier to our grasping (apprehending, understanding) any of those properties. (There is, of course, a related range of views about the *knowledge* that it is
possible for people to have concerning which properties are, in fact, possessed by the
being who bears the name. Naturally, it should be borne in mind that it is one question
whether we can [fully] understand [grasp, apprehend] a property, and quite another
question whether we have what it takes to be able to determine whether or not God in
fact *possesses* that property.)

The range of views concerning what it is possible for us to *say* about the properties
that God possesses depends in part on the views that we take about the understanding
of God's properties and in part on the theory of predication – and, in particular, on the
theory of the relationship between predicates and properties – that we adopt. Among
theories of properties and predication, the most important distinction for us – for present
purposes – is between *luxuriant* theories that suppose that every well-functioning
predicate expresses a property (or universal) and *sparse* theories that suppose that
there are many perfectly well-functioning predicates that fail to express properties
(or universals), even though these predicates can be truly predicated of at least some
objects. If we adopt a luxuriant theory of properties and predication, then we shall
suppose that whenever we make a true claim of the form "God is F," the predicate "F"
expresses a property that is possessed by God. However, if we adopt a sparse theory
of properties and predication, then we can suppose that, at least sometimes, when we
make a true claim of the form "God is F," there is no property that is expressed by the
predicate "F" that is possessed by God.

If we suppose that we cannot grasp (apprehend, understand) *any* of the properties
of God, and if we adopt a luxuriant theory of properties and predication, then it surely
follows that we cannot say anything at all about God. Indeed, this combination of
views seems incoherent, for, in order to fix the reference of the name "God," we need
to make use of some predicates that we take to be true of that which bears the name.
If we claim that those predicates express properties and yet also claim that we cannot
grasp (apprehend, understand) any of the properties of God, then we have lapsed into
self-contradiction.[3] Here I assume that one does not understand a predicate unless one
grasps (apprehends, understands) the property that is expressed by that predicate; if I
don't know which property is expressed by a predicate, then I cannot make meaningful
use of that predicate to express my own thoughts. (Note, by the way, that I am not
here assuming that the property that is expressed by a predicate is required to be the
literal content of the predicate. It could be that, in the case in question, the use of the
predicate is *metaphorical* or *analogical*. However, I am assuming that one does not
grasp [apprehend, understand] a metaphorical or analogical use of a predicate unless
one understands *which* property is being attributed to the subject of the predication

---

[3] Many complex issues – concerning, in particular, the doctrine of divine simplicity – arise here. If we suppose
that there must be unity of truth-makers for claims involving simple predicates, then it seems to me that the
doctrine of divine simplicity stands refuted unless we allow that we can grasp *the* divine property. However, if
we allow that there can be diversity of truth-makers for claims involving simple predicates, then perhaps we
can allow that there is a sense in which we understand simple predications that are made of God – given that
we understand the words that are used in making these predications, even if there is another sense in which
we don't understand these predications – because we cannot fully understand that in virtue of which these
predications are true. In the remainder of this chapter, I shall ignore considerations about divine simplicity.

by the metaphor or analogy in question. This is not quite a rejection of the view that there can be irreducible [essential] metaphors or analogies; however, it is the view that, where there are irreducible [essential] metaphors or analogies, these arise because of limitations on our powers of *representation* and *expression*, and not because of limitations on our powers to grasp [apprehend, understand] the properties that are possessed by things.)

If we suppose that we cannot grasp (apprehend, understand) any of the properties of God, and if we adopt a sparse theory of properties and predication, then there will be things that we can say truly of God. Perhaps, for example, we can truly say that God is self-identical, while denying that there is any such thing as the property (universal) of "being self-identical." Although this view does not collapse quite so immediately into self-contradiction, it is not clear that this view can be seriously maintained. In particular, it seems doubtful that one can plausibly allow that the predicates that are used in the kinds of reference-fixing descriptions mentioned earlier – and the predicates that are entailed by those predicates that are used in the kinds of reference-fixing descriptions mentioned earlier – fail to express properties. Consider Swinburne's definition. Any being that is omnipotent, omniscient, perfectly good, and creator of the world ex nihilo will be good, powerful, possessed of knowledge, creative, and so forth (i.e., will be such that the predicates "good," "powerful," "possessed of knowledge," "creative," and so forth, can be truly predicated of it). But can it be plausibly maintained that *none* of these are properties (universals)? Sparse theories of properties (universals) must satisfy the constraint that, among the properties (universals) over which they quantify, there are those properties (universals) that constitute the basic building blocks for *our* world. It is not, I think, plausible to suppose that not one of the predicates that can be truly applied to God expresses a property (universal).

The view that, while we cannot *fully* grasp (apprehend, understand) any of the properties of God, we can have a *partial* or *incomplete* grasp (apprehension, understanding) of *some* of the properties of God seems to me to be subject to much the same kinds of difficulties as the view that we cannot grasp (apprehend, understand) any of the properties of God. On the one hand, if we adopt a luxuriant theory of properties and predication, then this view will have us saying that we have no more than a partial or incomplete grasp (apprehension, understanding) of such properties as self-identity, existence, uniqueness, and the like. On the other hand, even if we adopt a sparse theory of properties and predication, then this view will have us saying that we have no more than a partial or incomplete grasp of the properties that are expressed by predicates such as "is good," "knows," "is powerful," "is creative," and the like. Neither of these views seems to me to be at all attractive.

Once we proceed to views that allow that we can fully grasp (apprehend, understand) *some* of the properties of God, the kinds of difficulties that we have been exploring thus far lapse. So long as we allow that the properties that we appeal to – or that are entailed by those properties that we appeal to – in fixing the reference of the name "God" are among the properties that we can fully grasp (apprehend, understand), we have no (immediate) reason to fear that our theory of the fixing of the referent of the name "God" is self-contradictory, incoherent, or evidently inadequate. Certainly, this is clear if we allow that *all* the properties that we appeal to – *and* that are entailed by those properties that we appeal to – in fixing the reference of the name "God" are among the properties that we can fully grasp (apprehend, understand). But, plausibly, the consequence remains clear even on weaker readings of the condition: if all the properties that we appeal to – and all the properties that we *in fact* infer from those properties that we appeal to – in fixing the reference of the name "God" are among the properties that we can fully grasp (apprehend, understand), then we have no reason to fear that our theory of the fixing of the referent of the name "God" is self-contradictory, incoherent, or evidently inadequate. Indeed, it *may* even be plausible that the consequence remains on *much* weaker readings of the condition, for the most important constraint here is just that our theory of the fixing of the reference of the name "God" should not impute partial or incomplete grasping (understanding, apprehension) of predicates in cases in which we have good independent reason to insist that there is a full grasp (understanding, apprehension) of those predicates. Of course, *this* constraint can be satisfied even if some of the predicates that are used in the fixing of the reference of the name "God" are only partially or incompletely grasped, and even if many of the predicates that are entailed by the predicates that are used in the fixing of the reference of the name "God" are only partially or incompletely grasped, so long as there are some *other* predicates that can be truly applied to the object picked out by the reference-fixing description that are fully grasped.[4]

Of course, the discussion to this point does not exhaust the questions that arise concerning the views that those who believe that there is a unique "ultimate source of everything" take concerning the language that they use when talking about this ultimate source. In particular, I have said nothing thus far about the view that there are properties of God of which we can – as a matter of logical or metaphysical necessity – have no more than a partial or incomplete grasp. This view is the subject of the next section of my chapter.

11.3 Understanding Properties

Various foundational debates about properties and predicates have so far gone without mention in our discussion. Among these, the most important for present purposes are (1) the various debates about the tenability of quantification over predicate position and (2) the debates about whether there is a nonpleonastic sense in which predicates have properties as semantic values. In the previous section, the discussion takes it for granted that there is a nonpleonastic sense in which predicates have properties as semantic values – the idea that there is a distinction between luxuriant and sparse theories of properties lapses if this assumption is rejected – and also takes it for granted that we can make intelligible quantification over predicates, talking freely about the existence

---

[4] There are a number of interesting questions to be raised here about the propriety of using reference-fixing descriptions that contain predicates that one does not fully understand. There are also interesting questions here about the relationship between entailment and (full) understanding, as well as the relationship between devising analyses and possessing (full) understanding. Perhaps most importantly of all, there are fundamental questions to ask about what is involved in the full – and in the partial or incomplete – grasping (understanding, apprehension) of properties. Much work remains to be done to achieve clarity on all of the relevant issues that arise in connection with these questions.

of properties of various kinds, and so forth. If either or both of these suggestions are rejected, then we will need to seriously reconsider the terms of that previous discussion.

If we reject the claim that there is a nonpleonastic sense in which predicates have properties as semantic values, and if we adopt, instead, the proposal that properties are no more than the ontological shadows of meaningful predicates, then it is not clear that we can even make sense of the idea that there are properties of God of which we can, as a matter of logical or metaphysical necessity, have no more than a partial or incomplete grasp. If to be a property is just to be the ontological shadow of a meaningful predicate in a human language, then there are no properties that elude our understanding. While it is perhaps consistent with the suggestion that there are no properties that elude our understanding and that the expressive power of our language is susceptible of indefinite improvements, it is not clear that the idea that the expressive power of our language is susceptible of indefinite improvement of itself is sufficient to support the claim that, as a matter of logical or metaphysical necessity, we have only a partial or incomplete understanding of God. At the very least, it seems to me that some investigation is needed of the consequences of deflationary semantics for the claim that there are properties of God of which we can, as a matter of logical or metaphysical necessity, have no more than a partial or incomplete grasp.

If we reject the claim that there can be intelligible quantification over predicate position – or even if we insist on the claim that the best choices for canonical notation and logic are based on languages in which there is no quantification over predicate position – then, again, it is not clear that we can even make sense of the idea that there are properties of God of which we can, as a matter of logical or metaphysical necessity, have no more than a partial or incomplete grasp. If we cannot intelligibly quantify over predicate position, then we cannot make sense of any claim of the form "there are properties of God which . . . " and hence, in particular, cannot make sense of the claim that there are properties of God of which we can – as a matter of logical or metaphysical necessity – have no more than a partial or incomplete grasp. While there are reasons to think that we should allow that quantification over predicate position is not merely intelligible but actually acceptable and, indeed, required in order to allow us to say some of the things that we want to be able to say, a fully carried out project on the foundations of claims about God and infinity would need to include some investigation of these matters.

Suppose, however, that we allow that there is a nonpleonastic sense in which predicates have properties as semantic values, that there can be intelligible quantification over predicate position, and, perhaps, that the best choices for canonical notation and logic are based on languages in which there is quantification over predicate position. What should we then say about the claim that there are properties of God of which we can, as a matter of logical or metaphysical necessity, have no more than a partial or incomplete grasp? Although we cannot hope to adequately address this problem here, perhaps we can make a few useful preliminary observations.

From the outset, it is important to distinguish between the claim that it is logically or metaphysically necessary that there are properties of God of which we can have no more than a partial or incomplete grasp and the claim that there are properties of God of which it is logically or metaphysically necessary that we can have no more than a partial or incomplete grasp. The former claim could be true of things other than God if, for example, those things have infinitely many logically independent properties and we are only capable of fully and completely grasping a finite range of properties, and the former claim could be true of things other than God if there is no upper bound to the number of logically independent properties that are possessed by different things, but there is an upper bound to the number of properties that we can fully and completely grasp, and so forth. On the other hand, the latter claim can only be true if there is something about the nature of a particular property that causes it to be the case that that property is resistant to our full and complete understanding. I believe that it is the latter claim that is primarily of interest to us in the present context.

It is often said – or suggested, or implied – that there are properties of God that are resistant to our understanding, in the sense that we cannot understand what it would be like to possess those properties. For example, it is sometimes said that we cannot understand what it would be like to be omniscient. (We might think that Dennett uses this observation in order to undermine the knowledge argument against physicalism: because we don't know what it is like to be omniscient, we are in no good position to judge what Mary would know if it were true that she knew all the physical truths about the world.) However, I take it that this kind of ignorance – ignorance about what it would be like to possess a certain kind of property – is perfectly compatible with full and complete knowledge about which property it is that is in question. Supposing that, for example, it is true that to be omniscient is to know every proposition that it is logically possible for one to know, *given that* there are the weakest possible constraints on what it is logically possible for one to know, then one can have full and complete knowledge about what omniscience is even if one cannot even begin to imagine (picture, "understand from the inside") what it would be like to be omniscient.

Once we have the distinction between (1) the possession of full and complete knowledge of what a property F *is* and (2) the possession of full and complete knowledge of what it would be *like* to possess property F, we can apply this distinction to the question of whether we should want to assent to the claim that there are properties of God of which it is logically or metaphysically necessary that we can have no more than a partial or incomplete grasp. As noted in the previous paragraph, it seems quite reasonable to allow that there are properties of God of which it is logically or metaphysically necessary that we have no more than a partial or incomplete grasp of what it would be *like* to possess those properties. However, it is much less obvious that it is reasonable to allow that there are properties of God of which it is logically or metaphysically necessary that we have no more than a partial or incomplete grasp of what those properties *are*. At the very least, I think that it is clear that there is room for much further fruitful investigation of this issue.

11.4 Infinite Domains and Infinite Degrees

At the end of the first section of this chapter, I said that the primary question to be investigated herein is whether it is true – and, if so, in what ways it is true – that those who wish to investigate "the ultimate source of everything" need to equip themselves with the fruits of an investigation of "limits" and "bounds" as a subject matter in its own right. Prima facie, at least, there are various syntactically simple claims that many believers have been inclined to make that suggest that those who wish to investigate "the ultimate source of everything" *do* need to equip themselves with the fruits of an investigation of "limits" and "bounds" as a subject matter in its own right. On the one hand, believers often claim that God is infinite. On the other hand, believers often claim that God is omnipotent, omniscient, omnipresent, eternal, perfectly good, the sole creator of the universe, and so forth. All these claims, when interpreted in a straightforward and literal way, strongly suggest that believers must actually be relying on the results of investigations of "limits" and "bounds" as subject matters in their own right.

Now, of course, it might be said that these various claims should not be interpreted in a straightforward and literal way. However, I take it that the discussion in the previous two sections of this chapter strongly supports the view that believers ought not to take such a line. While believers can perfectly well maintain that a *complete* characterization of God is beyond our imaginative and conceptual capacities, such believers are obliged to allow that we have the capacity to provide an intelligible – literal, straightforward – description that fixes the referent of the name "God." Of course, some will not be persuaded that this is so. No matter; those not persuaded should think of this inquiry as conditional in form: what should those who suppose that it is straightforwardly and literally true that God is omnipotent, omniscient, and so forth, allow that investigations of "limits" and "bounds" as subject matters in their own right contribute to their understanding of these claims?

There are straightforward ways in which literal interpretations of the claims that God is omnipotent, omniscient, and so on, *suggest* involvement with investigations of "limits" and "bounds" as subject matters in their own right. To say that God is omnipotent is, at least roughly, to say that it is within God's power to do anything that it is logically possible for God to do. To say that God is omniscient is, at least roughly, to say that God knows the truth status of every proposition for which it is logically possible that God know the truth status of that proposition. To say that God is omnipresent is to say that every spatiotemporal location is present ("available") to God. To say that God is eternal is to say that every time is present ("available") to God. (Some say, rather, that God is sempiternal, i.e., that God exists at every time. My formulation is neutral on the question of whether God is *in* time.) To say that God is perfectly good is to say, inter alia, that there is no moral obligation, moral duty, or moral good to which God fails to pay due accord. To say that God is the sole creator of the universe is to say that God is the sole original creator of all contingently existing things. And so forth. In every case, the attribution of one of these properties to God brings with it quantification over a domain of objects, and, in each case, there is then a serious question to address concerning the measure or cardinality of that domain.

Consider the case of omniscience. If we suppose that God knows the truth status of every proposition for which it is logically possible that God know the truth status of that proposition,[5] then a natural question to ask is, how many propositions are

---

[5] Note that if we suppose that God knows the truth status of every proposition for which it is logically possible that God know the truth status of that proposition, we do not suppose – but rather take no stance on the claim – that there are propositions for which it is not logically possible that God know their truth status. My formulation here is meant to be neutral on, for example, the question of whether Gödel's limitative theorems would have application in the case of God.

there concerning which God knows the truth status? Before we try to investigate the question, we need to tighten it up a little. In the case of human beings, it is a reasonable conjecture that there is a quite small bound on the number of propositions that are *explicitly* represented by a human agent over the course of a typical human life. Of course, it *might* be that the finite number of propositions that are explicitly represented over the course of a typical human life *entails* an infinite number of propositions that *might* then be said to be implicitly represented over the course of a human life. However, at least on standard accounts of divine knowledge, there is no corresponding distinction in the case of God's knowledge: every proposition that God knows is a proposition of which God has explicit representation (or, perhaps better, direct acquaintance). However, if every proposition that God knows is a proposition of which God has explicit representation (or with which God has direct acquaintance), and if God knows the truth status of every proposition for which it is logically possible that God know the truth status of that proposition, then one might think that there is good reason to suppose that God has explicit representation of (or direct acquaintance with) infinitely many distinct propositions. At the very least, it seems *implausible* to suppose that there are only finitely many distinct propositions concerning which God can have knowledge of their truth value.

There are, of course, many subtleties here. While those of us of a Platonist bent may be inclined to suppose that even a natural language such as English has the capacity to represent infinitely many distinct propositions (consider, for example, the propositions expressed by the sentences $1 + 1 = 2$, $1 + 2 = 3$, $1 + 3 = 4$, etc.) there will be at least some radical finitists who deny that this is so. (Perhaps, they might say, there is no good reason to suppose that the operations that are invoked in the "specification" of infinite lists of well-formed sentences of English are total!) Moreover, while those of us of a Platonist bent may also be inclined to suppose that there *are* infinitely many distinct propositions that could be expressed by sentences of English, there will be at least some intuitionists and constructivists who deny, at least in the case of the initial example, that there are propositions that exist independently of the actual construction or tokening of the relevant sentences in some language. (Perhaps, that is, they might say that there is merely a potential infinity of propositions that can be expressed by sentences of English.) However, regardless of the correct position to take concerning the expressive capacities of natural human languages such as English, questions about the nature of the representational properties that are attributed to God also need to be taken into account. If we suppose, as standard Christian theology would have it, that there is nothing *potential* in God, then it seems that there is good reason to deny that it is possible to apply a constructivist or intuitionistic – or even radically finitist or formalist – account of mathematical truth and mathematical ontology to God's knowledge or to the propositions that are known by God. Of course, we might wonder whether it is appropriate to suppose that God has a language of thought – or, indeed, whether it is appropriate to suppose that God has beliefs or other representational states of that kind – but, no matter how these matters are resolved, it seems at least prima facie plausible to suppose that the attribution of omniscience to God will lead us to claim that there *are* infinitely many distinct propositions that are known to God. (I return to the consideration of some of the relevant subtleties that are raised by the discussion of omniscience in the next section.)

What goes for omniscience goes for the other properties that I listed in the previous section. On plausible interpretations of the simple subject-predicate sentences that I listed, it is highly natural to suppose that the truth of any one of those sentences brings with it a commitment to infinite domains of objects and/or infinite magnitudes of degreed properties. At the very least, it is prima facie plausible to suppose that there are infinitely many different possible actions that an omnipotent being can perform, and that there are infinitely many different tasks that have been carried out by a sole creator of all contingently existing things; it is prima facie plausible to suppose that a four-dimensionally omnipresent being is present to an infinite volume of spacetime (and this because it is plausible to suppose that the universe is open in the future); it is prima facie plausible to suppose that an eternal being is present to an infinite extent of times (again because it is plausible to suppose that the universe is open in the future); it is *prima facie* plausible to suppose that a perfectly good creator has created a world of infinite value (because it would be unworthy of such a being to create a world of lesser value than some other world that it might have created); and so forth. Moreover, of course, the commitment to infinite magnitudes of degreed properties seems evident on its face in the case of the claim that God is infinite (although see the following section for discussion of some of the difficulties that are raised by this claim).

Considerable recent philosophical activity has sought to apply recent mathematical discussions of infinity to the divine attributes that are currently under discussion. In particular, there is a considerable literature on omniscience that draws on Cantorian theories of the infinite (mostly drawing on or discussing ideas that were first canvassed by Patrick Grim [1991]). However, even in the case of omniscience, there has been no *systematic* study of the kind that would be needed to address the kinds of questions that I have been raising in this section. There is a large program of research here waiting to be carried out.

11.5 God and Infinity

Among the various claims listed for consideration in the previous section, the claim that *God is infinite* raises special difficulties. As we noted initially, some might suppose that this claim is only to be interpreted in a loose or metaphorical sense: what it really means to say is that God is imperishable, or unchanging, or the source of everything, or the like. Of course, if this is all that the claim that God is infinite is taken to really mean, then understanding of the claim that God is infinite will not be enhanced by considerations drawn from an investigation of "limits" and "bounds" as a subject matter in its own right. But I do not think that it is plausible to suppose that this is all that those who now claim that God is infinite mean to assert; certainly, it is not all that *many* of those who now claim that God is infinite mean to assert. From this point, I shall proceed under the assumption that those who claim that God is infinite mean to assert something that is susceptible of explanation in terms of considerations drawn from an investigation of "limits" and "bounds" as a subject matter in its own right.

Perhaps the most plausible way to interpret the claim that God is infinite is to take it to be the claim that God is infinite in certain respects. Some might think that it should be taken to be the claim that God is infinite in *every* respect, but unless we have some very subtle way of determining what counts as a respect, it seems likely that this further claim will have untoward consequences. For example, there are few who would wish to claim that God has infinitely many parts, or that God consists of infinitely many distinct persons, or that God has created infinitely many distinct universes, or the like. Surely there are *none* who would wish to make contradictory claims – for example, that God is infinitely small and infinitely large, or infinitely heavy and infinitely light, or infinitely knowledgeable and infinitely ignorant, and so forth. However, if we take the claim that we are interested in to be the claim that God is infinite in *certain* respects, then, of course, we shall naturally wish to inquire about the nature of those relevant respects.

One natural thought is that, for any degreed property that it is appropriate to attribute to God, God possesses that property to an infinite degree. However, there are reasons for thinking that this thought is not obviously correct. Suppose that God is three-dimensionally omnipresent, and that three-dimensional omnipresence is taken to be understood in terms of presence to every volume of space. It should not be a conse-quence of the claim that God is infinite that God is present to an infinite volume of space, for it may be that we want to deny that it is even possible for the volume of space to be infinite, and even if we do not wish to deny that it is possible for the volume of space to be infinite, we should surely allow that we do not currently have overwhelming reason to think that the spatial volume of our universe is infinite. Yet the property of being present to a volume of n m^3 *is* a degreed property: something could be present to a volume of 1 m^3, 2 m^3, 3 m^3, and so forth.

The fix here is not hard to see. Rather than suppose that, for any degreed property that it is appropriate to attribute to God, God possesses that property to an infinite degree, we should say rather that, for any degreed property that it is appropriate to attribute to God, God possesses that property to the maximal or minimal possible extent that is consistent with the obtaining of other relevant facts, or else God possesses that property to an infinite degree. In the case of presence to volumes of space, God is present to the most inclusive volume of space; if that volume of space happens to be finite, then God is present to a finite volume of space – N m^3 – but there is nothing objectionable about the fact that God does not possesses this degreed property to an infinite extent.

Some philosophers might have thought it preferable to try for a different fix. Suppose that we have some acceptable way of distinguishing between the intrinsic – or perhaps *nonrelational* – properties of God and the extrinsic – or perhaps *relational* – properties of God. Then, among the degreed properties, it may seem right, at least initially, to say that whether extrinsic properties of God are infinite in degree can depend on what it is that God is accidentally (contingently) related to under those properties. However, on this line of thought, it will then seem that whether the intrinsic properties of God are infinite in degree *cannot* depend on what it is that God is accidentally (contingently) related to under those properties – given that, by definition, those properties are nonre-lational – and so it will seem reasonable to insist that, in these cases, God must possess the properties to an infinite degree.

Consider, again, the example of God's knowledge. We have already seen that it is at least prima facie plausible to claim that there are infinitely many distinct propositions whose truth value is known to God. But it does not *immediately* follow from this prima facie plausible claim that it is prima facie plausible to attribute some kind of infinite faculty to God. For whether we should say that we are here required to attribute some kind of infinite faculty to God *might* be thought to turn up whether we are required to attribute knowledge of the truth value of infinitely many logically independent propositions to God. Given that the attribution of omniscience requires that God knows (more or less) every logically independent proposition, that in turn would invite assessment of exactly how many logically independent propositions there are. Various subtleties now arise. What, exactly, do we mean by logically independent propositions?[6] Are the propositions that it is possible that p and that it is possible that q logically dependent propositions for any propositions that p and that q? If we assume that the correct logic for modality is S5, then, for any proposition that p, if it is possible that p, then it is a necessary truth that it is possible that p. If we suppose that all necessary truths are logically dependent, then we shall arrive at the view that only some collections of contingent propositions are mutually logically independent. Yet, even if we accept the (controversial) assumptions required to arrive at this view, it is not clear whether we should then go on to draw the further conclusion that there are only finitely many logically independent propositions that are known by God. (Moreover, even if we do conclude that there are only finitely many logically independent propositions that are known by God, we might still think that, if there can be nothing potential in God, God is required to have explicit representations of infinitely many distinct propositions.)

If we suppose that God's omniscience requires that God is related to infinitely many contingently true propositions, and if we also suppose that this entails that God has infinitely many distinct explicit representations, then we might suppose that there are intrinsic properties of God that are infinite in degree, even though there is also a sense in which these intrinsic properties are dependent on the world in which God is located. It seems plausible to think that the counting of representational states is an intrinsic matter (how many distinct representational states one has at a given time supervenes merely on how one is at that time, and not at all on how the rest of the world is), even though there are causal relations that hold between representational states and the world that contribute to the determination of how the number of distinct representational states that one is in varies over time. Thus, whether God is related to infinitely many contingently true propositions can be both a question about an intrinsic property of God – how many distinct explicit representations does God have – and yet also a question about the world in which God is located (given that the number of God's distinct explicit representations of contingently true propositions simply reflects the complexity of the world in which God is located).

---

[6] Many important and interesting questions arise here. In particular, there are questions about whether we can understand talk of "logical independence" that is not tied to the specification of particular linguistic resources. If we specify a language and a proof theory (or model theory), then we can give an account of logical independence for the logical system thus specified. But what are we to make of talk of "logical independence" that is not thus tied to specification of a particular logical system? There are other places in these notes where some will suppose that what is said makes no sense because these kinds of foundational questions about languages, syntax, and interpretation have not been addressed. I take it that this points to yet another area of inquiry that cannot be avoided in a full examination of the implications for theology of investigations of "limits" and "bounds" as subject matters in their own right. (Some might think that there are also questions about the logic that is proper to discussion of God, just as some have supposed that there are questions about the logic that is proper to discussion of quantum mechanics. My own view is that classical logic is the proper logic for discussions of both quantum mechanics and God.)

If, then, we take the claim that God is infinite to be the claim that God is infinite in certain respects, then *perhaps* we can say something like the following: For any degreed property that it is appropriate to attribute to God, God possesses that property to the maximal or minimal possible extent that is consistent with the obtaining of other relevant facts, or else God possesses that property to an infinite degree.[7] If there is some sense in which a degreed property is relational, then it may be that whether that property is infinite in degree depends on what it is that God is accidentally (contingently) related to under that property, but where there is no sense in which a degreed property is relational, it cannot be that whether that property is infinite in degree depends on what it is that God is accidentally (contingently) related to under that property. Given that it is possible for an intrinsic property to nonetheless be, in some senses, relational, it should not be thought that, merely because a degreed property is intrinsic, it cannot be that whether that property is infinite in degree depends on what it is that God is accidentally (contingently) related to under that property.

Perhaps, however, we shouldn't take the claim that God is infinite to be the claim that God is infinite with respect to *all* of an appropriately restricted class of degreed properties; perhaps, rather, we should take the claim that God is infinite to be the claim that God is infinite in *certain very particular* respects. I don't have any clear suggestion to make about what these very particular respects in which God is infinite might be; perhaps, however, further investigation of this line of thought might turn up some interesting results.

Anselm refers to God by the formula "that than which no greater can be conceived." It is not impossible that one might think that the claim that God is infinite should be tied to the sense of greatness that is implicated in St. Anselm's formulation. Surely St. Anselm would have agreed with the claim that God is infinitely great, and surely it is not utterly implausible to think that modern theories of mathematical infinity might be pressed into service in the understanding of this claim. Alas, however, it is not clear what sense should be interpreted to "greatness" in Anselm's formulation. (Indeed, this is a much debated question in the recent literature on this topic.) Thus, although we might make progress on what is meant by "*infinitely* great," it is less clear that we will make progress on what is mean by "infinitely *great*." (I continue with this theme in the next section.)

Of course, it should not be thought that the preceding discussion exhausts the kinds of considerations – never mind the details of the considerations – that should be raised in the course of an examination of the claim that God is infinite, when that claim is given a straightforward, literal interpretation that draws on investigations of "limits" and "bounds" as subject matters in their own right. As in previous sections of this chapter, I claim to have done little more than indicate where further work needs to be done.

---

[7] Even this formulation is at best provisional. I have already noted that "consisting of N persons" is a degreed predicate. If we allow that it is possible that a being consists of three persons, how can we deny that it is possible that a being consists of four persons? Yet, somehow, the Christian theologian needs to be able to defend the claim that God consists of exactly three persons. I shall not speculate here about further refinements to the principle that I have begun to formulate.

11.6 God and the Transfinite

Throughout the chapter to this point, I have made free use of the expressions "infinite" and "infinity" in talking about domains of objects and magnitudes of degreed properties. Of course, even if one grants that we can make sense of talk of "infinite" domains of objects and "infinite" magnitudes of degreed properties, one might insist that, in the light of the development of Cantor's theory of the transfinite, one needs to bring far more precision to this kind of talk than our initial discussion has recognized. In particular, given that there is a hierarchy of infinite cardinals, it seems that we need to ask about the particular infinite cardinals that might be thought to be appropriate to the characterization of God's properties.

One way, among many, into this topic is by way of some reflection on the theory of numbers developed in Conway (1976).[8] In Conway's system, there is a "gap" – "**On**" – that lies at the end of the number line. Intuitively, **On** – that is, {**No** | }, where "**No**" is shorthand for the entire number line – is "greater" than all the numbers, including, in particular, all of Cantor's infinite cardinals. Anything that has a magnitude that is properly characterized by **On** will have a magnitude that is not properly characterized by any number, however large.[9]

In Conway's theory of numbers – as in Cantor's theory of ordinal numbers – we have a sequence of numbers ordered by the "greater than" relation, including a series of (special) limit ordinals that can be identified with the distinct infinite cardinals of Cantor's theory of cardinal numbers. Thus, if we are talking about "infinite" quantities in the context of Conway's theory of numbers – or Cantor's theory of infinite cardinals – the question will always arise about the size of the infinity under consideration. Moreover, if one wishes to "exceed" any limitations that might be placed on size, then one will be driven to talk about things that are not properly considered to be numbers at all. (I suspect that this point is linked to the idea – to be found in many versions of set theory – that there are collections that form only proper classes and not sets, because they are "too big" to be collected into sets.)

If, then, one is to say that God is "infinite" in such and such respects – or that God is "infinite" (*sans phrase*) – the question will always arise about the size of the infinity under consideration. Given the discussion in Section 11.5, one might think that the appropriate thing to say is something like this: for any degreed property that it is appropriate to attribute to God, God possesses that property to the maximal or minimal possible extent that is consistent with the obtaining of other relevant facts, or else God possesses that property to an unlimited (unquantifiable, proper-class-sized)

---

[8] For a reasonably brief exposition of Conway's theory, see Oppy (2006, pp. 42–4).

[9] There is at least a loose sense in which **On** can be identified with Cantor's "absolute infinity." Like Cantor's absolute infinity, **On** "lies beyond" all of the transfinite numbers. However, there are claims that are sometimes made about Cantor's absolute infinity that are clearly not true of **On**. In particular, it should be noted that there is nothing inconceivable about **On**; nor is it the case that **On** cannot be either uniquely characterized or completely distinguished from the transfinite numbers (see Rucker 1982, p. 53). On the contrary, **On** is a gap rather than a number, and it is distinguished, in particular, by the fact that it "exceeds" all of the transfinite numbers. (I think that the claims in question are no more plausible in the case of Cantor's absolute infinity, but it is, I think, even more clear that they are not plausible in the case of **On**.)

extent. Under this reformulation, we allow that it is possible that God's possessing a given property to a certain maximal extent forces us to say that God possesses that property to a given infinite cardinal degree. (If, for example, there are $\aleph_{15}$ true propositions, then God knows $\aleph_{15}$ true propositions; in that case, God's knowledge is infinite, but it is not unquantifiably infinite.) However, we also allow that, at least until further considerations are brought to bear, it remains an open question whether God possesses some properties to an unquantifiable extent. (If, for example, there are proper class many – **On** – true propositions, then God's knowledge is unquantifiably infinite.)

There may be some pressures that nudge theologians in the direction of saying that God does possess at least some properties to an unquantifiable extent. Suppose, for example, that we accept Anselm's formulation: "God is that than which no greater can be conceived." Because it seems that we can conceive of creatures who possess some great-making properties that are of unquantifiable extent, there is at least some reason to suspect that we will be driven to the conclusion that God possesses those properties to an unquantifiable extent. However, once again, this is a matter for more careful investigation.

11.7 Checking for Consistency

In the introduction to Oppy (2006), I hinted at the existence of an argument for the conclusion that "there is no conception of the infinite that can be successfully integrated into relatively orthodox monotheistic conceptions of the world" (p. xi). About this hint, I wrote: "Since all that this brief introduction aims to do is to make it seem plausible that there is a *prima facie* interesting question to address, I shall leave further discussion of this argument to the future" (p. xiii). In this section of the present chapter, I reproduce the earlier discussion and provide some further comments on it (although certainly not on the scale envisaged in the just-quoted remark).

Here's what I wrote:

> If we are strict finitists – and thus reject all actual and potential infinities – then we are obliged to say that God is finite, and that the magnitudes of the divine attributes are finite. But what reason could there be for God to possess a given magnitude to degree N rather than to degree N + 1? More generally, how could a finite God be the kind of endpoint for explanation that cosmological arguments typically take God to be?
>
> If we are potential infinitists – i.e. if we reject all actual infinities, but allow that some entities and magnitudes are potentially infinite – then it seems that we will be obliged to say that God is potentially infinite, and that the magnitudes of the divine attributes are potentially infinite. But what kind of conception of God can sustain the claim that God is susceptible of improvement in various respects? If God possesses a magnitude to degree N even though God could possess that magnitude to degree N + 1, surely God just isn't the kind of endpoint for explanation that cosmological arguments typically take God to be.
>
> If we are neither strict finitists nor potential infinitists, then it seems that we must be actual infinitists, i.e. we must suppose that God is actually infinite, and that the magnitudes of the divine attributes are actually infinite. But is there a conception of the infinite that can sustain the claim that God is actually infinite, and the claim that the magnitudes of the

divine attributes are actually infinite without undermining the kinds of considerations to which orthodox cosmological arguments appeal in attempting to establish that God exists? Indeed, more generally, are there conceptions of the infinite that can sustain the claim that God is actually infinite, and the claim that the magnitudes of the divine attributes are actually infinite *tout court*? Moreover, if there is a conception of the infinite that can sustain the claim that God is actually infinite, can this conception of the infinite also sustain the idea of an incarnate God, and the idea that there is an afterlife in which people share the same abode as God?

(Oppy 2006)

As I noted at the end of the previous section, it seems to me that there are pressures that drive theologians in the direction of claiming that God possesses some properties to an unquantifiable – "more than proper class many," "**On**" – extent. But, if that is right, then it seems that theologians should not look with any fondness on those philosophical views that deny that we can form a coherent conception of actually infinite domains and actually infinite magnitudes. Rather than side with formalists, or radical finitists, or constructivists, or intuitionists, or those who insist that there are none but merely potential infinities, believers in God should say instead that there can be domains and properties that are "unquantifiably infinite," that is, not measurable by any of the cardinalities that are to be found in Cantor's paradise. Thus, I take it, the direction of thought that is expressed in the first two paragraphs of the quotation is acceptable without qualification (although there is much more to be said in defense of the main theses outlined therein). Furthermore – and for the same reasons – I take it that the line of thought that is expressed in the first part of the third paragraph is also acceptable.

However, when we turn to the question of traditional arguments for the existence of God, matters are rather more interesting than the above compressed presentation allows. What is true is that there are *some* traditional arguments for the existence of God (e.g., one a priori version of the kalām cosmological argument) in which it is explicitly assumed that there can be no actual infinities. *Those* arguments cannot be defended consistently with the adoption of the conception of divine infinity articulated in the previous section. Of course, there are many other arguments, including many other cosmological arguments, that make no such (implicit or explicit) assumption about the impossibility of actual infinities. *These* arguments are not impugned by the considerations about infinity to which I have been here adverting. (They may be impugned by *other* considerations about infinity, but that's another story.) It would be an interesting project to run an inventory of arguments about the existence of God, to determine where considerations about infinity come up, and to check to see how those arguments fare under the kinds of considerations that were adduced earlier.

When we turn to matters such as the idea of an incarnate God and the idea that there is an afterlife in which people share the same abode as God, there is yet a further raft of considerations that comes into view. It is not easy to reconcile the suggestion that God is actually infinite with the idea that God took on a finite physical form. It is not easy to reconcile the idea that God's abode is infinite with the idea that that abode is inhabited by finite physical creatures (such as ourselves). At the very least, there is clearly an interesting possible project that investigates the ways in which particular Christian doctrines – concerning, for example, incarnation, trinity, atonement, and so forth – are affected by particular theories about the ways and respects in which God is actually infinite. As I suggested at the outset, a full investigation of the implications for theology of an investigation of "limits" and "bounds" as a subject matter in its own right is likely to be very prolonged indeed.

11.8 A Concluding Stocktaking

As I said initially, this chapter is intended to be a kind of prolegomenon to the discussion of infinity in theological contexts. What I have tried to do is to raise various kinds of issues in a preliminary way, without in any way supposing that my comments on these issues constitute decisive verdicts. Perhaps it will be useful, in closing, to provide a summary of the range [...] has been canvassed (and of the opinions that I have expressed).

First, there are iss[...] d the question of the *normative significance* that mathematical a[...] gations of the infinite have for one another. On the one hand, [...] highly plausible to think that mathematical investigations of i[...] ant consequences for theology that should be recognized on all [...] s claim relies on some contentious assertions about the proper [...] on of theological talk; more about this anon.) On the other ha[...] be equally plausible to suppose that *only* those who actually a[...] al presuppositions will suppose that theological investigations [...] nt implications for mathematics and the physical sciences. As [...] of Section 11.1, there are no substantive results in contempo[...], or physics that depend essentially on theological assumptions [...] o make theological assumptions in order to earn entitlement [...] ce, or the fundamental theory of calculus, or the theory of g[...] atever.[10]

Second [...] cluster around the *interpretation* of theological talk. There is [...] iming that much – or even all – theological talk is [...] at any rate, not susceptible of a straightforward realist constru[...] t it is highly plausible to think that mathematical inves-tigatio[...] ignificant consequences for theology only if theological talk is [...] dly realist construal. At the very least, it is very hard to see h[...] t mathematical analyses of the infinite bear at all on theo-logic[...] te if the latter talk is all taken to be merely metaphorical, or ana[...]

[...] that cluster around the *limits* that one might wish to impose on [...] listic theological talk because of alleged limitations in our ca[...] completely understand the central objects of theological talk.

[...] ent expressions of interest in, and support for, the notion of "theistic science." I take it that the key [...] s that there might be significant scientific – or logical, or mathematical – results that depend essentially on theological assumptions. If, for example, the fine-tuning of the cosmological constants is best explained – and only explainable – by the hypothesis of intelligent design, then that explanation *might* count as an example of "theistic science." Thus, the claim that I have made in the main text is not entirely uncontroversial.

I have suggested that it is not at all obvious that it is reasonable to allow that there are properties of God of which it is logically or metaphysically necessary that we have no more than a partial or incomplete grasp of what those properties *are*. However, if there are properties of God of which it is logically or metaphysically necessary that we have no more than a partial or incomplete grasp of what those properties *are*, then that suggests one kind of limitation on the application of mathematical investigations of infinity in theological contexts that will need to be respected.

Fourth, there are issues that cluster around the *identification* of those parts of theology where it is plausible to suppose that mathematical investigations of infinity will have significant consequences. I take it that the obvious place to look is the discussion of divine attributes. There are many divine attributes that seem to involve some kind of imputation of infinite magnitude to properties or infinite domains of entities. While I mentioned a few plausible candidates – omniscience, omnipotence, omnipresence, eternity, perfect goodness, sole creation of the world *ex nihilo* – I don't pretend that this list is either systematically generated or exhaustive. However, it should not be supposed without further investigation that there are no other parts of theology – that is, apart from discussion of the divine attributes (and, of course, the arguments for and against the existence of God) – where mathematical investigations of infinity will have significant consequences.

Fifth, there are issues that cluster around the *application* of the results of mathematical investigations of infinity to those parts of theology where it is plausible that mathematical investigations of infinity will have significant consequences. How exactly can or do mathematical accounts of infinity contribute to the analysis, understanding, or explanation of particular divine attributes (or of particular arguments concerning the existence of God)? What commitments to infinite magnitudes of properties or infinite domains of entities are plausibly incurred by way of the attribution of particular divine attributes (or the adoption of particular arguments concerning the existence of God)? What kinds of infinities are involved in those cases in which there are commitments to infinite magnitudes of properties or infinite domains of objects incurred by way of the attribution of particular divine attributes (or the adoption of particular arguments about the existence of God)?

Sixth, there are issues that cluster around the *consistency* or stability of uses of the results of mathematical investigations of infinity in theology. Once we have in view a map of the ways in which the results of mathematical investigations of infinity have been – or could be – applied across a range of theological domains, we are then in a position to ask whether those results have been – or would be – applied in a consistent manner across those domains. I have suggested, for example, that there are serious questions to be asked about the consistency of the treatment of infinity in some of the standard arguments for the existence of God with the treatment of infinity in some of the standard analyses of the divine attributes.

Seventh, there are issues that cluster around the question of the *normative significance* that philosophical and theological investigations of the infinite have for one another. I take it that theological hypotheses can have significant consequences for philosophical debates about the ways (if any) in which infinity is present in the world. I sketched an argument that suggests that standard theological hypotheses bring with them a range of commitments to actual infinities and to Platonist interpretations of contested philosophical domains. If this is right, then, for example, the adoption of standard theological hypotheses has important consequences for the debates about formalism, finitism, intuitionism, and Platonism in the philosophy of mathematics for those who take these theological hypotheses seriously.

Eighth, there are issues that cluster around the *application* of the results of mathematical investigations of infinity to specific parts of Christian theology and doctrine – for example, to discussions of trinity, incarnation, immortality, and so forth. To the extent that there has been prior discussion of the *application* of the results of mathematical investigations of infinity to theology, this discussion has tended to focus on questions about generic divine attributes, that is, divine attributes as these are conceived on most monotheistic conceptions of God. However, it seems to me that there are bound to be questions that are quite specific to Christian theology and doctrine for which investigation of "limits" and "bounds" as a subject matter in its own right has important consequences.[11]

Although the examination of infinity in theological contexts is doubtless not itself an *infinite* task, it is abundantly clear – even from this relatively superficial and incomplete overview – that there is *plenty* of work to be done.

Acknowledgments

I am grateful to all of the participants in the "New Frontiers in Research on Infinity" Conference, San Marino, August 18 to 20, 2006. In particular, I am grateful to Wolfgang Achtner, Denys Turner, and Marco Bersanelli for their detailed comments on an earlier draft of this paper; to Ed Nelson, Hugh Woodin, Anthony Aguirre, and Michael Heller for discussion of relevant points in mathematics, set theory, and theoretical physics; and to Charles Harper, Bob Russell, and Melissa Moritz for their feedback, enthusiasm, and extraordinary organizational skills.

References

Barrow, J. 2005. *The Infinite Book*. London: Vintage.

Benardete, J. 1964. *Infinity: An Essay in Metaphysics*. Oxford: Oxford University Press.

Conway, J. 1976. *On Numbers and Games*. London: Academic.

Grim, P. 1991. *The Incomplete Universe: Totality, Knowledge and Truth*. Cambridge: MIT Press.

Lavine, S. 1994. *Understanding the Infinite*. Cambridge: Harvard University Press.

Lewis, D. 1991. *Parts of Classes*. London: Blackwell.

Meyer, R. 1987. God exists! *Nous* 21: 345–61.

Moore, A. W. 1990. *The Infinite*. London: Routledge.

Moore, A. W. 1998. Infinity. In *Routledge Encyclopaedia of Philosophy*, E. Craig (ed.), pp. 772–78. London: Routledge.

---

[11] Perhaps because of the nature of my own interests, I have focused here particularly on considerations from logic, philosophy of language, and metaphysics. There are also interesting *epistemological* issues that are raised by questions about the infinite (for some introduction to these considerations, see, e.g., Thomson 1967; Lavine 1994) and, in particular, by questions about the infinite in the context of theology.

Oppy, G. 2006. *Philosophical Perspectives on Infinity*. Cambridge: Cambridge University Press.
Owen, H. 1967. Infinity in theology and metaphysics. In *The Encyclopaedia of Philosophy*, P. Edwards (ed.), pp. 190–93. London: Collier-Macmillan.
Rucker, R. 1982. *Infinity and the Mind: The Science and Philosophy of the Infinite*. Sussex: Harvester Wheatsheaf.
Thomson, J. 1967. Infinity in mathematics and logic. In *The Encyclopaedia of Philosophy*, P. Edwards (ed.), pp. 183–90. London: Collier-Macmillan.

CHAPTER 12

Notes on the Concept of the Infinite in the History of Western Metaphysics

David Bentley Hart

12.1 The Infinite as a Metaphysical Concept

1. There is not – nor has there ever been – any single correct or univocal concept of the infinite. Indeed, the very word by which the concept is named typically – and appropriately – possesses a negative form and is constructed with a privative prefix: a[-peiron, aj-perivlhpton, aj-ovriston, aj-pevranton, aj-mevtrhton, *in-finitum*, *Un-endliche*, and so on. In order, therefore, to fix on a proper conceptual "definition" of the infinite, it is necessary to begin with an attempt to say what the infinite is not.

2. Before that, however, one ought to distinguish clearly between the "physical" (or mathematical) and "metaphysical" (or ontological) acceptations of the word "infinite." The former, classically conceived, concerns matters of quantitative inexhaustibility or serial interminability and entered Western philosophy at a very early date. Even in pre-Socratic thought, questions were raised – and paradoxes explored – regarding such imponderables as the possibility of infinite temporal duration, or of infinite spatial extension, or of infinite divisibility, and mathematicians were aware from a very early period that the infinite was a function of geometric and arithmetical reasoning, even if it could not be represented in real space or real time (that is to say, a straight line must be understood as logically lacking in beginning or end, and the complete series of real, whole, even, odd, etc., numbers must be understood as logically interminable). The paradoxes of Zeno, for example, apply entirely and exclusively to this understanding of the infinite and concern the apparent conceptual incompatibility between the logical reality of infinite divisibility and the physical reality of finite motion (inasmuch as the infinite divisibility of space would seem to imply the necessity of an infinite, ever more "local" seriality within all actions in space or time).

3. The metaphysical concept of the infinite is rather more elusive of definition and, as a rule, must be approached by a number of elliptical and largely apophatic paths. Perhaps its most essential "negative attribute" is that of absolute indeterminacy: the infinite is never in any sense "this" or "that"; it is neither "here" nor "there"; it is unconditioned; it is not only "in-finite" but also "in-de-finite." Granted, in the developed

speculative traditions of the West, pagan and Christian alike, this indeterminacy – when ascribed to the transcendent source of being – came to be understood also as a kind of "infinite determinacy"; but even here this determinacy consists in an absolute transcendence of finite determination. The rule remains indubitable: only that which is without particularity, definition, limit, location, nature, opposition, or relation is "infinite" in the metaphysical sense. Hence, "matter" – in the sense of u{lh or *materia prima* – is understood by classical and mediaeval philosophy as infinite precisely because it is utterly devoid of the impress of morfhv or *forma*, and not because it is in any sense limitlessly extended or divisible. Neither extension nor divisibility, in fact, applies in any way to prime matter; in Aristotelian terms, prime matter belongs entirely to the realm of toV dunatoVn, the possible, so long as no ejnevrgeia supervenes on it to grant it actual form; its infinity, therefore, is purely privative. Thus, space and time are not "metaphysically" infinite, despite their interminabilities. Neither, moreover, is any arithmetic series: the set of even numbers, for instance, while endless, is nevertheless definite: it includes and excludes particular members; it is a particular kind of thing; it is bounded by its own nature. In the purely metaphysical sense, in fact, the mathematical infinite remains within the realm of the finite.

4. This distinction between the mathematical and metaphysical concepts of the infinite was not obvious to the earliest thinkers of the Greek tradition and was often at best only a tacit distinction within their reflections. Even in the classical age of Greek thought, the difference between the interminability of "number" and the indeterminacy of "possibility" was not much remarked, although it is obviously implicit in Aristotle's observations (in *Physics* III, 6–7) regarding, on the one hand, spatial extension and divisibility and, on the other, the purely *potential* existence of the infinite in the realm of discrete substances.

5. Aristotle's distinction – however undeveloped it may be – leads toward the rather striking (but more or less inevitable) conclusion that the infinite, metaphysically conceived, is invariably and necessarily related to the question of being: how is it that anything – any finite thing, that is – exists? To resort to a somewhat Heideggerean idiom, whereas the mathematical infinite is an essentially "ontic" category, the metaphysical infinite is thoroughly "ontological." It is that over against which the finite is (phenomenologically) set off, or out of which it is extracted, or on which it is impressed, or from which it emanates, or by which it is given. This is true at both extremes of the metaphysical continuum: at the level of prime matter or at the level of the transcendent act or source of being; at the level of absolute privation or at the level of absolute plenitude.

6. From this one can draw a very simple metaphysical distinction between the finite and the infinite. The realm of the finite is that realm in which the *principium contradictionis* holds true. In an utterly vacuous sense, the more "positive" *principium identitatis* holds true in all worlds and at every level of reality, but the more "negative" *principium contradictionis* describes that absolute limit by which any finite thing is the thing it is. In a very real sense, finite existence *is* noncontradiction. Neither infinite potentiality nor infinite actuality excludes the coincidence of opposites, inasmuch as neither in and of itself *posits* anything; only when, on the one hand, potentiality is realized as a single act (and so becomes "this" thing rather than "that" thing) and, on the other hand, actuality limits itself by commerce with potentiality (and so becomes

the particular existence of "this" thing rather than "that" thing) does any particular thing come into being. Only that which is posited – only that which has form and can be thought – "exists." Or, to phrase the matter purely in phenomenological terms, existence is manifestation, and manifestation is finitude; the *Sache* of thought is that object in which a certain set of noetic intentions are realized and by which an endless multitude of other intentions are frustrated.

7. Thus, we can say that the "infinite" – at least metaphysically speaking – is not merely that which lacks boundary or end (despite the word's etymology), but is rather that wherein the *principium contradictionis* does not hold true. In a very real sense, therefore, the infinite does not "exist." It is devoid of morfhv, it has no mode, it cannot be thought. This, in fact, is essentially what Aristotle says in the *Physics*, although there he is still thinking principally in terms of potentiality and unformed u{lh. One may also, however, say that what later Christian metaphysics would call toV o[ntw" o[n or *actus purus* is infinite in much the same sense: it too does not have a finite mode; it too does not, in the common sense, exist. Of *potentia pura*, one may say that it does not *exist* but nevertheless *is* insofar as it *is* possible; of *actus purus*, one may say that it does not *exist* but nevertheless *is*.

8. We see here, then, that between the mathematical and the metaphysical senses of "infinite" there exists not merely a distinction, but very nearly an opposition. In the realm of the mathematical infinite, the *principium contradictionis* remains of necessity inviolable. Thus, between the two acceptations of the word "infinite" any apparent univocity is illusory, and any possible analogy is at best pictorial, affective, and immeasurably remote.

9. The two models under which the metaphysical infinite was "classically" conceived – before the Christian period, at least – were those of "indeterminacy" and "totality." And in both the pre-Christian and post-Christian metaphysical epochs, these two models functioned often each to the other's exclusion, or were only imprecisely distinguished from one another, or were implicitly at odds with one another. In Hegel's thought, however, they were placed in explicit opposition to one another, as (on the one hand) the "bad" infinite of endless and meaningless particularity and (on the other) the "good" infinite of dialectical sublation. It is tempting to describe this entire tradition of reflection on the infinite as "idealism," although its most fundamental premises were present in Western thought from the very beginning, well before any "idea" ever floated free of the cosmos.

12.2 The Metaphysical Concept of the Infinite in Pre-Christian Thought

1. These two "masks of the infinite" appear with almost archetypal perfection at the very dawn of the Western philosophical tradition, in the two great "systems" of the pre-Socratic age, the Heracleitean and the Parmenidean. Neither Heracleitus nor Parmenides enunciates any particular metaphysics of the infinite as such, of course, but each provides a conceptual form for such a metaphysics: in the thought of the former, it is that of indeterminacy, the fecundity of chaos, the sheer boundless inexhaustibility of becoming and perishing; in the latter, it is that of the totality, absolute closure and

enclosure, the simple and eternal fixity and fullness of the whole, the eternal actuality of all possibilities.

2. For pre-Socratic thought as a whole, however – if indeed any general categorization is possible here – the "infinite" is conceived principally as the antinomy of order, and as therefore both unthinkable and dangerous. For Anaximander the a[peiron is a kind of eternal and limitless (although perhaps in some sense spherical) elemental plenum from which the finite and bounded cosmos has been extracted and against which the cosmic order is continuously preserved. For Empedocles, also, the universe is a kind of small, fragile, local island of order amidst the boundless flux of material being, a sort of city liberated and walled off from chaos. For the Pythagoreans it is pevra" – born of number – that subdues the limitless and gives dimensions (and thereby existence) to finite things. At this point, however, the distinction between the metaphysical and the mathematical infinite has not been clearly drawn. The finite is set off against the infinite, but almost entirely in the fashion of a stable concrete object set off against an unstable and fluid, but nonetheless material, "first element." The Pythagorean mysticism of number is still not the abstract Aristotelian metaphysics of form, or of the actual and the potential.

3. For Plato – and, really, for the entire classical philosophical tradition of Greece, including Stoicism – the infinite was solely a negative concept. Words like a[peiron, ajperivlhpton, ajovriston, and so forth, were more or less entirely opprobrious in connotation; they were used to designate that which was "indefinite" or "indeterminate" and, hence, "irrational" or "unthinkable." The infinite is that which lacks form, that which reflects no ei\do" and receives the impress of no morfhv. As such, it is pure deficiency. Hence, Plato would never have called the Good beyond being "infinite." Aristotle's ontology was entirely concerned with finite substances (a category that included even God, the supreme substance) and had no room for anything like a concept of "infinite being."

4. Of the entirety of classical Greek thought, from Plato onward, it may fairly be said that this "Platonic" (or perhaps "Pythagorean") prejudice remains constant: the highest value and the only ground of rational meaning is intelligibility, and so the highest good can never be conceived as lacking rational limit. Only that which possesses eidetic dimensions and boundaries is thinkable; all else is chaos, formlessness, pure irrationality, and therefore malign. One sees this not only in Plato's epistemology, but in his cosmology. In the *Philebus*, for example, the universe is considered as the product of four primordial forces: limit, the limitless, the mixture of these two, and the first cause of this mixture; the infinite in itself is an aboriginal tumult of oppositions – such as the dry and the moist or the hot and the cold – that can come to constitute a living world only when it is brought under the governance of number and harmony. In the *Timaeus*, it is only in imposing the limiting proportions of distinct ideas on chaos that the demiurge brings about a world.

5. Stoic metaphysics, needless to say, is a tradition unto itself, and although there was a constant cross-pollination of Platonic, Aristotelian, and Stoic thought over the centuries before and after the rise of Christianity in the Empire, whatever distinctive concept of the infinite one might ascribe to Stoicism surely cannot be said to be "idealist" in its guiding premises. Nevertheless, the model of reality peculiar to Stoicism

is, if anything, even more intransigently, aboriginally Greek than that of the other developed schools; it might almost be described as a remarkably refined and embellished, but essentially pure, expression of a certain pre-Socratic vision of the whole. The "Stoic infinite" obeys the logic of totality rather than that of indeterminacy. The entire cosmos, which is, viewed transcendentally, convertible with the divine mind, is an enclosed and finite order, spatially and temporally determined, endlessly recurrent, and yet invariable in its parts and processes. It is a perfect plenum, possessed of perfect order, a "cosmopolis" in no respect deficient and in every respect admirable; it is at once both the material plenitude and the harmonious city of cosmic order described in the thought of Anaximander and Empedocles. As such, its "infinitude" is also its limitation; as both an ideal order and the whole of all that is, it can contain its plenitude only in the form of a distribution of parts and succession of events. In this perfectly sealed order of arising and perishing, advancement and retreat, the whole must always and again "make room" for what it contains. Death is necessary so that new beings can arise within the divine plenitude; and, in its fullness, the whole of the spatial and temporal order is destined for ecpyrosis, a return into the primordial fire of divine mind, only that it might arise once more and repeat the same circuit again and again, identical in every detail, throughout eternity. No more astonishing, sublime, or terrible vision of the world as totality – and none more perfectly sealed within itself – can be found or imagined. However, in this system, the concept of the infinite in a truly transcendent metaphysical sense – as that which is absolved of all finite determination while giving being to the finite – remains unthought.

6. The first evidence of a purely "positive" metaphysical concept of the infinite in Western thought might be found in Plotinus. At least, one should note his willingness to speak of the One as infinite, or of Nous as infinite and the One as inexhaustible. According to W. Norris Clarke, at least, this is not mere apophasis: by "infinite" Plotinus means, inter alia, limitless plenitude as well as simplicity, absolute power as well as absolute rest. Plotinus even asserts that love for the One must be infinite because its object is (*Enneads* VI.7.32). The infinity of Plotinus's One, then, is no longer mere indeterminacy, as the infinite was for earlier Greek philosophy, but is rather the indeterminate but positive wellspring of all of being's virtues, possessing those virtues in an indefinite (and so limitless) condition of perfection and simplicity. Yet, even so, for Plotinus the infinity of the One is still in some sense only dialectically related to the totality of finite beings; it is the metaphysical reverse of the realm of difference, its abstract and formless "super-essence," at once its opposite and its substance, the absolute distinction and absolute unity of being and beings. The infinite, then, is the ground of the finite precisely in that it is "limited" by its incapacity for the finite; the One's virtues are "positive" only insofar as they negate, and so uphold, the world. Thus, Plotinus's thought comprises, if only implicitly, a kind of diremption and recuperation (to speak in Hegelian terms): the ambiguous drama of *egressus* and *regressus*, *diastole* and *systole*, a fortunate fall followed by a desolate recovery. As the world's dialectical counter, its "credit" or "treasury," sustaining its totality, the One is of necessity the eternal oblivion of the here below; it is not mindful of us, and it shows itself to us only in the fragmentation of its light, shattered in the prism of Nous, dimly reflected by Psyche in the darkness of matter. Hence, the bounty of being is pervaded by a tragic

truth: the diffusiveness of the good is sustained only by the absolute inexpressiveness of its ultimate principle. At the last, the "infinite" in Plotinus is still a concept governed by the α-privative.

7. There was no reason for any of the classical pagan schools of thought to move beyond this concept of the infinite; no principle indigenous to pagan thought demanded that they do so.

12.3 The "Christian Infinite"

1. In the intellectual world of the first three centuries before the Council of Nicaea, especially in the eastern half of the empire, something like a "Logos metaphysics" was a crucial part of the philosophical lingua franca of almost the entire educated class, pagan, Jewish, Christian, and even Gnostic (even though the term generally preferred was rarely "logos"). Certainly, this was the case in Alexandria: the idea of a "derivative" or "secondary" divine principle was an indispensable premise in the city's native schools of Trinitarian reflection and in the thought of either "Hellenized" Jews like Philo or of the Platonists, middle or late. Furthermore, one could describe all of these systems, without any significant exception, pagan and Jewish no less than Christian, as "subordinationist" in structure. All of them attempted, with greater or lesser complexity, and with more or less vivid mythical adornments, to connect the world here below to its highest principle by populating the interval between them with various intermediate degrees of spiritual reality. All of them, that is, were shaped by the same basic metaphysical impulse, one sometimes described as the "pleonastic fallacy": the notion that, in order to overcome the infinite disproportion between the immanent and the transcendent, it is enough to conceive of some sort of *tertium quid* – or of a number of successively more accommodating quiddities – between, on the one hand, the One or the Father or oJ QeoV" and, on the other, the world of finite and mutable things.

2. In all such systems, the second "moment" of the real – that which proceeds directly from the supreme principle of all things (*logos*, or *nous*, or what have you) – was understood as a kind of economic limitation of its source, so reduced in "scale" and nature as to be capable of entering into contact with the realm of discrete beings, of translating the power of the supreme principle into various finite effects, and of uniting this world to the wellspring of all things. This derivative principle, therefore, may not as a rule properly be called oJ QeoV", but it definitely is qeoV": God with respect to all lower reality. (This, of course, is why Augustine, in *Confessions* VII.9, credits the Platonists with having taught him the truths revealed in John 1:1–3.) Moreover, this inevitably meant that this secondary moment of the real was understood as mediating this supreme principle in only a partial and distorted way, for such a Logos (let us settle on this as our term) can appear within the totality of things that are only as a restriction and diffusion of – even perhaps a deviation or alienation from – that which is "most real," the Father who, in the purity of his transcendence, can never directly touch this world. For Christians who thought in such terms, this almost inevitably implied that the Logos had been, in some sense, generated *with respect to* the created order, as its most exalted expression, certainly, but as inseparably involved in its existence nonetheless.

Thus, it was natural for Christian apologists of the second century to speak of the Logos as having issued from the Father in eternity shortly before the creation of the world.

3. It was also thus that the irreducibly Alexandrian theology of Arius inevitably assumed the metaphysical – or religious – contours that it did: the divine Father is absolutely hidden from and inaccessible to all beings, unknowable even to the heavenly powers, and only through the mediation of an inferior Logos is anything of him revealed. What, of course, was distinctive in Arianism was the absence of anything like a metaphysics of participation that might have allowed for some sort of real ontological continuity (however indeterminate) between the Father and his Logos; consequently, the only revelation of the Father that Arius's Logos would seem to be able to provide is a kind of adoring, hieratic gesture toward an abyss of infinitely incomprehensible power, the sheer majesty of omnipotent and mysterious otherness. The God (oJ QeoV") of Arius is a God revealed *only* as the hidden, of whom the Logos (qeoV" oJ lovgo") bears tidings, and to whom he offers up the liturgy of rational creation. However, as the revealer of the Father, his is the role only of a celestial high priest, the Angel of Mighty Counsel, the coryphaeus of the heavenly powers; he may be a kind of surrogate God to the rest of creation, but he too, logically speaking, cannot attain to an immediate knowledge of the divine essence.

4. Even, however, in late antique metaphysical systems less ontologically austere than Arius's, in which the economy of divine manifestation is understood as being embraced within some sort of order of metochv or metousiva, the disproportion between the supreme principle of reality and this secondary principle of manifestation remains absolute. Hence, all revelation, all disclosure of the divine, follows on a more original veiling. The manifestation of that which is Most High – wrapped as it is in unapproachable darkness, up on the summit of being – is only the paradoxical manifestation of a transcendence that can never become truly manifest – perhaps not even to itself, as it possesses no Logos immanent to itself. It does not "think"; it cannot be thought. This, at least, often seems to be the case with the most severely logical, and most luminously uncluttered, metaphysical system of the third century, that of Plotinus. For the One of Plotinus is not merely *a* unity, not merely solitary, but is oneness as such, that perfectly undifferentiated unity in which all unity and diversity here below subsist and are sustained, as at once identity and difference. Plotinus recognized that the unity by which any particular thing is what it is, and is at once part of and distinct from the greater whole, is always logically prior to that thing; thus, within every composite reality, there must always also be a more eminent "act" of simplicity (so to speak) that makes its being possible. For this reason, the supreme principle of all things must be that One that requires no higher unity to account for its integrity, and that therefore admits of no duality whatsoever, no pollution of plurality, no distinction of any kind, even that between the knower and the known. This is not, for Plotinus, to deny that the One is in some special and transcendent sense possessed of an intellectual act of self-consciousness, a kind of "superintellection" entirely transcendent of subjective or objective knowledge (*Enneads* VI.vii.37.15–38.26; ix.6.50–55). However, the first metaphysical moment of *theoria* – reflection and knowledge – is of its nature a second moment, a departure from unity, Nous's "prismatic" conversion of the simple light of the One into boundless multiplicity; the One itself, possessing no "specular" other within itself, infinitely exceeds all reflection.

5. Nor did philosophy have to await the arrival of Hegel to grasp that there is something fundamentally incoherent in speaking of the existence of that which is intrinsically unthinkable, or of "being" in the absence of any proportionate intelligibility: for in what way is that which absolutely – even within itself – transcends intuition, conceptualization, and knowledge anything at all? Being *is* manifestation, and to the degree that anything is *wholly* beyond thought – to the degree, that is, that anything is not "rational" – it does not exist. Thus, it was perhaps with rigorous consistency that the Platonist tradition after Plotinus generally chose to place "being" second in the scale of emanation: for as that purely unmanifest, unthinkable, and yet transfinite unity that grants all things their unity, the One can admit of no distinctions within itself, no manifestation *to* itself, and so – in every meaningful sense – *is* not (although, obviously, neither is it not *not*).

6. In truth, even to speak of an "ontology" in relation to these systems is somewhat misleading. Late Platonic metaphysics, in particular, is not so much ontological in its logic as "henological," and so naturally whatever concept of being it comprises tends toward the nebulous. "Being" in itself is not really distinct from entities, except in the manner of another entity; as part of the hierarchy of emanations, occupying a particular place within the structure of the whole, it remains one item within the inventory of things that are. Admittedly, it is an especially vital and "supereminent" causal liaison within the totality of beings, but a discrete principle among other discrete principles it remains. What a truly ontological metaphysics would view as being's proper act is, for this metaphysics, scattered among the various moments of the economy of beings. One glimpses its workings now here and now there: in the infinite fecundity of the One, in the One's power to grant everything its unity as the thing it is, in the principle of manifestation that emanates from the One, in the simple existence of things, even in that unnamed, in some sense *unnoticed,* medium in which the whole continuum of emanations univocally subsists.

7. Ultimately, the structure of reality within this vision of things is (to use the fashionable phrase) a "hierarchy within totality," held together at its apex by a principle so exalted that it is also the negation of the whole, in all of the latter's finite particularities (*Enneads* VI.vii.17.39–43; ix.3.37–40; cf. V.v.4.12–16; 11.1–6.; etc.). What has never come fully into consciousness in this tradition is (to risk anachronism) the "ontological difference" – or, at any rate, the *analogy* of being. So long as being is discriminated from the transcendent principle of unity, and so long as both figure in some sense (however eminently) within a sort of continuum of metaphysical moments, what inevitably must result is a dialectic of identity and negation. This is the special pathos of such a metaphysics: for if the truth of all things is a principle in which they are grounded and by which they are simultaneously negated, then one can draw near to the fullness of truth only through a certain annihilation of particularity, through a forgetfulness of the manifest, through a sort of benign desolation of the soul, progressively eliminating – as the surd of mere particularity – all that lies between the One and the noetic self. This is not for a moment to deny the reality, the ardor, or the grandeur of the mystical elations that Plotinus describes, or the fervency with which – in his thought and in the thought of the later Platonists – the liberated mind loves divine beauty (*Enneads* VI.vii.21.9–22.32; 31.17–31; 34.1–39; ix.9.26–56; etc.). The pathos to which I refer is a sadness residing not within Plotinus the man, but within any logically dialectical metaphysics

of transcendence. For transcendence, so understood, must also be understood as a negation of the finite, as well as a kind of absence or positive exclusion from the scale of nature; the One is, in some sense, *there* rather than *here*. To fly thither one must fly hence, to undertake a journey of the alone to the alone, a sweetly melancholic departure from the anxiety of finitude, and even from being itself, in its concrete actuality: self, world, and neighbor. For so long as one dwells in the realm of finite vision, one dwells in untruth.

8. One must keep all of this in mind when considering the metaphysical implication of Nicene theology. For the doctrinal determinations of the fourth century, along with all of their immediate theological ramifications, rendered many of the established metaphysical premises on which Christians had long relied in order to understand the relation between God and the world increasingly irreconcilable with their faith, and at the same time, suggested the need to conceive of that relation – perhaps for the first time in Western intellectual history – in a properly "ontological" way. With the gradual defeat of subordinationist theology, and with the definition of the Son and then the Spirit as coequal and coeternal with the Father, an entire metaphysical economy had implicitly been abandoned.

9. These new theological usages – this new Christian philosophical grammar – did not entail a rejection of the old Logos metaphysics, perhaps, but certainly did demand its revision, and at the most radical of levels. For not only is the Logos of Nicaea *not* generated with a view to creation, and *not* a lesser manifestation of a God who is simply beyond all manifestation; it is, in fact, the eternal reality whereby God is the God that He is. There is a perfectly proportionate convertibility of God with his own manifestation of himself to himself; and, in fact, this convertibility is nothing less than God's own act of self-knowledge and self-love in the mystery of his transcendent life. His being, therefore, is an infinite intelligibility; his hiddenness – his transcendence – is always already manifestation; and it is this movement of infinite disclosure that is his "essence" as God.

10. Thus it is that the divine Persons can be characterized (as they are by Augustine) as "subsistent relations" – meaning not that, as certain critics of the phrase hastily assume, the Persons are nothing but abstract correspondences floating in the infinite simplicity of a logically prior divine essence, but that the relations of Father to Son or Spirit, and so on, are not extrinsic relations "in addition to" other, more original "personal" identities, or "in addition to" the divine essence, but are the very reality by which the Persons subsist; thus, the Father is eternally and essentially Father *because* he eternally has his Son, and so on (Augustine, *De Trinitate* VII.i.2; John of Damascus, *De Fide Orthodoxa* I.14). God *is* Father, Son, and Spirit; nothing in the Father "exceeds" the Son and Spirit. In God, to know and to love, to be known and to be loved are all one act, whereby he is God and wherein nothing remains unexpressed. Furthermore, if it is correct to understand "being" as in some sense necessarily synonymous with manifestation or intelligibility – and it is – then the God who is also always Logos is also eternal Being: not *a* being, that is, but transcendent Being, beyond all finite being.

11. Another way of saying this is that the dogmatic definitions of the fourth century ultimately forced Christian thought, even if only tacitly, toward a recognition of the full mystery – the full transcendence – of Being within beings. All at once the hierarchy of hypostases mediating between the world and its ultimate or absolute principle had

disappeared. Herein lies the great "discovery" of the Christian metaphysical tradition: the true nature of transcendence, transcendence understood not as mere dialectical supremacy, and not as ontic absence, but as the truly transcendent and therefore utterly immediate act of God, in his own infinity, giving being to beings. In affirming the consubstantiality and equality of the Persons of the Trinity, Christian thought had also affirmed that it is the transcendent God alone who makes creation to be, not through a necessary diminishment of his own presence, and not by way of an economic reduction of his power in lesser principles, but as the infinite God. In this way, he is revealed as at once *superior summo meo* and *interior intimo meo*: not merely the supreme being set atop the summit of beings, but the one who is transcendently present in all beings, the ever more inward act within each finite act.

12. This does not, of course, mean that there can be no metaphysical structure of reality, through whose agencies God acts, but it does mean that, whatever that structure might be, God is not located within it, but creates it, and does not require its mechanisms to act on lower things. As the immediate source of the being of the whole, he is nearer to every moment within the whole than it is to itself, and he is at the same time infinitely beyond the reach of the whole, even in its most exalted principles. In addition, it is precisely in learning that God is not situated within any kind of ontic continuum with creation, as some "other thing" mediated to the creature by his simultaneous absolute absence from and dialectical involvement in the totality of beings, that we discover him to be the *ontological* cause of creation. True divine transcendence, it turns out, transcends even the traditional metaphysical divisions between the transcendent and the immanent. God is at once superior *summo meo* and interior *intimo meo*.

13. This recognition of God's "transcendent immediacy" in all things, it should also be said, was in many ways a liberation from a certain sad pathos native to metaphysics, for with it came the realization that the particularity of the creature is not in its nature a form of tragic alienation from God, which must be overcome if the soul is again to ascend to her inmost truth. If God is himself the immediate actuality of the creature's emergence from nothingness, then it is precisely through becoming what it is – rather than through shedding the finite "*idiomata*" that distinguish it from God – that the creature truly reflects the goodness and transcendent power of God. The supreme principle does not stand over against us (if secretly within each of us) across the distance of a hierarchy of lesser metaphysical principles, but it is present within the very act of each moment of the particular. God is truly Logos, and creatures – created in and through the Logos – *are* insofar as they participate in the Logos's power to manifest God. God is not merely the "really real," of which beings are distant shadows; he is, as Maximus the Confessor says, the utterly simple, the very simplicity of the simple (*Ambigua*, PG 91:1232BC), who is all in all things, wholly present in the totality of beings and in each particular being, indwelling all things as the very source of their being, without ever abandoning that simplicity (*Ambigua*, PG 91:1256B). This he does not as a sublime unity absolved of all knowledge of the things he causes, but precisely *as* that one infinite intellectual action proper to his nature, wherein he knows the eternal "*logoi*" of all things in a single, simple act of knowledge (*Capita Theologica* II.4, PG 90:1125D–1128A).

14. In this theology, God in himself is an infinite movement of disclosure, and in creation, rather than departing from his inmost nature, he discloses himself again by

disclosing what is contained in his Logos, while still remaining hidden in the infinity and transcendence of his manifestation. To understand the intimacy of God's immediate presence *as God* to his creatures in the abundant givenness of this disclosure is also – if only implicitly – to understand the true difference of Being from beings.

15. One consequence of all of this for the first generations of Nicene theologians was that a new conceptual language had to be found, one that could do justice not only to the Trinitarian mystery, nor even only to the relation between this mystery and finite creation, but to our knowledge of the God thus revealed. For, in a sense, the God described by the dogmas of Nicaea and Constantinople was at once more radically immanent within and more radically transcendent of creation than the God of the old subordinationist metaphysics had ever been. He was immediately active in all things, but he occupied no station within the hierarchy of the real. As Augustine says, he is manifest in all things and hidden in all things, and none can know him as he is (*Enarrationes in Psalmos* LXXIV.9). He was not the Most High God of Arius, immune to all contact with the finite, for the Logos in whom he revealed himself as creator and redeemer was his own, interior Logos, his own perfect image, his own self-knowledge and disclosure; nor certainly was his anything like the paradoxical transcendence of the One of Plotinus, "revealed" only as a kind of infinite contrariety. In fact, the God who is at once the Being of all things and beyond all beings, and who is at once revealed in a Logos who is his coequal likeness and at the same time hidden in the infinity of his transcendence, is immeasurably *more* incomprehensible than the One, which is simply the Wholly Other, and which is consequently susceptible of a fairly secure kind of *dialectical* comprehension (albeit, admittedly, a comprehension consisting entirely in negations).

16. The Christian God, by contrast, requires one to resort to the far more severe, far more uncontrollable, and far more mysterious language of analogy – to indulge in a slight terminological anachronism – in the sense enunciated in Roman Catholic tradition by the Fourth Lateran Council: a likeness always embraced within and exceeded by a greater unlikeness. In the terms of Gregory of Nyssa, however much of God is revealed to the soul, God still remains infinitely greater, with a perfect transcendence toward which the soul must remain forever "outstretched." Or, as Maximus says, Christian language about God is a happy blending of affirmation and negation, each conditioning the other, telling us what God is not while also telling us what he is, but in either case showing us *that* God is what he is while never allowing us to imagine we comprehend *what* it is we have said (*Ambigua*, PG 91:1288C).

17. That said, in a very significant way, the fully developed Trinitarianism of the fourth century allowed theologians to make real sense of some of those extravagant scriptural claims that, within the confines of a subordinationist theology, could be read only as pious hyperbole: "we shall see face to face," for instance, and "I shall know fully, even as I am fully known" (I Cor. 13:12); or "we shall see him as he is" (I John 3:2); or "who has seen me has seen the Father" (John 14:9); or "the Son . . . is the exact likeness of his substance" (Heb. 1:3); or even "blessed are the pure of heart, for they shall see God" (Matt. 5:8) – makavrioi oiJ kaqaroiV th/' kardiva/, o{ti aujtoiV *toVn* qeoVn o[yontai (note the definite article). In considering the God of Nicene theology, we discover that the knowledge of the Father granted in Christ is not an external apprehension of an unknown cause, not the remote epiphenomenon

of something infinitely greater than the medium of its revelation, and not merely a glimpse into the "antechamber of the essence"; rather, it is a mysterious knowledge of the Father himself within the very limitlessness of his unknowability. Then again, the God who is the infinite source of all cannot be an object of knowledge contained within the whole; furthermore, if the Logos is equal to the Father, how can he truly reveal the Father to finite minds? And if – as became clear following the resolution of the Eunomian controversy – the Spirit, too, is not the economically limited medium of God's self-disclosure, but is also coequal with the Father and the Son, and is indeed the very Spirit by which God's life is made complete as knowledge and love, power and life, how can he reveal to us the Father in the Son?

18. These questions are made all the more acute, obviously, by the quite pronounced apophatic strictures that all post-Nicene theologians were anxious to impose on their language. As Augustine repeatedly affirms, every kind of vision of God in himself is impossible for finite creatures (see, e.g., *De Trinitate* II.16.27; *In Ioannis Evangelium tractatus* CXXIV.3.17; *Contra Maximinum Arianorum Episcopum* II.12.2), none can ever know him as he is (*Enarrationes in Psalmos* LXXIV.9), nothing the mind can possibly comprehend is God (Sermon 52.6; Sermon 117.5), God is incomprehensible to anyone except himself (Epistle 232.6), we are impotent even to conceive of God (Sermon 117.5), and so when speaking of God we are really able to do so properly only through negation (*Enarrationes in Psalmos* LXXX.12). And yet he also wants to say that, even in failing to comprehend God in himself, we are led by the Spirit truly to see and know and touch God.

19. Similarly, Gregory of Nyssa denies that any creature is capable of any qewriva of the divine essence, and yet he wants also to say that, in stretching out in desire toward God, the soul somehow sees God and attains to a qewriva tw'n ajqewrhvtwn, a vision of the invisible (*De Vita Moysis* II). Gregory even speaks of David as going out of himself really to *see* the divine reality that no creature *can* see, and remaining ever thereafter unable to say how he has done this (*De Vita Moysis* II). Maximus, who raises the "Greek" delight in extravagant declarations of apophatic ignorance to its most theatrical pitch, nonetheless makes it clear that his is an apophaticism of intimacy, born not from the poverty of the soul's knowledge of God, but from the overwhelming and superconceptual immediacy of that knowledge. The mind rises to God, he says, by negating its knowledge of what lies below, in order to receive true knowledge of God as a gift, and to come ultimately – beyond all finite negations – to rest in the inconceivable and ineffable reality of God (*Ambigua*, PG 91:1240C–1241A). When the mind has thus passed beyond cognition, reflection, cogitation, and imagination and discovers that God is not an object of human comprehension, it is able to know him directly, through union, and so rushes into that embrace in which God shares himself as a gift with the creature (*Ambigua*, PG 91:1220BC), and in which no separation between the mind and its first cause in God can be introduced (*Ambigua*, PG 91:1260D).

20. It was in terms such as these that the metaphysical reserve of older Greek thought regarding the "rationality" of the infinite was ultimately overcome. At first, even Christian thought was reluctant to depart very far from antique philosophy on the matter of the unintelligibility of the infinite. Origen actually denied that God could be called "infinite," arguing that an infinite divine essence would simply be an essence without definition and would, in consequence, be incomprehensible even to itself (*De*

Principiis II.9.1). It is arguable – and indeed has been argued – that the first "Greek" thinker either to attribute to God, or to develop a philosophical description of, positive infinity was St. Gregory of Nyssa, in the fourth century.

21. As a Christian, Gregory was bound to conceive of God's infinity as a love that gives without need, rather than as the unresponsive speculative completion of the world's necessity; as Trinity, the divine a[peiron is already determinate, different, related, and sufficient; it freely creates and wholly exceeds the world, being neither the world's emanative substance nor its dialectical consummation. Whereas the One spends its inexhaustible power in a ceaseless disinterested overflow into being, the Christian God creates out of that *agape* that is the life of the Trinity: God not only gives being to difference but elects each thing in its particularity, is turned toward and regards it, and takes it back to himself without despoiling it of its difference. Thus, no mere metaphysical prescinsion from multiplicity can lead thought up to being's highest truth. To pass from the vision of the world to the *theoria* of the divine is not simply to move from appearance to reality, from multiplicity to singularity, but rather to find the entirety of the world in all its irreducible diversity to be an analogical expression (at a distance, in a different register) of the dynamism and differentiation that God is. Difference within being, that is, corresponds precisely *as difference* to the truth of divine differentiation. Thus, the energy of desire drawing creation to God is not a recoil back from finitude, toward an unexplicated and disinterested simplicity whose "eyes" are forever averted from the play of being and its deficiencies, but answers – *corresponds to* – God's call to what he fashions for himself, and what is in itself nothing but an ontic ecstasy *ex nihilo* and *in infinitum*.

22. All that said, what Gregory understands "infinity" to mean when predicated of God is very much (at least on the face of it) what Plotinus understood it to mean in regard to the One: incomprehensibility, absolute power, simplicity, eternity. God is uncircumscribable (ajperivlhpton), elusive of every finite concept or act, boundless, arriving at no terminus: in Gregory's idiom, "that which cannot be passed beyond" (ajdiexivthton) (*Contra Eunomium* III.vii). God is the perfect completeness of what he is; the boundaries of bounty, power, life, wisdom, and goodness are set only where their contraries are encountered (*Contra Eunomium* I; *De Vita Moysis* I.5), but God is without opposition, as he is beyond nonbeing or negation, transcendent of all composition or antinomy; it is in this sense of utter fullness, principally, that God is called simple. To say, moreover, that he is eternal is to say not only that he is without beginning or end, but that he is without extension or succession at all (*Contra Eunomium* III.vii); the divine nature knows no past or future, no sequence, but is like an endless ocean of eternity (*Contra Eunomium* I); it is not time, although time flows from it (*Contra Eunomium* I). Extension, whether of time or of space, belongs exclusively to the created order and distinguishes it from the unimaginable infinity of God, who contains beginning and end at once in his timeless embrace (*Contra Eunomium* I; III.vi). This is hardly an extraordinary aspect of Gregory's thought, but it must be kept in view nonetheless, inasmuch as the difference between creaturely diachrony and divine eternity is also, for Gregory, the condition that makes union between the creature and God possible. Gregory's understanding of the infinite is one of utter positivity and utter plenitude, in which the finite can participate without in any way departing from its finitude. The insuperable ontological difference between creation and God – between the dynamism

of finitude and an infinite that is eternally dynamic – is simultaneously an implication of the infinite in the finite, a partaking by the finite of that which it does not own, but within which it moves – not dialectically, abstractly, or merely theoretically, but through its own endless growth in the good things of God (*In Canticum Canticorum* VI). Creation's "series," its ajkolouqiva, is at an infinite distance from the "order" and "succession" of the trinitarian *taxis* (*Contra Eunomium* II), but that distance is born of God's boundlessness: the Trinity's perfect act of difference also opens the possibility of the ontico-ontological difference, as the space of the gift of analogous being, imparted to contingent beings who then receive this gift as the movement of an ontic deferral.

23. God's transcendence is not absence, that is, but an actual excessiveness; it is, from the side of the contingent, the impossibility of the finite ever coming to contain or exhaust the infinite; the soul must participate in it successively or endlessly traverse it, "outstretched'" by a desire without surcease, an "infinition" of love; but God pervades all things, and all is present to his infinite life (*Oratio Catechetica* XXXII). Because the difference between God and creation is not a simple metaphysical distinction between reality and appearance, but the analogical distance between two ways of apprehending the infinite – God being the infinite, creatures embracing it in an endless sequence of finite instances – the soul's ascent to God is not a departure from, but an endless venture into, difference. The distance between God and creation is not alienation, nor the Platonic *chorismos* or scale of being, but the original ontological act of distance by which every ontic interval subsists, given to be crossed but not overcome, at once God's utter transcendence and utter proximity; for while the finite belongs to the infinite, the converse cannot be so, except through an *epektasis,* an everlasting "stretching out" of the soul, toward more of the good, which can be possessed only ecstatically – possessed, that is, in dispossession.

24. Without rehearsing the whole history of Christian ontology, one can note that, from the high patristic period onward, Christian thought developed an ontology of the infinite, which St. Thomas Aquinas (inspired in great measure by the Pseudo-Dionysius) brought to its most lucid Latin expression: being is not reducible either to essence or to finite existence; "infinity," which for Aristotle was to be ascribed only to the inchoate potentiality of matter, became now a name for the fullness of *esse*, which is the transcendent act – *actus purus*, the *actus essendi subsistens* – in which essence and existence alike participate; thus, being differs from beings more or less as does goodness from anything good. It is this approach to ontology alone that made it possible within the Western tradition to conceive of the distinction between being and beings as something other than an ontic economy, and to make "infinity" the proper name for ontological transcendence (albeit a transcendence analogically reflected in the finite world that participates in God) and a term more or less convertible with "being." Even if, in the realm of the ontic, the possible is in some sense the fountainhead of the actual, this obviously finite order still must be conceptually inverted (which involves an appeal to the infinite) if one is to be able to think of being as such, for possibility – however one conceives of it – must first *be*. This is the Christian "metaphysics of supereminence," which brought (one might argue) the entire classical tradition to its supreme expression, by uniting it to a Trinitarian understanding of God and a doctrine of creation that in a sense liberated antique metaphysics from dialectic, from a tragic

understanding of the relation between finitude and the infinite, and from the rule of fate.

25. Nicholas of Cusa understood, perhaps better than any other Christian philosopher, the remarkable consequences of imagining the infinite's freedom from the *principium contradictionis* as a positive good in the realm of act. For him, God is the *coincidentia oppositorum*, but not in the sense of a coincidence of opposed possibilities such as one could ascribe to the realm of pure potential. Rather, God – as infinite actuality – is the one in whom the supereminent fullness of being *simply* comprises the totality of what is, such that even the polarities of finite existence are present in his essence as a single act. The fetching images he uses to illustrate his insight – such as the actual identity of an "infinite circle" and a straight line – are wonderfully evocative. More powerful and more illuminating, however, are the various names he applies to God in an attempt to unfold the ontological implications of his vision of the divine: God as the absolute maximum who is also (by virtue of comprising nothing in a composite way or as a series) the absolute minimum; God as the infinite "complication" of all things who is at once implicated and explicated in all things; and God as the *non aliud* who, as the transcendent act of the infinite, is not any thing over against other things, but is the ever more inward act of every act of being.

26. In Christian thought, then, the difference between the infinite and the finite is the difference between being and beings. In Thomistic terms, it is the difference between an infinite actuality, in which essence and existence are one and the same, and that condition of absolute contingency in which there is a constant dynamic synthesis of essence and existence, and in which the act of existence is limited by essence and essence is actualized by existence, and in which both essence and existence are imparted to the finite being from a transcendent wellspring of being. Whether one uses Thomistic language or not, however, is unimportant. For Christian philosophy, the thought of the utter qualitative difference between infinite and finite is one with the thought of what Heidegger calls "the ontico-ontological difference."

27. In the twentieth century, the remarkable Erich Przywara used the term *analogia entis* as a convenient description of this entire ontological tradition. His arguments defy any easy summary, but one can say that, at the most fundamental level, the term *analogia entis* means only that being can be neither univocal between God and creatures (which would reduce God to a being among beings, subject to a higher category) nor equivocal (which would, curiously enough, have precisely the same result), and so the only coherent understanding of the relation of created being to uncreated must be analogical. *In him* we live and move and have our being. Every creature exists in a state of tension between essence and existence, in a condition of absolute becoming, oscillating between what it is and that it is, striving toward its essence and existence alike, receiving both from the movement of God's grace while possessing nothing in itself, totally and dynamically dependent, sharing in the fullness of being that God enjoys in infinite simplicity, and so infinitely other than the source of its being. Thus, the analogy of being does not analogize God and creatures under the more general category of being, but it is the analogization of being in the difference between God and creatures; for this reason, it is quite incompatible with any naive natural theology: if being is univocal, then a direct analogy from essences to "God" (as the supreme substance) is possible, but if the primary analogy is that of being itself, then an infinite analogical

interval has been introduced between God and creatures, precisely because God is truly declared in both the essence and the existence of all things. Conversely, the rejection of analogy, far from preserving God's transcendence, can actually only objectify God idolatrously as that which is "over against" creation: such a duality inevitably makes God and creation balancing terms in a dialectical opposition, thus subordinating God to being after all. The strident mysticism of the "Wholly Other" that one finds in, say, the theology of Karl Barth – both early and late – represents a "Christian" assault on the ontology of the *analogia entis*, but this may safely be ignored: the sort of critique Barth mounts against the *analogia entis* is so contaminated by misunderstandings, logical errors, and thoroughly modern prejudices that it is better to dismiss it as mere confusion, or to treat it as a fundamentally post-Christian phenomenon.

28. In the *analogia entis*, as Przywara describes it, the term above all terms is God, the full act of being, in whom all determinacy participates, but himself beyond all finite determination, negation, or dialectic: not one of two poles, not the infinite "naught" against which all things are set off (which would still be a "finite infinite"). Furthermore, it is not possible to regard this transcendent act as a primordial convertibility of being and nothingness, requiring its tragic solution in the finite – by way of Hegel's "becoming," or Heidegger's "temporalization," or what have you – as such a convertibility would already comprise an ontic opposition, a finite indetermination subordinate to its own limits, and so would still be in need of an ontological explanation, some account of the prior act of simplicity in which its unresolved and essential contradiction would have to participate in order to constitute a unity (in order, that is, to be). Being can be neither reduced to beings nor negated by them; it is peacefully expressed and peacefully withheld in its prismation in the intricate interweaving of the transcendentals. Even the transcendental moments of "this" and "not this" have their source in God's triune simplicity, his coincidence within himself of determinacy (as Trinity) and "no-thing-ness" (as "the all" – Sirach 43:27 – in whom we live, move, and are). The analogy thus permits the very difference of creatures from God, their integrity as what they are, and their ontological freedom to be understood as manifestations of how the Trinitarian God is one God. The analogy allows one to see being as at once simplicity and yet always already difference, not as a result of alienation or diremption, but because the fullness of being is God's one movement of being, knowing, and loving his own essence; to be is to be manifest; to know and love, to be known and loved – all of this is the one act, wherein no essence is unexpressed and no contradiction awaits resolution. The analogy thus stands outside the twin poles of the metaphysics of the necessary: negation and identity. After all, purely dialectical and purely "identist" systems are ultimately the same; both confine God and world within an economy of the absolute, sharing a reciprocal identity. If God is thought as either total substance or total absence, foundation or negation, "ground of Being" or static "Wholly Other," God is available to thought merely as the world's highest principle rather than as its transcendent source and end.

12.4 The Decline of the Christian Infinite

1. The fully positive metaphysics of the infinite began its decline arguably as early as the fourteenth century, in certain new movements within scholastic thought. It had

effectively vanished, definitely, by the time of the birth of modern philosophy, understood as an autonomous discipline, entirely severed from theology. For Descartes, the idea of the infinite that appears in the ego's cogitations, but not in its cognitions, is proof of God's existence, but is no more than that; because Descartes's "ontology" is more or less entirely confined to his division between thinking and extended substances, there is nothing in his thought remotely approaching the concern with the difference between being and beings that first became possible in Christian metaphysics. Thereafter, the more philosophy asserted its autonomy from theology, the more the concept of the infinite was forced to retreat to its "negative" form.

2. One could adduce Spinoza, of course, as evidence to the contrary, and as a counterweight to the more exclusively "transcendental" tradition of modern philosophy. For him, divine infinity – infinite substance expressing itself in infinite and infinitely many modes – could scarcely be more concrete, dynamic, or actual. And yet, one must note, the infinite is not really an ontological category at all in Spinoza's thought; it is, rather, a description of what is, essentially, sheer ontic magnitude. The endless diversity of divine modalities, the limitless extension and activity of divine substance, the absolute plenitude of the divine reality – all of this is, at the end of the day, a matter of mere quantity. The infinite as truly absolute – as truly absolved of becoming, as immanent through being transcendent, as that against which the finite is set off – appears nowhere in this philosophy.

3. Descartes did not intend to confine thought to an "anti-metaphysical metaphysics"; indeed, he found one of the surest confirmations of the trustworthiness of his perceptions in this presence within his mind of the thought of the infinite (and of the infinite God), which must, he reasoned, have been placed in him *by* God, inasmuch as it cannot in any way be abstracted from finite experience, and which must then be a proof of God's existence. But it is just this reduction of the infinite to what is effectively no more than a suggestively extra-empirical concept that makes it obvious that, from the transcendental vantage, immanent and transcendent truth are dialectically rather than analogically related; the former concerns a world of substances that exhaust their meaning in their very finitude, while the latter can appear among these substances only in the form of a paradox – a "knowledge" whose only evidence and condition is itself. God's infinity, thus conceived, is not truly the infinite; it is not qualitatively other than every finite thing precisely by being "not other," the possibility and "place" of all things, the unity and fullness of being in which all beings live and move and are; rather, it is merely the negation of finitude, the contrary of limit, found nowhere among my finite cognitions.

4. A new conceptual pattern becomes visible here. This "modern" infinite is not only beyond finite vision; it is without any analogous mediation – any *via eminentiae* – within the visible. A world certified by my founding gaze, and then secondarily by God (the postulated *causa efficiens et causa sui*), can admit the infinite into its calculus only as the indivisible naught that embraces every finite quantity, the "not-this" that secures every "this" in the poverty of its particularity. But the power of such an "infinite" to disrupt the ordered internal universe of the ego, as Descartes thought it must, is an illusion. The distinction between God's featureless, superempirical infinity and the palpable limits of the perceiving ego functioned for Descartes as proof that the "I" does not constitute its experience in any original way, but rather is itself constituted by the creative will of an invisible (but presumably undeceiving) God; however, it required

only the next logical step of distinguishing the empirical ego from the transcendental ego to collapse the distinction between the infinite and subjectivity altogether.

5. This was one of Kant's singular contributions to the dismantling of the old metaphysical categories. The transcendental project in its inchoate, Cartesian, insufficiently "critical" form could escape the circularity of knowledge – the ceaseless oscillation of epistemic priority between understanding and experience, between subject and object – only by positing, beyond empirical ego and empirical data alike, a transcendent cause. But for Kant this was simply to ground the uncertifiable in the unascertainable. Moreover, he certainly felt no impulse to retreat to an alternative anything like the premodern language of illumination, which would resolve the question of the correspondence between perception and the perceived by ascribing that correspondence to the supereminent unity in which the poles of experience – phenomena and gaze – participate. Now that every standard of validity had been definitively situated within the knowing subject, within the self's assurance of itself, such a metaphysics was a critical impossibility, as it could never be established by the autonomous agency of reason; indeed, it could be seriously entertained as a satisfactory answer only by a mind resigned to a certain degree of passivity, a trust in a transcendent source of truth that is, by definition, elusive of the scrutiny of the independent ego. Thus, Kant grounded the circle of knowledge in a transcendental capacity behind empirical subjectivity, the "transcendental unity of apperception" that accompanies all the representations available to the empirical ego. In a sense, Kant thus gained for subjectivity not only a mastery over the realm of the "theoretical" but a "supersensible freedom" so profound that even the moral law – once the exclusive preserve of God, the gods, the Good beyond being – could be regarded as its achievement.

6. For Kant, moreover, the transcendental ego is capable of the thought of the infinite – despite the absence of any corresponding sensuous intuition, or any datum more probative than the purely suggestive affective experience of the "dynamical sublime" – and this is proof of the self's inherent rational freedom. The infinite is no longer necessarily associated with God, however, who is himself now only a regulative idea in the realm of ethical reflection. Nothing here is left of the Christian concept of a positive infinity, hospitable to participation by the finite; rather, thought has arrived at a "finitist" ontology, to master which philosophy requires no theological "supplement."

7. Hegel was in some obvious sense a rebel against this transcendentalism. He recognized even better than Fichte (or the Kant of the *Opus Postumum*) that a liberty that consisted in resting secure in the impregnable citadel of the self, never hoping to take possession of the lands beyond its walls, was scarcely distinguishable from incarceration. He understood that philosophy would always be in retreat if it could not actually overcome theology's metaphysical tradition, provide an account of theology that could cogently portray the Christian story as an unrefined foreshadowing of a story that only philosophy would be able to tell fully, and so assume theology into itself. He was willing, therefore, to surrender some real measure of the isolated ego's independence because he was engaged in the near-impossible task of bringing under speculative control the radical contingency that Christian theology had introduced into the "absolute" and, moreover, in an attempt to overcome a final irreducible difference that Christian thought had imagined within God. By historicizing Spirit, Hegel sought

at once to spiritualize history and to idealize the Trinity, making God and history a mutually sustaining process.

8. What Hegel could not tolerate was the notion either of a God who possesses in himself difference, determinacy, plenitude, and perfection independent of any world or of a world thus left devoid of meaning in the ultimate speculative sense of "necessity" and so reduced to the status of something needless, something thoroughly aesthetic, not accomplishing – but merely expressing – a love God enjoys in utter self-sufficiency. Divine infinity, conceived as Trinitarian, always infinitely "determined" toward another, does not require time's tragic probations and determining negations; all created being is an unnecessary, *excessive* display of God's glory, and thus the world of things is set free in its "aimless" particularity. The Christian infinite is its own "exteriority," without need of another, negative exteriority to bring it to fruition. And without the majestic mythology of necessity governing the realm of the absolute, philosophy enjoys no sure authority over the contingent.

9. When one takes into account Hegel's distinction between the bad infinite and the "pleromatic" infinite of synthesis (which makes the true – the philosophical – concept of the infinite a "finite infinity," so to speak), one realizes that, even for him, the infinite is no longer a properly ontological category. It is either indeterminacy or totality (although a totality from which a certain kind of "mere particularity" has been excluded), but it is not convertible with the indeterminate "that it is" of the whole; it is not that transcendent act of being that imparts being out of its fullness, ever "absolved" of the finite.

10. Nor is this any less the case for any of modernity's or postmodernity's "post-metaphysical" forms of philosophy. The most radically anti-idealist thinkers, from Nietzsche to Derrida, tend to possess an inflexible conviction that being is finite, that what is must be, that death is the possibility of life and absence the possibility of presence, and that the world's most terrible limits are being's indispensable conditions. Of course, at the level of finite act and finite potentiality, this is undeniably the case; but it is no longer possible, it seems – or no longer conceivable – for the philosopher to abstract from the finite priority of possibility over actuality to the necessary priority of pure act over becoming. Heidegger, especially, insists on the limitation of being's event, its necessary confinement within the dispensations of its epochal sending, its finitude. Being *must* be thus: manifestation as obscuration, truth as duplicity, mission as "errance." Of course, Heidegger must approach ontology this way, as he is convinced that the philosopher must confine himself to the immanent.

11. Furthermore, in all its manifestations, the "higher nihilism" of contemporary continental thought consists in an unyielding insistence on the finite economy of being, the "nihilation" of being in beings, the passage from nothingness to nothingness that "eventuates" in the world of finitude. Jean-Luc Nancy, for instance, refuses to speak of the limitless that lies beyond figuration, but will speak only of the "movement of unlimitation," which is another name for the passage of beings from the nothing and back to the nothing, without transcendent source.

12. The story of the infinite as a metaphysical concept can be brought to an end here. More could be said, but it is perhaps enough to make two final observations. The first is that, speaking in the broadest and most impressionistic terms, the course of this history could be described almost as a kind of palindrome: both beginning and ending

with an unresolved dialectic (or opposition) between indeterminacy and totality, the interval between beginning and end spanned by the rise and fall of a fully ontological metaphysics of the infinite. The second is that, once the thought of the infinite has been thought – and with it a fully coherent concept of the difference between being and beings – it cannot be conjured away again. As that devoutly Christian philosopher Leibniz asked, "Why is there anything at all, and not – much rather – nothing?" This question having appeared on the philosophical horizon, we can never cease to recognize it as an enigma that cannot be reduced to an "analytic" collapse of being into beings, or to a "finitist" dialectic of the "nothing nothinging" (to use the Heideggerean term). It is a question that renders all philosophy henceforth questionable. It is, so to speak, an infinite question.

References

Aristotle. 1951. Physica. Ed. Sir David Ross. Oxford: Oxford University Press.

Augustine. 1845. Contra Maximum Arianorum Episcopum. In *Opera Augustini, Patrologia Latina*, Vol. 42. Paris: Migne.

Augustine. 1968. De Trinitate. Ed. Pieter Smulders. In *Corpus Christianorum series Latina*, Vols. 50–51. Turnhout: Brepols Publishers.

Augustine. 1956. Enarrationes in Psalmos. Eds. Eligius Dekkers and Johannes Fraipont. In *Corpus Christianorum series Latina*, Vols. 38–39. Turnhout: Brepols Publishers.

Augustine. 1954. In Iohannis evangelium tractatus CXXIV. Ed. Nicol Willem. In *Corpus Christianorum series Latina*, Vol. 36. Turnhout: Brepols Publishers.

Gregory of Nyssa. 1997. Contra Eunomium. Ed. Werner Jaeger. In *Gregorii Nysseni Opera*, Vols. 1–2. Leiden: Brill Academic Publishers.

Gregory of Nyssa. 1991. De Vita Moysis. Ed. Herbert Musurillo. In *Gregorii Nysseni Opera*, Vol. 7. Leiden: Brill Academic Publishers.

Gregory of Nyssa. 1997. In Canticum Canticorum. Ed. H. Langerbeck. In *Gregorii Nysseni Opera*, Vol. 8. Leiden: Brill Academic Publishers.

Gregory of Nyssa. 1996. Oratio Catechetica. Ed. Ekkehard Mühlenberg. In *Gregorii Nysseni Opera*, Vol. 3.4. Leiden: Brill Academic Publishers.

John of Damascus. 1857. De fide Orthodoxa. In *Opera Iohannis Damasceni, Patrologia Graeca*, Vol. 94. Paris: Migne.

Maximus the Confessor. 1858. Ambigua. In *Opera Maximi, Patrologia Graeca*, Vol. 91. Paris: Migne.

Maximus the Confessor. 1858. Capita Theologica. In *Opera Maximi, Patrologia Graeca*, Vol. 90. Paris: Migne.

Origen. 1836. De Principiis. Ed. E. R. Redepenning. Leipzig: In Bibliopolio Dykiano.

Plotinus. Enneads. 1964-1983. Eds. H. R. Schwyzer and Paul Henry. In *Plotini Opera*, Vols. I–III. Oxford: Oxford University Press.

CHAPTER 13

God and Infinity: Theological Insights from Cantor's Mathematics

Robert John Russell

13.1 Introduction

Western monotheism begins with the fundamental assertion that God is Absolute Mystery, the incomprehensible ground and source of being itself, and that we can speak about God only because God first speaks to us in revelation received in faith and understood always inadequately through reason, tradition, and experience. Theology as self-critical reflection on revelation and religious experience starts with what we do not understand before it seeks to say something about that which we understand at least in part. In the long traditions of Western monotheism, the way of unknowing leads, and the way of knowing follows behind. These ways have come to be called the *via negativa* (the negative way of denial) and the *via positiva* (the positive way of affirmation), respectively. This means, in turn, that when we talk about God we should begin by attempting to say something about those divine attributes that we only know by negation, that is, by their utter contrast with our experience of ourselves and our universe. We should begin with what are called "apophatic" statements, from the Greek αποφατικος for "negative" or "denial." With this we return full circle to the fundamental assertion that the most inclusive divine attribute is the incomprehensibility of God. But this incomprehensibility, too, tells us something inestimably important about God's relation to the world: it is only by being incomprehensibly different from this world that God can be the source of this world and its final home. Thus, the way of unknowing is, paradoxically, a way of knowing and a source of confident hope.

With this in mind, we may also attempt to speak about what God has made known to us about God's nature and purposes, particularly through that dimension of culture shaped by the great discoveries of the natural sciences and modern mathematics in the past four centuries. Here we can talk about God not by sheer contrast but by making a positive analogy with key experiences in our own lives, with key events in history, and

This paper is a revised and expanded version of chap. 2, "The God Who Infinitely Transcends Infinity: Insights from Cosmology and Mathematics," in Russell (2008, pp. 56–76). The latter is a slightly revised version of a chapter with the same title in Templeton (1997).

with the world in which we live, particularly as it is illuminated by science. This way of speaking is called "kataphatic," from the Greek κατάφασις, meaning "according to" or by analogy. Thus, when we experience wonder and awe at the immensity and beauty of the universe, we are led to think of God as utterly wondrous, terribly awesome, the source of ecstatic beauty. When we experience love in our lives, when we are forgiven our iniquities by those we have wronged, when we know the goodness of home, hearth, health, and family, we can be led to speak of God as perfect love, unconditional mercy, the source of all that is good, our final home beyond death and grief. Most importantly, when we look up from the daily routine of life and witness the sacred in our midst, as Moses did when he turned aside from tending his sheep to go to see the burning bush (Exod. 3:1–6), we are compelled to confess God as utterly holy. Thus, we are enveloped by the mystery of a God who surpasses all knowing, yet we know that this God seeks us and would be known by us, and so we move ahead in the light of this knowledge. This requires us to remember the poverty of faith in light of the surpassing mystery of God and cloak all that we wish to affirm in the spreading folds of our unknowing. The way of faith is always the disposition of humility in which the kataphatic undergirds the apophatic.

In this chapter I want to explore ways in which modern mathematics bears important implications for our theological conversation about God. The key mathematical piece will be Georg Cantor's work on infinity in relation to the ways in which divine revelation both reveals and veils God and in relation to the concept of the divine eternity in the theology of Wolfhart Pannenberg. This choice to focus on Cantor's mathematics arises because we normally think of infinity in relation to theology through the *via negativa*: by way of utter contrast with the finite and temporal world that God created. This reflects a view of infinity that dates back to the ancient Greeks, where infinity is defined apophatically through its contrast with the finite: the infinite is the *apeiron* – the unbounded, the unlimited, or the formless. Recent developments in mathematics since the nineteenth century may shed new light on this issue. Georg Cantor, in particular, has given us a new conception of infinity that is much more complex than previously thought, with layers of infinities leading out endlessly to an unreachable Absolute. In effect, we now can say a lot more about infinity than merely that it contrasts with the finite – indeed, an infinite amount more! These revolutionary discoveries in mathematics might, then, lead to new insights into the ways we can use the concept of infinity in thinking theologically about God.[1]

13.2 A Note on Infinity in Mathematics, Philosophy, and Theology[2]

The history of the concept of the infinite in mathematics, philosophy, and theology from early Greek thought through the European Enlightenment is both extraordinarily

---

[1] For a readable and insightful introduction I recommend Rucker (1983). For a paper that explores themes in common with the present paper, see Pennings (1993). For online resources see http://en.wikipedia.org/wiki/Infinity and http://plato.stanford.edu/entries/set-theory.

[2] Sections 2 and 3.1 draw in part from Rucker (1983).

complex and endlessly fascinating, but even a short treatment is beyond the possibilities of this limited chapter. Fortunately, this volume contains several chapters that together offer a wealth of insight and references to this history; I would call particular attention to the chapters by Wolfgang Achtner (Chapter 1) and David Bentley Hart (Chapter 12).[3] Here Hart begins by stating that "there is not – nor has there ever been – any single correct or univocal concept of the infinite." Granted Hart's point, what I want examine here is a very basic distinction between that *aspect* of the concept of the infinite that by and large dominated this history, namely, the infinite as the negation of finite, and the new element in the concept of the infinite found in modern mathematics, specifically that of Georg Cantor. To portray this distinction in its simplest form, recognizing that this might, in fact, be an oversimplification, I touch briefly on key examples in this rich history.[4]

As is well known, the ancient Greeks defined the concept of infinity as the unbounded. In so doing, they unequivocally distinguished between the the finite: something is infinite if it has no boundary or end, or if it is lacks structure or order. Thus, the infinite is defined by contrast with infinite is totally different from, even opposed to, those things that m finite experience. In this view, underlying the world of finite entities indeterminate infinity, and finite entities are good precisely becaus bounded, and determinate in contrast with the infinite. Evidence of t in the form of paradoxes, such as the famous race between Achille date back at least to Zeno of Elea (490–430 BCE). These parad to arise precisely because of the sharp contrast between the fin

There are, in fact, roots in Greek thought on the infinite writings. Anaximander (610–547 BCE) conceived of the inf less substance that is eternal, inexhaustible, and lacks boun world of finite entities was seen as arising out of that w (570–500 BCE) rejected the infinite as having anything t contended that all actual things are finite and representa numbers. Pythagoras taught that the geometrical forms sions arise when a mathematical limit is imposed on t Plato (428–348 BCE) believed that the Good must be d therefore must be finite rather than infinite. Accordi imposes limitations (i.e., intelligible forms) on pree to the structured world around us as an ordered wh gible, infinite chaos. In all these cases, the infinite with the finite.

With the philosopher Aristotle (384–322 BC conception of infinity, and one that continue

---

3   See "Notes on the Concept of the Infinite in the History

4   Hart also makes a crucial distinction, at the outset, be
    ontological meanings of "infinity." While accepting t
    the ways in which the mathematical concept of infi
    implications of this change for the ways we impl
    metaphysical issues in the context of theology fo
    Pannenberg).

ics," p. 1.

thematical and the metaphysical/
pose of my paper is to begin with
he work of Cantor and explore the
tical sense of infinity in discussing
d on the divine attributes (following

challenge until the revolution in mathematics in the nineteenth century. Aristotle began, as those before him, by considering infinity as a negative quality, that is, as something lacking: "(I)ts essence is privation."[5] But he altered the concept of infinity to mean an unending and incomplete process: "(T)he infinite has this mode of existence: one thing is always being after another, and each thing that is taken is always finite, but always different."[6] Here the only infinity considered possible is potential infinity: something capable of being endlessly divided or added to, but never fully actualized as infinite. For example, consider the series of unending succession of natural numbers, the endless succession of the seasons, or the continuous division of a line interval into smaller and smaller portions.[7] Aristotle thought of these as potentially infinite, given that the series can be continued endlessly while remaining finite; these series are never actually infinite. Going through successive elements in a series, we never get beyond the finite steps in the series to reach their limit, the actually infinite. We never view the infinite series as a whole.[8]

Early Christian writers were informed by this more "positive" conception of infinity as they developed the doctrine of God.[9] In time the attribution of infinity to God became a standard cornerstone of Christian theology. Augustine (354–430), through his debt to Plotinus, believed that because God is infinite, God must know all numbers and must have limitless knowledge of the world.[10] Gregory of Nyssa (335–394) also argued that God is infinite in the context of his Christological debates against such Arians as Eunomius.[11] Writing a millennium after Augustine and drawing on the philosophy of Aristotle, Thomas Aquinas (1225–74) believed that only God is infinite and all creatures are finite. Yet even God, whose power is infinite, "cannot make an absolutely unlimited thing."[12] Turning to contemporary theology, most theists inherit this tradition, affirming that God creates and sustains the world in existence while utterly transcending it. God alone is infinite, and the world is finite. God's infinity is

[5] See *Physics*, III.7.207b35 in McKeon (1941).

[6] Aristotle, *Physics*, III.6.206a26-30.

[7] Aristotle, *Physics*, III.4.204b1-206a8.

[8] Aristotle, *Physics*, III.5.206a5b.

[9] In his very illuminating contribution to this volume, "A (Partially) Skeptical Response to Hart and Russell," Denys Turner offered the following criticism of my reference to a positive notion of the infinite: "I confess to not being clear about what is meant by a 'positive' notion of the divine infinity . . . (W)e need to distinguish between two sorts of predicates of God: those that can be said to be 'positive' (predicates) – such as 'goodness,' 'wisdom,' 'intelligence' – and are known 'by analogy' from what we know of such predicates as affirmed of creatures, and those that are . . . 'regulative' – such as 'infinity' and 'simplicity' – which are known to be true of God *only by denial* of what we know of creatures" (p. 4). I am appreciative of Turner's distinction between these two types of predicates of God. Using Turner's, together with Pannenberg's, terminology, the regulative predicates are appropriate for discussing those divine attributes that define the God who acts (i.e., tell us what "God" means), while the positive predicates are appropriate for discussing those divine attributes that describe God's actions (i.e., tell us what God does). See my discussion of Pannenberg in "The infinity of God in relation to the divine attributes," below. Regarding my comment on early Christian writers, what I meant to suggest is that these writers were informed by a positive concept of infinity through Aristotle's understanding compared with that of Plato. For the latter, God could never be said to be infinite. (See the chapter by Hart in this volume, sections 2 and 3.)

[10] Augustine, *City of God*, XII 18.

[11] Thomas Aquinas, *Summa* Ia, 7, 2–4.

[12] We shall see that Galileo's insight was to play a key role in the discovery of "transfinite" numbers by Georg Cantor in the nineteenth century (see Galileo 1954).

a mode of God's perfection even while God is incomprehensible. We can know *that* God is perfect, infinite self-existence, but we cannot conceive of *how* God is perfect, infinite self-existence.

Scholars trace the beginning of the modern understanding of the mathematical concept of infinity back to Galileo Galilei (1564–1642). In *Two New Sciences*, Galileo argued that, as far as the natural, or whole, numbers are concerned, infinite sets do not obey the same rules that finite sets obey. Consider the unending sequence of whole numbers, $1, 2, 3, \ldots$, and the unending sequence of their squares, $1, 4, 9, \ldots$. Intuitively, we think that there must be more whole numbers than their squares given that the set of whole numbers *contains* the squares along with other numbers (e.g., $2, 3, 5, 6, 7, 8, 10, \ldots$). Yet the whole numbers, $1, 2, 3, \ldots, n$, can be put into a one-to-one correspondence with their squares, $1, 4, 9, \ldots, n^2$, by the simple rule $n \rightarrow n^2$. Two sets of numbers whose elements can be put in a one-to-one correspondence such as this are said to be of the same size (or "equinumerous"). Adopting this definition, we find that the series of squares $1, 4, 9, \ldots$ is *equivalent* to the series of whole numbers $1, 2, 3, \ldots$. This seems to mean that there are as many squares as there are whole numbers! Given this paradoxical result, Galileo recognized that, unlike finite quantities, "the attributes 'larger,' 'smaller,' and 'equal' have no place . . . in comparing infinite quantities with each other."[13]

It took centuries for the insights of Galileo to prove their fruitfulness. While the simultaneous discovery of the calculus in the seventeenth century by Sir Isaac Newton (1642–1727) and Gottfried Wilhelm Leibniz (1646–1716) involved complex questions about the status of infinitesimals as both minute extensions in space and durations in time, even Newton, Leibniz, and Carl Friedrich Gauss (1777–1855) adhered to the view that such infinities and infinitesimals are only abstractions or potential infinities. It was only with the discoveries of Bernard Bolzano (1781–1848), Richard Dedekind (1831–1916), and Georg Cantor (1845–1918) that a radically new mathematical conception of infinity was born. My focus here is on Cantor, who discovered that there are many kinds of mathematical infinities and who explored their structure in detail.[14]

A biographical caveat is in order here to give us some sense of the person behind the discoveries. Georg Ferdinand Ludwig Philip Cantor, often called the creator of set theory and the discoverer of transfinite numbers, was born in St. Petersberg, Russia, on March 3, 1845. His mother was a devout Roman Catholic, his father a Jew who had converted to Lutheranism. Raised in a musical family that relocated to Germany and converted to Lutheranism, Cantor was an accomplished violinist. Cantor received his doctorate at age 22 on number theory and accepted a position at the University of Halle two years later. In 1873 he proved that the rational numbers are countable and, a year later, that the real numbers are uncountable. Starting in 1883 he began a series of attempts to prove the continuum hypothesis. In 1895 and 1897 he made his last major contributions to set theory, including finding the first of a series of paradoxes in set theory. During much of his lifetime Cantor's views were supported by mathematicians such as J. W. Richard Dedekind (1831–1916), but they were opposed

---

13 William Lane Craig (1979) gives a helpful presentation of Cantor's transfinite numbers. See also Hahn (1956) and Rucker (1983, chaps. 1 and 2).

14 For a detailed biography see http://www-groups.dcs.st-and.ac.uk/~history/Biographies/Cantor.html.

by many others, notably Leopold Kronecker (1823–91). This opposition may have contributed to Cantor's declining health in later years, including a series of nervous breakdowns that began in 1884. Cantor died on January 6, 1918 in a mental institution.[15]

It is thanks to Cantor that we know how to apply basic mathematical operations, which we use with finite sets, to infinite sets, including addition, multiplication, exponentiation, the relation "greater than," the equivalence of sets by counting, and so on. This, in turn, meant that he could give an explicit procedure for constructing different kinds of infinity and for testing their internal consistency. Cantor's fundamental claim is that these operations can be transferred from their foundations in finite sets to infinite sets. Cantor coined the term "transfinite" for the cardinal number of infinite sets. Rather than assuming that the infinite is in direct *contrast* with the finite, Cantor treats an infinite set in a direct *analogy* with how he treats finite sets.

Let's start with the simplest example of an infinite set, the set of natural numbers $\{1, 2, 3, \ldots\}$. Cantor chose $\aleph_0$ ("aleph-null") to represent its transfinite cardinal number. He thought of this set as a given whole, and not just an incomplete sequence of unending finite numbers ascending in scale. In other words, Cantor actually distinguished between an unending but finite sequence of elements, such as the sequence $1, 2, 3, \ldots$, a sequence that is potentially infinite but always, in fact, finite, and the complete infinite sequence of these numbers thought of as a whole, that is, the set $\{1, 2, 3, \ldots\}$. He called the potential infinite a "variable finite" and symbolized it as ∞; the actual infinite he symbolized by $\aleph_0$, as we saw earlier. Thus, ∞ never reaches completion, never becomes $\aleph_0$. To think about ∞ is to think of an ever-increasing series of numbers continuing forever without reaching an end. To think about $\aleph_0$ is to stand outside this series *sub specie aeternitatis* and to consider them as a single, unified, and determinate totality.

Cantor then extended what we know about counting finite sets to infinite sets: all infinite sets whose elements can be put in a one-to-one correspondence with the natural numbers will have the *same* cardinal number, $\aleph_0$. We call these sets *denumerably or countably infinite*. This leads to some surprising results. For example, recall Galileo's paradox about square numbers. Because there is a one-to-one correspondence between the natural numbers and the square numbers (technically a bijection), it means that the set of square numbers is countably infinite;[16] it has the same cardinal number, $\aleph_0$, as the set of natural numbers. Similarly, the set of even numbers is equivalent to the set of natural numbers, as is the set of odd numbers.[17]

Let's go further. We can generate infinities that are "bigger" than the set of natural numbers, although all of these still have cardinal number $\aleph_0$. In the case of finite sets, the ordinal and the cardinal numbers of a set are the same. It turns out, however, that they are *not* the same for infinite sets! We start, as before, with the infinite set of

---

[15] Note: Unfortunately there is a textual error in previous publication of portions of this essay in which the text here reads, erroneously, "uncountably infinite."

[16] While still a student, Cantor discovered that the set of all rational fractions (i.e., the quotient of two natural numbers such as 4/7) is denumerably infinite, and the same holds for the set of all "algebraic" numbers, such as $\sqrt{3}$. All these sets have the same transfinite cardinal number, $\aleph_0$.

[17] $\omega + 1$ is *not* equal to $1 + \omega$; the latter is equal to ω. That is, $1 + \omega = \omega$ because all the symbol $1 + \omega$ means is that we add one element to the elements that taken endlessly but thought of as a whole are the set of natural numbers whose ordinal number is ω. The former, $\omega + 1$, means adding to a *given* infinite whole, namely, $\{1, 2, 3, \ldots\}$, a new element 1, thus forming the set $\{1, 2, 3, \ldots, 1\}$. This means that $\{1, 2, 3, \ldots, 1\}$, taken as a whole, is not equivalent to the set $\{1, 2, 3, \ldots\}$ taken as a whole. Thus, $\omega + 1$ is not equivalent to ω.

natural numbers, $\{1, 2, 3, \ldots\}$; following mathematical custom, we designate its ordinal number as ω. Now let us think of this set as a whole, as complete in itself. If we do so, then it is possible to conceive of adding 1 to the set, forming a new set, $\{1, 2, 3, \ldots\}$; the ordinal number of this set would be $\omega + 1$. (Note: $\omega + 1$ is *not* equal to $1 + \omega$. The latter is equal to ω.)[18] Adding 1 to this set yields another new set, $\{1, 2, 3, \ldots, 1, 2\}$, whose ordinal number is $\omega + 2$. As we continue the process, we generate a whole ladder of increasingly complex infinities. For example, if we add the set of natural numbers to the set of natural numbers, we obtain the set $\{1, 2, 3, \ldots, 1, 2, 3, \ldots\}$, whose ordinal number is $\omega + \omega$, which we notate as $\omega \times 2$. Continuing from here, we can consider $(\omega \times 2) + 1$, $(\omega \times 2) + 2$, $(\omega \times 2) + 3$, and so on, until we reach $\omega \times \omega$, which we can write as ω^2. Again we continue adding to this set until we conceive of ω^3, ω^4, and so on. This in turn points toward its goal – ω^ω – but there's still more, in fact, infinitely more! We can think of an infinite series of exponential powers, raising ω to the ω power infinitely many times. What is even more astonishing is the fact that the elements in any of these transfinite sets can be put in a one-to-one correspondence with the elements in the set of natural numbers, $\{1, 2, 3, \ldots\}$. This means that all of these sets, even though differing in their ordinal number, are *denumerably* infinite: *they have the same cardinal number,* $\aleph_0$. Mathematicians express this fact by noting that $\aleph_0 + \aleph_0 = \aleph_0$, even though it remains true that $\omega + \omega \neq \omega$. Apparently the rules that infinite sets obey are *both like and unlike* the rules that finite sets obey!

Still, the story of Cantor's discoveries is not over – indeed it is hardly begun. We can now imagine sets whose "infinity" *cannot* be put into a one-to-one correspondence with the set of natural numbers: they are "uncountably infinite." Cantor proved this fact for the set of real numbers in 1874.[19] Because they can be put in one-to-one correspondence with the points of a straight line, Cantor called the cardinal number of the set of real numbers "the power of the continuum," designated by the letter "c." In 1877 Cantor proposed the "continuum hypothesis": the real numbers form the *first* uncountable infinity, that is, the first infinity whose cardinality is "greater than" the cardinality of the set of natural numbers, and he therefore denoted it $\aleph_1$. It is the first of David Hilbert's famous list of unanswered questions (1900). Even today, the continuum hypothesis remains controversial, as evidenced by the ongoing research of University of California, Berkeley, mathematician Hugh Woodin.[20]

The properties of $\aleph_1$ are startling. For example, the cardinal number of points in all line segments is the same, regardless of their length. More surprising is the one-to-one correspondence between the set of points in a plane and the set of points in a line, given that one might well have thought the former to be infinitely greater than the latter. Cantor then extended this result to the points in a three-dimensional space, and then to a space of any number of dimensions. All these mathematical objects – the

---

[18] Real numbers are composed of the natural numbers, the fractions, and the irrational numbers.

[19] In 1963 Paul Cohen seemed to establish that the continuum hypothesis is neither provable nor disprovable in terms of the axioms of stardard Zermelo-Fränkel Set Theory together with the Axiom of Choice. Hugh Woodin, however, has given reasons for believing that the continuum hypothesis might be false (see Woodin 2001). For a readable overview see http://www.phschool.com/science/science_news/articles/infinite_wisdom.html. For a recent technical report see http://www.math.unicaen.fr/~dehornoy/Surveys/DgtUS.pdf.

[20] In set theory a reflection principle states, roughly, that the collection (i.e., class) of all sets "resembles" the sets in it. Specifically, every property of the class of all sets is shared by at least one of its sets. Of course, the exact meaning of "resemble" requires significant clarification if contradictions are to be avoided.

line, the plane, the volume, the volume in four or more dimensions – have the *same* cardinal number, c. We can go further still. Cantor showed that one can construct a whole series of transfinite cardinal numbers starting with $\aleph_0$ and $\aleph_1$ and leading to $\aleph_\omega$ and beyond this to $\aleph_{\omega+1}, \aleph_{\omega+2}, \ldots, \aleph_{\omega^\omega}, \aleph_{\aleph_0}$, and so on. In fact, there is no end to the kinds of transfinite infinity we can construct. At the same time, these infinite sets share an important feature with finite sets because, no matter how complex they seem, they are conceivable by construction; hence, the reason for Cantor to call them "transfinite."

What lies beyond even the transfinite numbers? According to Cantor, lying beyond the transfinites is "Absolute Infinity," symbolized as Ω. In one sense Absolute Infinity is inconceivable. Yet in another sense we can know something about Ω, namely, we can know something about its properties! Does this lead us into a contradiction, namely, that Ω is both conceivable and inconceivable?

To see that it is not a contradiction, consider the converse. Suppose that Ω is as conceivable as the transfinites described earlier. Then there must be some property P that is exclusively a property of Ω, a property in terms of which we can conceive exclusively of Ω. Now in order to make Ω inconceivable, we merely have to stipulate that every such property P is, in fact, shared by both Ω and some transfinite ordinal: there is no property P that is unique to Ω. This means that we can consistently assert the conceivability and the inconceivability of Ω. On the one hand, we can conceive of Ω because each of its properties P is shared by some transfinite ordinal. Yet because of this, we can never differentiate Ω completely from the transfinite ordinals, given that we can never describe Ω as possessing a property P that it does not share with some transfinite ordinal. In essence, we can never distinguish between Ω and the transfinite ordinals because of the fact that all its properties are shared with each of the transfinites. We can never know if we are conceiving of Ω and not some transfinite ordinal. In short, we can never conceive of Ω as unambiguously distinct from a transfinite ordinal. Ω is thus inconceivable because Ω can never be uniquely characterized or completely distinguished as distinct from some transfinite ordinal. Instead, the transfinite numbers lead endlessly toward Absolute Infinity Ω but never begin to reach it, since Ω lies, absolute and unapproachable, beyond all comprehension. This argument is often referred to as Cantor's reflection principle,[21] and it led Cantor to set up a threefold distinction regarding the infinite:

> The actual infinite arises in three contexts: *first* when it is realized in the most complete form, in a fully independent other-worldly being, *in Deo*, where I call it the Absolute Infinite or simply Absolute; *second* when it occurs in the contingent, created world; *third* when the mind grasps it *in abstract* as a mathematical magnitude, number, or order type. I wish to mark a sharp contrast between the Absolute and what I call the Transfinite, that is, the actual infinities of the last two sorts, which are clearly limited, subject to further increase, and thus related to the finite.[22]

---

[21] From Cantor (1980, p. 378). This translation is from Rucker (1983, p. 10).

[22] Before exploring the potential usefulness of Cantor's ideas on infinity for theology, I must acknowledge and briefly respond to two challenges: (1) Do mathematicians agree with Cantor that infinity is a consistent and coherent concept in mathematics? (2) Is Cantor correct in believing that the world is actually infinite?

   Regarding the first challenge, some mathematicians accept Cantor's articulation of the transfinites and defend Platonic realism. Others point to antinomies in this work, including those of Burali-Forti, Cantor

13.3 Enriching Our Theological Understanding:
Two Examples[23]

13.3.1 Infinity in the Context of Divine Revelation: "The Transfinite as the Veil that Discloses"

Previously I suggested that the early Christian world helped to transform the meaning of infinity from its negative connotations, involving an unlimited chaos or a gnawing privation, to more positive connotations, suggesting ultimate reality as the ground of being, the highest good, and the source of the world. At the same time, however, theologians retained the classical Greek distinction of the infinite as wholly different from, and in contrast to, the finite. Thus, to say that God is infinite is to say that God is incomprehensible.

This distinction, inherited from the Greek philosophical culture by the early church, has predominated through the centuries in Christian theology as it seeks to speak about God. We see this most clearly in the distinction theologians make between the *apophatic* and the *kataphatic*, which we discussed in a previous section. To repeat, we start with acknowledging the apophatic, that is, how *little* we know of the unseen, incomprehensible mystery that is God. Then we move to the kataphatic, that is, we seek to express something that we *do* know about God by analogy with our experience of the world. The key point for our purposes here is that the term "infinity" has played a pivotal role in this history: it has been used almost exclusively to express the stark difference between the apophatic and the kataphatic. God as infinite and we as finite means that God is other than, unlike, wholly different from us. Now we are in a position to appreciate the importance of Cantor's work for the meaning of revelation. The key will be the way the reflection principle in mathematics points to a connection between what we know and what we do not know about Absolute Infinity. In theological language, we can say that the reflection principle relates what we have

himself, and most prominently Bertrand Russell, which have forced major revisions in set theory. These revisions lead into such new areas as logicism, axiomatization, and intuitionism, and at the same time they tend to undercut the Platonist/realist interpretation of infinity. My response is that these differences do not directly hamper the fruitfulness of using Cantor's work theologically. The issue of the consistency and coherence of Cantor's set theory does not speak directly for or against its applicability to the doctrine of God, where the notion of the infinite versus the finite has been, for far too long, embedded in Greek thought of the infinite as the *apeiron*.

Regarding the second challenge, if any property that we think of as a property of the finite world would turn out to be a property of the infinite, it might seem to challenge our understanding of God: either the world would *be* God, thus leading potentially to pantheism, or the world would simply be infinite and there would be *no need* for God, thus leading in principle to atheism. The importance of the finitude of the world has resided in the fact that it has been seen as a key defense against unbelief as well as a key constituent to the Christian distinction between Creator and creation, or more generally the Christian definition of the God-world relation in terms of Creator-creation. I think that these concerns, too, were a factor motivating the detailed arguments by Bill Stoeger and George Ellis against multiverses (see Stoeger, Ellis, and Kirchner 2005). My response to Stoeger, Ellis, and Kirchner is that the God-world distinction is founded on an ontological distinction that does not require the traditional finite/infinite distinction as its basis. Even a world filled with actual transfinites could still be considered a created world and not divine.

[23] The following section is drawn in part from a much larger project I am pursuing called *Time in Eternity: Physics and Eschatology in Creative Mutual Interaction*, funded by a Metanexus/Templeton Foundation grant.

kept separate, the apophatic and the kataphatic, the way of negation and the way of affirmation.

We begin with Cantor by way of defining Absolute Infinity as beyond comprehension, that is, as lying beyond the unending ladder of the transfinite numbers. Recall that the transfinites are infinities, but they are infinities that can be fully comprehended in the sense that their properties can be formally described and these descriptions can be used to distinguish one transfinite from another. Now the way we ensure that the Absolutely Infinite is beyond comprehension is by claiming that if it were not so, then it, too, could be described uniquely, that is, we could state at least one property that it alone possesses and that allows us to single it out from all the other transfinites that do not share this property. We therefore reverse this and insist, via the reflection principle, that all of the properties of Absolute Infinity must be shared by the transfinites. Because none of its properties are unique to itself, we can never point unequivocally to Absolute Infinity by pointing to that unique property. Given that Absolute Infinity cannot be uniquely described, it is, in this sense at least, incomprehensible.

But this, in turn, means that we do, in fact, know something about the Absolutely Infinite: all of the properties it possesses must be shared with and disclosed to us through the properties of the transfinites. The Absolute Infinite is in this sense knowable, comprehensible; each of its properties must be found in at least one transfinite number. The Absolute is disclosed through the relative, Absolute Infinity through the transfinite, and yet it is precisely through this same disclosure that Absolute Infinity remains hidden, ineffable, and incomprehensible.

Another way of putting this is that the incomprehensibility of Absolute Infinity is manifested by its partial comprehensibility. What we know about the Absolute Infinite is never more than partial knowledge, for it is the knowledge we get from knowing the properties of the transfinites. What is truly unique about Absolute Infinity is never disclosed, but forever hidden. To speak somewhat metaphorically, it is as though the transfinites form an endless veil surrounding Absolute Infinity. The veil is all we can ever know; what lies behind it, Absolute Infinity, is hidden by that veil and is forever beyond our knowledge. Yet, genuine knowledge about Absolute Infinity is forever revealed in the veil that hides it, for we can endlessly learn more and more about Absolute Infinity as we continue to discover more and more about the transfinites; thus, we can move endlessly to ever more knowledge of what can never be known exhaustively.

This dynamic relationship between comprehensibility and incomprehensibility in Cantor's mathematics leads me to explore a more subtle interpretation of the relationship between the apophatic and kataphatic in theology, one in which they mutually imply each other. On the one hand, we want to affirm that God is Absolute Mystery, the ineffable source of knowledge, wisdom, and existence lying forever beyond human comprehension. Yet we also want to affirm that we can know this God as our Creator and Redeemer. How can we affirm both of these seemingly contradictory views? It turns out that the standard theological response is already somewhat analogous to Cantor's reflection principle. In essence, the traditional meaning of divine revelation is that the God who is incomprehensible is the God who makes Godself known to the world. The God who we know as Creator and Redeemer is inherently incomprehensible Mystery.

The God who is known through special revelation (word and scripture) and general revelation (nature) is known as unknowable.

This brings us to a deeper analogy between the mathematics of infinity and the theology of revelation. Just as the incomprehensibility of Absolute Infinity in mathematics is safeguarded by its reflection in the transfinites, so too the mystery of God is vouchsafed by the revelation of God in our lives and in the universe that God creates and science explores. Mathematically, Absolute Infinity is known through the transfinites, and yet being so, it remains unknown in itself. Theologically, the God who is known is the God who is unknowable. What God has chosen to disclose to us, the kataphatic – God's existence as Creator, God's goodness, love, and beauty – is a veil behind which the reality of God is endlessly hidden precisely as it is endlessly revealed. To capture this theological understanding of God's hiddenness in God's self-disclosure, I suggest the following metaphor for revelation: "revelation is the veil that discloses."

13.3.2 Infinity in the Context of Wolfhart Pannenberg's Doctrine of God: From Hegel to Cantor[24]

Given its culminating role in the development of his lifetime of theological research, my focus here will be on Wolfhart Pannenberg's massive *Systematic Theology* (1991, 1994, 1998). The concept of infinity arises in a variety of places in the *Systematics*, where Pannenberg describes it in words such as these: "The thought of the true Infinite . . . demands that we do not think of the infinite and the finite as a mere antithesis but also think of the unity that transcends the antithesis" (Pannenberg 1991, p. 446). This concept of infinity plays a crucial role in Pannenberg's interpretation of the doctrine of God, particularly in his treatment of the divine attributes (Pannenberg 1991, chap. 6). Pannenberg attributes his understanding of infinity to Hegel, and he contrasts it with its meaning in the traditional Greek thought, where infinity as *apeiron* is seen as merely the antithesis of the finite. What I will explore here is the way insights from Cantor's work on the transfinites can build on Pannenberg's use of Hegel in the doctrine of God.

13.3.2.1 The Role of Infinity in Pannenberg's Doctrine of the Divine Attributes and His Reliance on Hegel

It is with Pannenberg's move to the section on "The Infinity of God: His Holiness, Eternity, Omnipotence, and Omnipresence" that we clearly find the central role of the concept of infinity in the doctrine of God (Pannenberg 1991, pp. 392–448). Pannenberg begins by telling us that there are two kinds of divine attributes: (1) those attributes that define the God who acts (i.e., what "God" means) – God as infinite, omnipresent, omniscient, eternal, omnipotent, and holy; and (2) those attributes that describe God's

---

[24] According to Pannenberg, God's holiness and infinity are directly related, while eternity, omnipotence, and omnipresence are "concrete manifestations of his infinity" under the categories of time, power, and space (Pannenberg 1991, p. 397).

actions (i.e., what God does) – God as kind, merciful, faithful, righteous, patient, good, gracious, and wise.

> When we say that God is kind, merciful, faithful, righteous, and patient, the word "God" is the subject of the descriptions . . . But what does it mean to say all these things of "God"? The answer lies in terms that explain the word "God" as such, e.g., terms like infinite, omnipresent, omniscient, eternal, and omnipotent.
>
> (Pannenberg 1991, p. 392)

He first discusses the pivotal role of the concept of the Infinite in the definitional attributes. He then turns to the second set of attributes, which, while "structurally" related to the concept of the Infinite found in the first set, are most clearly connected with the Johannine proclamation that the essence of God is love (1 John 4:8; Pannenberg 1991, pp. 396, 432).

Pannenberg notes that while the Bible does not refer directly to God's infinity, it is clearly implied in its descriptions of God's attributes: "Even if the Infinite is not the essential concept of God from which all the qualities of his essence are to be derived, it is still to be viewed as the initial concept of the divine essence to which all other statements about God's qualities relate as concrete expressions of the divine nature" (Pannenberg 1991, p. 396).[25] By "infinity" Pannenberg at first means "that which stands opposed to the finite, to what is defined by something else." It only secondarily means that which is endless. Freedom from limitation "is a consequence of negation of the finite." He tells us that in mathematics, this meaning of freedom underlies the idea of an unlimited progress in a finite series. In this way, the infinity of God is radically distinct from everything finite, limited, and transitory. When it is linked to the holiness of God, the concept of the infinite distinguishes God from all that is profane (Pannenberg 1991, pp. 397–98).

Still, when Pannenberg develops this conception of the infinity and holiness of God, he adds a crucial new component to the previous, sharp distinction between the infinite and the finite. God's holiness not only opposes the profane world but "embraces it, bringing it into fellowship with the holy God." For Pannenberg this reflects a "structural affinity" between God's holiness and God's infinity. Finally, when Pannenberg summarizes his understanding of the concept of infinity in the doctrine of God, he specifically refers to Hegel:

> The Infinite that is merely a negation of the finite is not yet truly seen as the Infinite (as Hegel showed), for it is defined by delimitation from something else, i.e., the finite. Viewed in this way the Infinite is a something in distinction from something else, and it is thus finite. The Infinite is truly infinite only when it transcends its own antithesis to the finite. In this sense the holiness of God is truly infinite, for it is opposed to the profane, yet it also enters the profane world, penetrates it, and makes it holy. In the renewed world that is the target of eschatological hope the difference between God and creatures will remain, but that between the holy and the profane will be totally abolished (Zech. 14:20–21).
>
> (Pannenberg 1991, p. 400)

---

[25] This is Pannenberg's translation of Boethius's definition of eternity, who wrote: "*interminabilis vitae tota simul et perfecta possessio.*" Its standard translation is: "Eternity is the complete, simultaneous and perfect possession of everlasting life."

Pannenberg notes that Hegel's abstract concept of the Infinite actually contains a logical paradox: the Infinite, while being the negation of the finite, also comprehends this negation in itself. "There appears to be no way of showing how we can combine the unity of the infinite and the finite in a single thought without expunging the difference between them. We cannot solve this problem, as Hegel thought, by the logic of concept and conclusion" (Pannenberg 1991, p. 446). Instead, to resolve this paradox, Pannenberg turns to theology's understanding of God's holiness, the Spirit of God, and the Johannine proclamation of God as love. Here he believes we can see the paradox as resolved, for we come to understand that "[On the one hand] God gives existence to the finite as that which is different from himself, so that his holiness does not mean the abolition of the distinction between the finite and the infinite ... (while on the other hand) the unity of God with his creature ... is grounded in the fact that the divine love eternally affirms the creature in its distinctiveness and thus sets aside its separation from God but not its difference from him" (Pannenberg 1991, pp. 400, 446).

With Pannenberg's addition to Hegel's concept of the true Infinite in place, we can return briefly to his discussion of the divine attributes, in which this revised concept plays such a crucial role. To illustrate this role, let us focus on the definitional attribute, eternity, because it leads directly to Pannenberg's argument for the doctrine of the Trinity. Pannenberg starts by adopting the definition of eternity that Boethius took up from Plotinus: eternity is "the simultaneous and perfect presence of unlimited life."[26] Pannenberg insists that eternity must not be reduced to either timelessness or everlasting, unending time. A timeless concept of eternity would assume an "improper" concept of the infinite as defined simply by its opposition to the finite and would therefore be itself merely finite (Pannenberg 1991, pp. 408, 412). An everlasting concept of eternity would never allow for the simultaneous possession of all of its endless string of isolated "present" moments. Instead, eternity must be such that all created things are simultaneously present to God in a way that preserves their intrinsic temporal differences. Pannenberg then makes a crucial move: such a view of eternity requires a Trinitarian doctrine of God. Specifically, the presence of all things to God with their intrinsic temporality preserved is only possible "if the reality of God is not understood as undifferentiated identity but as intrinsically differentiated unity. But this demands the doctrine of the Trinity" (Pannenberg 1991, p. 405).

13.3.2.2 Cantor's Concept of the Transfinites and Absolute Infinity as Fruitful Resources for Pannenberg's Conception of the Infinite

I believe that Cantor's mathematical understanding of the relation between the finite and the infinite, in turn, provides certain advantages for Pannenberg's project over his use of Hegel, particularly in the way Cantor displays the analytic logic underlying both the similarities and the differences between the finite, the transfinite, and Absolute Infinity.

---

[26] For his development of the doctrine of the Trinity, see Pannenberg (1991, chap. 5).

Let us recall Pannenberg's own assessment of the importance of Hegel for his project:

> The Infinite that is merely a negation of the finite is not yet truly seen as the Infinite (as Hegel showed), for it is defined by delimitation from something else, i.e., the finite... The Infinite is truly infinite only when it transcends its own antithesis to the finite... We (must) combine the unity of the infinite and the finite in a single thought without expunging the difference between them.

We can reformulate Pannenberg's statement in the following way:

1. The Infinite is the negation of the finite. Yet if it is nothing more than this negation, the Infinite too is finite.
2. To avoid being merely finite through this negation, the Infinite transcends the negation by uniting itself with the finite without destroying their difference.

My suggestion is that Cantorian mathematics provides a way to restate these claims as follows:

1. Cardinality: The cardinality of the transfinites and the cardinality of the finites represent one aspect of the radical difference between the infinite (i.e., the transfinites) and the finite. This difference is analogous to Pannenberg's understanding of the relation of negation between the infinite and the finite.
2. Ordinality: The ordinality of the transfinites and the ordinality of the finites represent one aspect of the striking similarity between the infinite (i.e., the transfinites) and the finite without undermining their radical difference, as represented by their differing forms of cardinality. This similarity is analogous to Pannenberg's understanding of the idea of the infinite being united with the finite.

Now if we follow Cantor in using the concept of "Absolute Infinite" in reference to God, we can consider viewing Cantor's transfinites as aspects of the world that traditionally have been seen as fully finite without in any way challenging the radical distinction between God and creation formalized as the doctrine of creation *ex nihilo*. Doing so would suggest that the relation between the transfinites and the finite, differing in cardinality and similar in ordinality, serves as a metaphor, and perhaps even an analogy, for Pannenberg's assertion that God's holiness not only opposes the profane world but "embraces it, bringing it into fellowship with the holy God... In this sense the holiness of God is truly infinite, for it is opposed to the profane, yet it also enters the profane world, penetrates it, and makes it holy. In the renewed world that is the target of eschatological hope the difference between God and creatures will remain, but that between the holy and the profane will be totally abolished."

13.4 Conclusion

Cantor's contributions to the mathematics of infinity offer a relatively unexplored and immensely promising topic for theological reflection. This brief chapter has pointed to two broad topics for reflection: the meaning of revelation as apophatic and kataphatic modalities of hiddenness and disclosure ("the veil that reveals"), and the conception of the divine attributes in Pannenberg's theology. Further discussions about infinity in the topics discussed, particularly the unity of God and the distinction between the divine essence and existence, seem to be particularly promising for future theological research into the implications of the mathematics of Cantor.

References

Cantor, G. 1980. *Gesammelte Abhandlungen Mathematischen und Philosophischen Inhalts.* Berlin: Springer.

Craig, W. L. 1979. *The Kalam Cosmological Argument.* New York: Barnes & Noble.

Galileo Galilei. 1954. *Dialogues Concerning Two New Sciences.* Translated by H. Crew and A. de Salvio. New York: Dover.

Hahn, H. 1956. Infinity. In *The World of Mathematics,* vol. 3, J. R. Newman (ed.), pp. 1593–1611. New York: Simon & Schuster.

McKeon, R. (ed.). 1941. *The Basic Works of Aristotle.* New York: Random House.

Pannenberg, W. 1991. *Systematic Theology,* vol. 1. Translated by G. W. Bromiley. Grand Rapids, MI: Eerdmans.

Pannenberg, W. 1994. *Systematic Theology,* vol. 2. Translated by G. W. Bromiley. Grand Rapids, MI: Eerdmans.

Pannenberg, W. 1998. *Systematic Theology,* vol. 3. Translated by G. W. Bromiley. Grand Rapids, MI: Eerdmans.

Pennings, T. J. 1993. Infinity and the Absolute: Insights in our world, our faith and ourselves. *Christian Scholar's Review* 23 (2): 159–80.

Rucker, R. 1983. *Infinity and the Mind: The Science and Philosophy of the Infinite.* New York: Bantam Books.

Russell, R. J. 2008. *Cosmology from Alpha to Omega: Theology and Science in Creative Mutual Interaction.* Philadelphia: Fortress Press.

Stoeger, W. R., Ellis, G., and Kirchner, U. 2005. Multiverses and cosmology: Philosophical issues. astro-ph/0407329v2.

Templeton, J. M. (ed.). 1997. *How Large Is God? The Voices of Scientists and Theologians.* Philadelphia: Templeton Foundation Press.

Woodin, W. H. 2001. The continuum hypothesis, part I. *Notices of the American Mathematical Society* 48 (6): 567–76.

CHAPTER 14

A (Partially) Skeptical Response
to Hart and Russell

Denys A. Turner

I want to suggest a way into the theological notion of infinity via the notion of "otherness," or, if you prefer a more negative and restrained proposition, I want to suggest that there is a problem about how to speak of the divine infinity, one that connects with a problem about how to speak of the divine "otherness." The chapters by Hart and Russell in this volume have ably shown how the "infinity" of God was historically a problem for Christian theologians, inheriting as they did Greek notions of the infinite as formless and vacuous "indeterminacy," and both are right to emphasize the crucial role of Gregory of Nyssa in generating a notion of the divine infinity that allows us to speak non-oxymoronically of God as "infinite perfection." Not wishing to rehearse that historical question – I could neither want nor hope to match the brilliant lucidities of either's chapter – my concern is rather with how that same problem with which early Christian theologians were faced about divine infinity recurs for us today in connection with general notions of "otherness." Furthermore, we can see in what way there is a problem – as in general terms arising out of our ordinary conceptions of "otherness" – from the following objection to the proposition that God is "infinite" found in Thomas Aquinas's *Summa Theologiae* (ST Ia, q7, a1, obj3). The objection goes like this:

> What exists in such a way as to be "here" and not "there" is finite in respect of place. Hence, what so exists as to be "this somewhat" rather than "another somewhat" is finite as "things" go [*secundum substantiam*]. But God is "this" and not something else: for he is not stone or wood. Therefore God is not an infinite "somewhat" [*non est infinitus secundum substantiam*].

One further disclaimer: I am not concerned in this chapter with how Thomas resolves this problem so much as with the way in which it arises for him. The question of how to characterize the divine "otherness" is, we may say, a question about how to characterize the divine "transcendence." But if we try to set out that transcendence in terms of "otherness," we immediately run up against Thomas's objection. A "this" that is other than a "that" is necessarily finite; indeed, *any* "this" other than a "that" would have to be a "somewhat" (identifiable existent), so if we can give an account of the "other than" in such relational propositions, then we have, as it were, drawn the terms so related into a finite community of difference that contains them. As Thomas himself often says, reporting Aristotle, *eadem est scientia oppositorum* (*Peri Hermeneias*, 6, 17a, 31–33): to know what counts as not-p is the same knowledge as knowing what counts as p. If we can say in what God differs from creatures (from "stone" or "wood") or, which is to say the same, if we can characterize how "other" God is, then, as Thomas puts it, we end up having to deny the divine infinity and the divine transcendence altogether. God becomes reduced to just another existent alongside our finite, creaturely existence, a "this" or "that," because there would have to be something in common between God and creatures that they differ *as*. Nor would it help to say that God differs from this and that in this way and that, only *infinitely* so. For, as Hart rightly says, it is only as *finite* that things can differ in one way or another, by howsoever much. Any infinity defined in terms of contrast with finiteness would have to be a *created* infinity, such as that, as Russell explains, of Cantor's transfinites. Moreover, as Hart puts it (in very different terms), it is for want of seeing this that the Heideggerian "ontico-ontological difference" between God and creatures is obliterated in onto-theological idolatry (Hart, Chapter 12).

Recognizing this, some theologians have sought a way out of the dilemma of the divine transcendence by saying that God is not "other" than creatures in any particular respect (in "this" way rather than "that") but is "wholly other," or, as one might say, "infinitely other," "other" in *every* respect. But this way out is no way out at all – not, at any rate, as it stands, and certainly not for Christian theologians in the classical traditions. Such a notion of "otherness" is entirely vacuous. To be "other" in every respect is to be nothing in any respect, and to be "other" in no particular respect is not to be "other" at all. Moreover, the "wholly other" is not just contentless in itself. Any notion of the divine "infinity" that is connected with it will itself be wholly vacuous, and we are back, once again, with that classical Greek notion of infinity, as of "otherness," as pure indeterminacy. Being back to that problem, we are back where Gregory of Nyssa started: how do you get a notion of an "infinite *being*," or of a being that is "infinitely *good*," from a notion of infinity as absolute indeterminacy? In short (no doubt too short), it would seem that you can buy the divine infinity at the price of evacuating it of all determinate content. But if you want determinate content to the divine "difference," you must sell off your interest in the divine infinity. To put it another way, if you want to say that God contains all "perfections," you will have to sacrifice his/her infinity – or vice versa. Either way, on a purely negative concept of infinity, for example, on that of Plato, who thought of the infinite as formless imperfection, it would seem that you can make no sense of an infinite *existent* – hence, none of an infinite God.

I am not at all sure that the chapters by Hart and Russell – Hart proposing that a way out of this dilemma was found at least in some classical theologies, and Russell suggesting that the problem is soluble on an analogy provided by Cantor's distinction between the Absolute Infinite and the transfinite – are, as they stand, entirely convincing. Certainly both are right that Christian theologians, or at least some of them, did acknowledge the problem and faced up to it. Apart from Gregory of Nyssa, Thomas faces up to it squarely, as we have seen. So, in their different but related ways, do Meister Eckhart and Nicholas of Cusa. On the other hand, I am not entirely sure that Hart is right in saying that from Descartes, through Kant and Hegel, the dilemma is eliminated because one of its horns is removed, by means of retreat into a purely negative conception of the infinite (i.e., I am not sure that *that* is what is wrong with their accounts of infinity, although what *is* wrong with them is another story). In any case, my main problem with Hart's chapter is that what he identifies as the "historical" solution in patristic and medieval theology (requiring the simultaneous affirmation of "infinity" and "perfection," and so of "infinity" and "determinacy") seems not to amount to anything much more than simply attempting to have it both ways, without formally removing what is prima facie the inconsistency of so doing. Although I think that the inconsistency can be removed, I doubt whether it can be removed just by working with what Hart calls a "positive" notion of infinity.

Russell's chapter, on the other hand, while developing what seems to me to be a most illuminating *parallel* between Cantor's account of the relationship of the Absolutely Infinite to the transfinite and that of the divine infinity to the creaturely finite, does not, in the end, demonstrate what he claims for this parallel, namely, that there is any real *analogy* to be found there. For, as Hart shows, an analogical relationship between p and q depends on a relationship of participation one in the other, and I cannot see how sense can be made of any created infinity's *participating* in the divine infinity. However, given that both Hart and Russell rest their distinct cases on the theological necessity of a "positive" notion of infinity, I will direct my comments principally to the discussion of that notion. I think that such a notion is neither possible logically nor necessary theologically in the defense of the divine infinity.

A single example will illustrate what I mean. Hart says, rightly, that Christians will want to say things like the following (this is Hart in explication of Gregory's "solution"):

> The insuperable ontological difference between creation and God – between the dynamism of finitude and an infinite that is eternally dynamic – is simultaneously an implication of the infinite in the finite, a partaking by the finite of that which it does not own, but within which it moves – not dialectically, abstractly, or merely theoretically – but through its own endless growth in the good things of God (*In Canticum Canticorum* VI).
>
> (Hart, Chapter 12)

However, my problem with being able to say such things (and as a Christian theologian I, like Hart, want to be able to say them) is that the defensibility of such theological talk requires the demonstration that the following two propositions are not inconsistent: (a) "God is without opposition, as he is beyond nonbeing or negation, transcendent of all composition or antinomy" (Hart, Chapter 12); and (b) "Difference within being, that is, corresponds precisely as difference to the truth of divine differentiation" (Hart, Chapter 12).

The problem is that (a) is the proposition that God is "wholly other," that is, a being beyond "difference," "transcendent of all composition and antinomy," as Hart says. To be "beyond antinomy," of course, is to be "beyond" any relation of opposition with anything else. To escape the antinomies of "this" rather than "that" is to escape the clutches of the Aristotelian *eadem est scientia oppositorum*. It follows that such a notion of the "otherness" of God has therefore to be that of the "*wholly* other." On the other hand, (b) seems to set creaturely differences in at least an "analogical" relation with the "divine differentiation," so that our creaturely "differentiations" in some way participate in the "divine differentiation" that is the plenitude of all perfections. Hart's case for a "positive infinity" rests on showing how (a) and (b) are consistent, or at the very least on demonstrating that the question of formal consistency is undecidable. But you could not demonstrate that "undecidability" if you had not first seen off any arguments purporting to demonstrate formal inconsistency, such as that which Thomas offers. What I am not sure about, then, is whether Hart's version of how historically theologians have dealt with the problem – as Thomas raises it – shows that they have, in fact, seen the objection off.

The grounds of my skepticism lie in the proposition, claimed by both Hart and Russell, that Christian theology not only historically originated but also logically requires a "positive" notion of infinity. I confess to not being clear about what is meant by a "positive" notion of the divine infinity. The best I can make of this notion is that of an actually existent infinite being, or of an infinitely good existent, and so forth. Although I am far from having any quarrel with Hart and Russell over those propositions – after all, they are central to any Christian theology – I do not see how, just by virtue of that certain truth, it follows that "infinity" is, as they want to say, a "positive" attribute of God.

At any rate, my uncertainty about this (perhaps it is just a failure of clarity on my part) arises from a feeling that we need to distinguish between two sorts of predicates of God: those that can be said to be "positive," such as "goodness," "wisdom," "intelligence," and are known "by analogy" from what we know of such predicates as affirmed of creatures, and those that are, as I shall say, "regulative," such as "infinity" and "simplicity," which are known to be true of God *only by denial* of what we know of creatures. Let me explain.

When we say that God is "simple," we mean to say of God something we know we cannot say of any creature at all: that everything true of God *is* God. If I am good, there is an I and there is my goodness: these are distinct. I could not come to be good, nor could I cease to be good, if I and my goodness were identical, for were I, being once good, to become less so, then the I who was less good would not be the same I who was once better. A being identical with an attribute cannot cease to possess that attribute without ceasing to be that being. Now when we say that God is "simple," we mean to say that any of the attributes we affirm positively of God belong to God in such a way that God would cease to be God were s/he to lose them. And when I say that a predicate affirmed of God in the way that "simple" is has a "regulative," rather than an "attributive," character, I advert to this fact of the logic of such terms: the divine "simplicity" refers not to some substantive attribute, on a par with "goodness," "beauty," and so forth, but to *how* such positive attributes are predicated of God; that is to say, they are predicated of God *minus* any implication of complexity carried over with such positive attributes from their applications to creatures.

We can see this if we note how the logic of "complex" as predicated of creatures parallels that of "simple" as predicated of God. Of course, any creature is, by contrast with the simplicity of God, "complex." However, there is a sort of category mistake in supposing that the following is a logically homogeneous list of substantive attributes of a creature: "Peter is good, strong, handsome, intelligent, and complex." For whereas "good," "strong," and so forth, are "positive" attributes of Peter, having "positive content," Peter's being "complex" is not another attribute of Peter alongside the others and on all fours with them logically. Peter's being "complex" is rather a "second-order" property of *predications of* Peter, not a first-order "attribute" of Peter, for "Peter is complex" is rather a remark about the *list* itself of first-order substantive attributes, informing us, namely, that its members are not identical with one another, nor the whole list with Peter. It is not, of course, that "complex" tells you nothing about Peter: it certainly does, for it tells you that Peter is a creature, that Peter who is strong can become weak, that Peter who is good can become bad, that Peter who is handsome can become wrinkled and ugly. But what it tells you is something about *how* Peter possesses the attributes we positively ascribe to him; it is not to add another attribute to the list logically on a par with the others. In the same way, "simple" is predicated of God not in the way that "good" is predicated of God, but rather *as governing the way in which* "good" is predicated of God: namely, anything truly predicated of God *is* God, God *is* his goodness. Equivalently the same, I propose, holds for "infinite" as predicated of God. It is not a "positive attribute."

Now if we take the line that broadly – only "broadly," because their accounts differ – is taken by Gregory of Nyssa and Thomas Aquinas, but also broadly proposed, likewise in different ways, by Hart and Russell, that positive attributes of God are predicated "by analogy" from creatures, we can see how and why it is *precisely because* that is so that "infinite" cannot itself be a positive attribute of God, itself known "by analogy." Following that line, we need to make a distinction concerning our attribution of names to God, between the *res significata* and the *modus significandi*. As Thomas Aquinas says (*Summa Theologiae*, 1a q13 a3 *corp.*), *what* is signified by "good" (the *res significata*) belongs properly and primarily to God, because God is the source of all goodness in creatures, and what belongs to the cause belongs to it in a manner prior to its manner of existence in the effect. But so far as *our understanding* of goodness goes, and so far as the logic goes that governs our use of the word (our *modus significandi*), goodness belongs to creatures primarily. Hence, we have somehow to signify by means of our limited conceptions of goodness in creatures that goodness that in God is without creaturely limitation.

This account of how theological language works, that is, "by analogy," is an attempt to hold together in tension the two features of (a) language as *creaturely* and (b) the reality of the *Creator* that lies unutterably beyond the reach of language. As we might say (a little simplistically), there are "positive" and "negative" elements involved, and Russell is exactly right to say (and Hart, too, in his extensive demonstration) that these must be seen as implying one another within a single structure of analogy: the *via negativa* and the *via affirmativa* are not independent routes to God, and analogy is nothing but their conjunction. They do in this way reciprocally imply, again as Hart and Russell rightly emphasize, through the core Christian doctrine of creation *ex nihilo*. For insofar as God is the creator of all things "out of nothing," all things in some way reflect their origin in the divine creative act; hence, our language descriptive of all created things becomes an inexhaustible repertoire of talk about God. Because Hart seems to think of infinity as in the same way predicated positively of God, and so by analogy from creaturely infinity, it is thus, as he says, that the "finite infinities" of the created world participate in the absolute infinity of God. Likewise Russell, relying on an analogy with the way in which Cantor's transfinites "participate" in, "reveal and veil," Absolute Infinity construes created infinities as participating analogically in the divine Infinity. It is precisely here that the need arises to distinguish between the way in which positive attributes are predicated of God – namely, by analogy – and the way in which regulative attributes are predicated – namely, by negation. Both forms of utterance, we may say, "fail" of God, but they fail differently.

As for the positive attributes, we can truly predicate goodness of God, because all creaturely realities reflect God, their Creator. But just because of that, just insofar as what creatures reflect is the Creator of all things out of nothing, and just insofar as that notion of *creatio ex nihilo* utterly defeats our powers of comprehension, all that positive talk fails, and is *known* to fail, of God. It fails, that is to say, not because we are short of things to say about God, for we are far from being short of affirmative theological utterance if *all* creation in some way speaks God: theology is notoriously and rightly garrulous. Nor does our speech about God fail because we are short of the epistemological instruments for dissecting true from false propositions about God, for we are not. Our speech about God fails because all of what we can say about God, even truly, falls short of him – infinitely. And that is precisely the point of difference between the predication of positive attributes, which by analogy affirm some positive content of God, and the predication of infinity. The predication of infinity is simply the measure of the shortfall of the positive predications; or perhaps it would be more proper to say that the predication of the infinity of God is required because there is *no* measure of that shortfall. The falling short is infinite, because God is without limit, and language is essentially limited, designed for the discrimination of the limits of a "this" that mark it out from a "that."

Coming to see that God is infinite is not, then, to acquire some new piece of information about God further to our knowledge of God's goodness, or beauty, or truth, or whatever. God's infinity refers to the manner in which these positive attributes hold true of God; hence, secondly, it refers to the failure of thought and speech to capture the *how* of their existence in God. So within this "affirmative/negative" complex that is theological language, it would seem that the "infinity" of God functions somewhat in the manner of an operator in mathematics and logic. The operator "square of *x*" is an incomplete expression governing a mathematical procedure, whose "content" can be known only when a value is substituted for the variable *x*. The operator is not, of course, itself a number, but if you know how to conduct the operation of squaring a number, then you know what value the whole expression "square of *x*" has when you know what value to substitute for *x*. It is in the same way that God's infinity is not a substantive attribute of God, as are "goodness" and "wisdom"; instead, it functions in relation to such substantive attributes in the manner in which an operator does over the values of its variables. More simply, the concept of the "infinite" itself has only "negative" meaning, as the denial of all creaturely limitation: if, indeed, it is the nature of the creaturely existence to be limited, then the denial of *all* creaturely limitation could not possibly have any "content" of its own; at any rate, it could not have any content if, as the doctrine of analogy supposes, language about God gets its content from its creaturely reference. For, again, language is *essentially* thus limited, finite, in its *modus significandi*. Nonetheless, combined with a substantive attribute, such *complete* expressions as "God is infinitely good, wise . . . " yield just that conjunction of affirmativity and negativity that, as both Hart and Russell rightly say, characterizes our language about God. Furthermore, that *conjunction* is what we call "analogy." As for "infinite," that we do *not* know "by analogy," but only by negation.

Otherwise, we could never get analogy going at all. If, as Russell suggests in his chapter, we knew God's "absolute" infinity by analogy from creaturely infinity, on a model of the way in which Cantor's "transfinites" enable knowledge of Absolute Infinity, then it could not serve, in the manner just described, as a quasi-operator within language about God, to contribute to the structure of analogical utterance about God. The infinite/finite distinction *defines* the gap that analogy has to cross; it is not an analogical means of crossing it. If the absolute "infinity" of God were known by analogy from any creaturely "finite" infinity, then we would need another term, not predicated by analogy and defined only negatively, as a "qualifier" over the supposedly "substantive" notion of "infinity," and so on. Such a procedure would be like trying to *define* "square of . . . " by means of the values it generates. Just as "square of . . . " operates *over* numbers, and so itself cannot be a number, so "infinite," it seems to me, is a second-order qualifier, operating *over* our affirmations of substantive attributes of God, but is not itself a substantive "attribute" in the way that " . . . is good" and " . . . is wise" are.

However, if the divine "absolute" infinity is not a positive attribute, known by analogy from the more limited infinities of creaturely experience, but only by negation of those more limited infinities, it is important to distinguish, as the Pseudo-Dionysius does, between "ordinary negation" and "negation by transcendence" (*Mystical Theology*). Thus, it is important in itself to distinguish types of "negativity," but it is also important if we are to be precise about the nature of the disagreement I have with Hart and Russell. I suspect that the reason why both are resistant to conceding that "infinity" as predicated of God is a negative concept is because they suppose that the negativity involved could be only of the Pseudo-Dionysius's "ordinary" kind. Now as the Pseudo-Dionysius says, there is all the difference in the world between saying "God is not good," in the sense entailing the "ordinary" negation of "God is good," namely, "God is evil," and saying "God is not good," as meaning "God is not one of the good things that there are, for God is the *cause* of all the good things that there are, and what is the cause of all cannot be one of the things thus caused." Now, if you supposed that to say that the infinity of God is "only negative" meant nothing more than is entailed by "ordinary negation," then, of course, you would have every reason for denying that the infinity of God is but a negative concept. To adopt Russell's comparison, this would reduce the theological notion of divinity to nothing more than the equivalent of Cantor's transfinites, which may, of course, be serially endless, but being only the endless extrusions of series of finite numbers, or else of sets of finite numbers, are each finite as a whole. To get at the infinity of God, we would have to "get at" some intelligible content to the notion of an infinite *being*, and that we cannot get at. We possess no concepts and no language, except that of negation, for doing so. Such a notion is utterly incomprehensible, beyond any power of expression, negative "by transcendence."

How, then, does "negation by transcendence" differ logically from "ordinary" transcendence – in what way, that is, relevant to the logic of predicating infinity of God? If we are to speak of the divine infinity, it seems necessary to distinguish between infinity understood as the simple negation of the finite such as yields its corresponding contradictory, namely, that of the infinite understood as endlessness – whether of mathematical or of temporal seriality – and the infinity that is yielded by the *negation of the negation* between the finite and the infinite as so understood in mathematics.

Logically, the difference consists in that, in the case of "ordinary" negation, the notion of infinity amounts but to the negation of serial finiteness, an endlessly extruded series each of whose parts is finite, because mutually exclusive, each being a "this" rather than a "that." In the case of the divine being, however, we have to speak of an infinity that transcends that known by force of "ordinary" negation of the finite, because the negation required is the negation of that "ordinary" negation itself. As such, this "transcendent" notion of the infinite is such as to *exclude all exclusion*, not only, then, the exclusions of one finite by another, but also the exclusion of the finite by the transfinite. To put it more positively, God's not being any sort of "this" rather than a "that" entails not maximum vacuousness and indeterminacy but, on the contrary, maximum *inclusiveness* – God's being "all in all." When, therefore, the Pseudo-Dionysius says that "there is no kind of thing that God is," it is not to proclaim a God of absolute emptiness, but, on the contrary, it is *so as* to be able to say "there is no kind of thing that God is not." God is maximally inclusive: is this what Hart and Russell mean when they speak of a "positive" notion of infinity? Perhaps. But if so, it is at least misleading so to say, for what we have here is no contentful notion of the divine infinity, but a *double* negation, the negation of exclusiveness, of which conception there is no possible "positive" understanding.

Hence, to return to the beginning of this chapter, if such is true of "infinity," then the same will be true of "difference" and "otherness," for to say that God is "wholly other" is to say no more than that God escapes every form of finite difference, that is, *every* form of utterable difference. The sense in which God is "wholly other" is not one that possesses any positive content, as if it were an *extension* "by analogy" of our creaturely conception of "otherness" in this or that respect. When we say, rightly, that God is "wholly other," we do not mean that there is some measure of how "other" God is, but that, as of God, *all* conceptions of "otherness" fail – or, as the Pseudo-Dionysius famously said, God is "beyond *both* similarity *and* difference" (*Mystical Theology*, 1048A). That failure of the language of "otherness" to capture the transcendence of God is exhibited precisely in the oxymoronic character of the phrase "wholly other," for no "otherness" can be "total." If "totally," then not "other." If "other," then not "totally." *That* was the point of Thomas's objection quoted at the beginning of this chapter: "otherness" is always *other in a certain respect*. It was because of this that Nicholas of Cusa thought it just as well to describe God as "ly *non*-aliud" (the "*not-other*") as to describe God as "*totally* other." Furthermore, it was for the same reason that Meister Eckhart said that God is distinct precisely in virtue of being the one and only being that is *not* distinct, an *unum indistinctum*. Both propositions amount to the same as that of the Pseudo-Dionysius mentioned earlier: all creatures fall within *some* family, whether of "difference" or "sameness," and God does not. And *that* – not being distinct – is the distinction between God and creatures.

For these reasons, then, it seems wrong, or at least misleading, of Hart to say that "difference within being . . . corresponds precisely *as difference* [my italics] to the truth of the divine differentiation." On the account I have given of the purely negative character, whether of infinity or of the divine difference, there is no relation of analogy, nor therefore of participation, between creaturely finitude and the divine infinity on the one hand, nor between creaturely difference and the divine differentiations on the other. Once again, this is not to say that finite *creatures* do not participate in the infinite

Godhead, or that creatures, inserted as they are into families of "difference," do not participate in the differentiated trinitarian Godhead: they do, by virtue of their own goodness and wisdom, participate in the divine goodness, wisdom, and so forth. What seems misleading is to say, along the lines of Russell's chapter, that creaturely infinities *as such* form an analogy with the divine infinity, or, as Hart does, that difference among creatures "corresponds precisely to the truth of the divine differentiation," for it is not *qua* creaturely and finite, or even *qua* any creaturely, and so finite, infinity that creatures participate in the divine attributes. As the medieval theologians used to say, we creatures are indeed all that God is, but what we are in a created way, God is as uncreated. It is only by distinguishing, in some such way as I have done in what precedes, between, on the one hand, the "substantive attributes" of God (goodness, wisdom, and so forth), which we know of by "analogy" from creatures and as such participate in those divine attributes, and, on the other, the "regulative properties" of simplicity and infinity, which we know of only by negation, that we can get the two sides of the "affirmative/negative" polarity to match up without internal incoherence. Hence, then, we can indeed say "God is *infinitely* good, beautiful . . ." as meaning "good, beautiful as we indeed are, but *incomprehensibly so*."

That said, my disagreement is *only* with the proposition that you need a "positive" notion of God's infinity to get going all the rest of what Hart and Russell so convincingly argue for – the substantive doctrine of the divine transcendence as being precisely that which permits the intimacy of the divine immanence to creatures, as well as the Christological necessity that requires and imposes that doctrine, the doctrine that the divine existence is infinite not by virtue of a maximally vacuous indifference but, on the contrary, by a maximized richness of differentiation. My argument is, rather, that it is only as *negatively policing* our substantive talk about God that it can be said that the divine infinity allows us to affirm, without formal contradiction, all those things of God that the revelation of the Trinity in Christ requires us to say. For it is just insofar as the divine infinity is so utterly transcendent of all creaturely knowing and saying as to *exclude all exclusion* (all "difference") from within that maximized plurality of the divine attributes that we may defend, against the accusation of simple incoherence and plain contradictoriness, our talk of God as both utterly simple and maximally *inclusive*. This is because, as Thomas puts it, "what are diverse and exclusive in themselves pre-exist in God as one, without detriment to his simplicity" (*Summa Theologiae*, 1a, q4, a2, ad1). Hart, Russell, and I all want to say this, and I imagine for the same reason. Not one of the three of us could, on any other terms, envisage a remotely orthodox defense of the Christological doctrine that the human and divine natures in Christ remain distinct, unconfused, but united in the one and only divine person. We seem to differ only as to *how* one might say it, that is, as to what account of the divine infinity is required as a condition of Christological coherence. All three of us agree that, in the end, what for a Christian theologian must be said about the divine infinity is whatever is required of it to permit the Pseudo-Dionysius's obiter dictum, namely, both "there is no kind of thing which God is," God's being exclusive beyond any possible inclusion, and "there is no kind of thing that God is not," God's being inclusive beyond all possible exclusion (*Divine Names*, 817D).

Index

WinCo
FOODS
The Supermarket Low Price Leader

www.wincofoods.com
855 Davis St, Suite A
Vacaville, CA 95687
Store #0060

Cashier: Theresa A

05/18/17
                          14:46:47

Item	Code	Price
BAGELS,ASSORTED	2624	.50 FS
G/MILLS CEREAL	1600048366	2.98 FS
CHEERIOS MG	1600048769	2.48 FS
REAL WTR ALKLZ	66914910552	2.97 TF
3 @ .99		
+CRV		.30 FS
3 @ .10	40000000052	
O/VALLEY MILK	9396600441	3.98 FS
WOOLITE	6233877940	8.91 TX
PAC BROTH CHIX	5260305425	3.36 FS
PEARL SOYMILK O	4139006156	3.54 FS
3 @ 1.18		
DAVE'S KLR BRD	1376402808	4.33 FS
CHICKEN THIGHS	2410558405	5.08 FS
SWT MINI PEPPE	81642601044	1.98 FS
MT HIGH YOGURT	7527000160	2.78 FS
S/B SPONGE	5114125403	2.53 TX
PLAIN GOAT LOG	7055176104	2.98 FS
MARUCHAN SOUP I	4178900128	.56 FS
2 @ .28		
RSTTO BTRNT SQU	7341602005	2.25 FS
K/OSCAR SARDINE	3480000525	2.52 FS
AVOCADO,SMALL	4046	.88 FS
COLD LBSTR TAI	22640100000	9.96 FS
C/BELLS CRM MSH	5100001261	1.96 FS
2 @ .98		
HALIBUT,FZN ST	22609100000	3.83 FS
PORK TENDERLOI	20335800000	3.04 FS
DRY GR VAN BEAN	4154800335	2.98 FS
SPLIT PEA SOUP	7107904367	2.01 FS
INST LUNCH BEEF	4178900122	1.12 FS
4 @ .28		
MARACHUAN CHICK	4178900121	.28 FS
INST RST CHIX	4178900157	.56 FS
2 @ .28		
MC BEEF STEW	5210004360	1.07 FS
MCCORMK MTX	5210009040	3.00 FS